Joy Tivy
DEPARTMENT OF GEOGRAPHY
THE UNIVERSITY OF GLASGOW

THE EFFECT OF RECREATION ON FRESHWATER LOCHS AND RESERVOIRS IN SCOTLAND

Commissioned by the
COUNTRYSIDE COMMISSION FOR SCOTLAND

A

ISBN 0 902226 47 9

CONTENTS

CONTENTS *continued*

LIST OF FIGURES

LIST OF TABLES

The University of Glasgow was commissioned in 1974 by the Countryside Commission for Scotland to undertake a study of the effect of recreation on the margins of freshwater lochs and reservoirs in Scotland. For the purpose of the study, lochs comprise natural lochs, impounded lochs and man-made reservoirs. The mutually agreed aims of this study are listed below:

(a) To carry out (as a background to the following items) a systematic study of recreation relevant to lochsides in Scotland. The aim of the systematic survey was to assess the extent to which the effects of recreation constitute a problem on lochsides in Scotland as a whole. It was decided, however, that a more detailed study of impact would be undertaken by means of site studies at sample lochs (varying as to size, form, water quality, location, accessibility and range of recreation uses) initially within West Central Scotland, and subsequently outwith this area where it was deemed necessary for representative cover.

(b) To investigate the nature of the effects of recreation on selected lochsides; attention was to be drawn to other land uses having effects on lochshores that may be confused with the effects of recreation deriving from both the land and water side of the lochshore.

(c) To investigate the ability of these selected lochsides to withstand the effects so identified; for this purpose, physical and biological data was to be provided for the selected lochsides. In doing so, consideration was to be given to presenting data in such a way that subsequent detailed studies on the effects of recreation could be related to a recommended classification of lochside types, constructed for the purpose of assessing recreation effects.

(d) To describe resource and recreation management techniques relating to recreation on lochsides. For this purpose, a report will be made of selected management techniques that illuminate principles of management on the lochside zone, distinguishing between techniques aimed at managing the resources and those directed at managing users.

(e) To identify areas for research, including possibilities for field experiment, and to make recommendations for priorities for investigation; it was agreed that research in this context should be management-oriented.

(f) To identify situations where there is a need to promote among recreation managers and users an awareness of the effects of recreation on lochsides and to make recommendation as to priorities for action.

The desk and field work on which this Report is based was undertaken in the period July 1974 to September 1977 by a study team, each member of which had a particular, though not mutually exclusive area of responsibility:

Dr Joan Rees, BSc, PhD (Research Assistant): the characteristics and vulnerability of the lochside vegetation;

Miss May Johnstone, BSc (Research Student): the classification of lochside site-types and the evaluation of the lochside for recreational activities;

Mr Martin Orrick, BSc (Research Student): the nature of loch waves and lochshore beaches.

In addition three placement students from Middlesex Polytechnic at Enfield each spent a year working on the project: Miss Sheila Potter (1974/75) was responsible for the initial compilation of the Tourism and Recreation Information Package (TRIP) lochside data set; Miss Sheila Morgan (1975/76) and Mr Gary Mitchell (1976/77) maintained and updated the data set, and assisted in coding and analysing the material.

To all the members of the study team, I would like to express my very deep and sincere gratitude. Their unflagging enthusiasm for and commitment to all stages of the project made the production of this Report possible. Further, their co-operation as an integrated team made it a particularly enjoyable and mutually stimulating intellectual experience.

I would also like to express my appreciation to those regular visitors to our Lochshore seminars: Mr Jim Hamilton, Department of Biology, Paisley Technical College; Mr P. L. Pearson and Mrs V. Clements of the Countryside Commission for Scotland; and also Miss N. Ward (now of the Scottish Economic Planning Department) for their continuing interest and, more particularly, for their constructive criticism, catalytic ideas and objective views which helped to keep our heads above the flood of our own data.

Finally, I would thank the University of Glasgow and the Department of Geography who made available the time and facilities to produce the report; Mr Ian Gerrard (Chief Technician, Department of Geography); Mr Mike Shand (Cartographer, Department of Geography); Miss Katezyna Kram who produced all the figures and Mrs Lorna Peedle whose speed and accuracy at deciphering manuscripts was well nigh miraculous; thanks are also due to all those individuals and organisations too numerous to list here, but who are named in the Acknowledgments at the end of the Report.

Without these resources this Report could not have been produced. I must however, take full personal responsibility for any errors of fact and all subjective opinions expressed.

Professor Joy Tivy
Department of Geography
The University of Glasgow

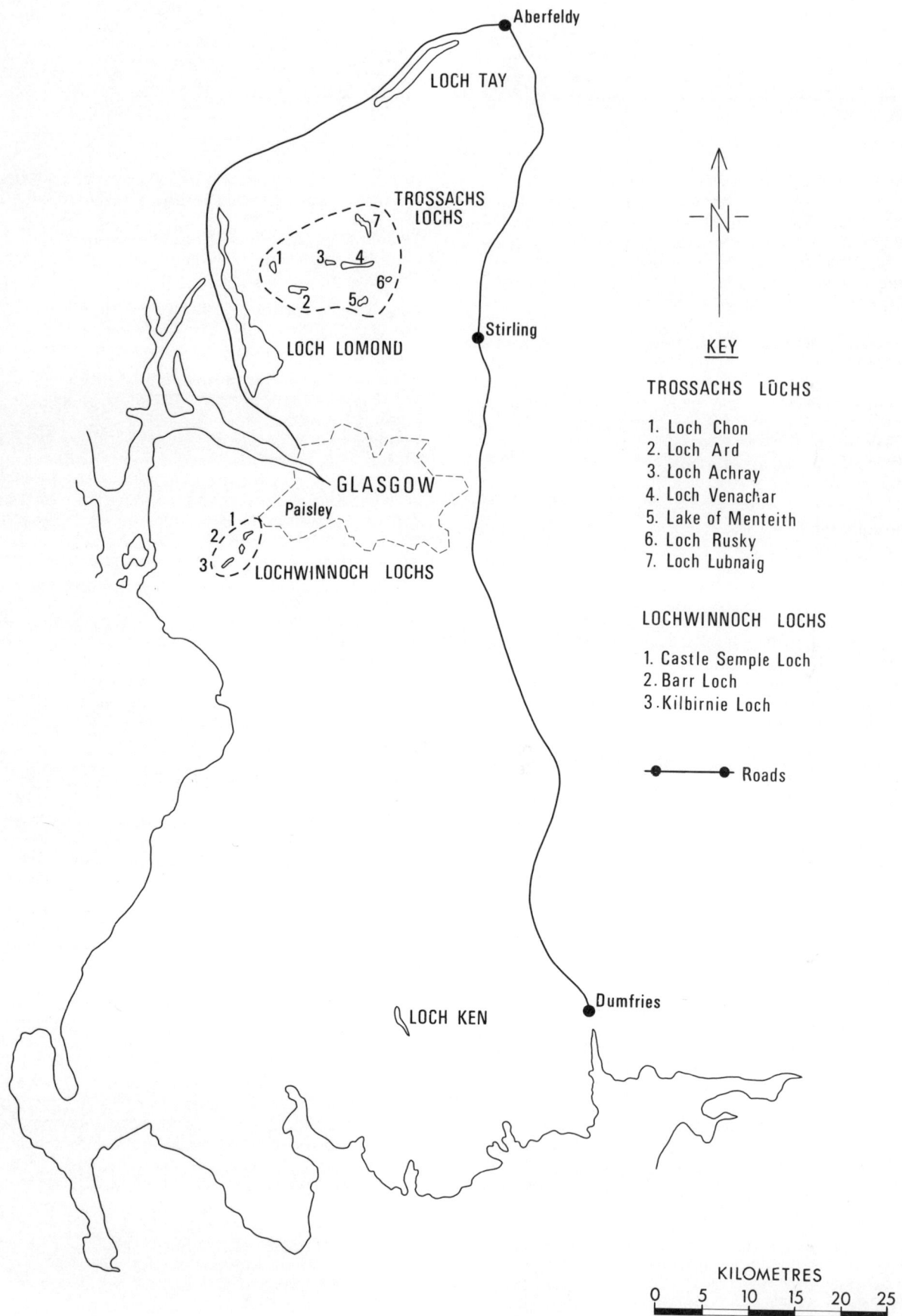

KEY

TROSSACHS LŪCHS

1. Loch Chon
2. Loch Ard
3. Loch Achray
4. Loch Venachar
5. Lake of Menteith
6. Loch Rusky
7. Loch Lubnaig

LOCHWINNOCH LOCHS

1. Castle Semple Loch
2. Barr Loch
3. Kilbirnie Loch

•————• Roads

Fig. 1   Special study area and lochs selected for detailed on-site surveys

1.1 Aims and Objectives

The aim of this study undertaken for the Countryside Commission for Scotland is to assess the nature, extent and the effects of outdoor recreational activities on the perimeter of natural and impounded freshwater lochs and reservoirs in Scotland. The Lochside is defined here as the zone from the water's edge to 100m inland, or to the roadside, where this was less than 100m from the summer water's edge. The object of the Report submitted is to provide data about:

(a) the characteristics of the lochside resource;

(b) the ways in which the lochside is used for land- and water-based recreational activities;

(c) the extent and physical impact of these activities;

(d) the vulnerability of the lochside to such impacts, and

(e) existing levels of lochside development and/or management techniques that are directly or indirectly relevant to recreational use and impact.

The Report also aims to provide information which will assist those concerned to develop policies for the planning and management of the recreational use of lochsides in Scotland. However, while it considers ways in which the resources involved may determine the range of possible management options, the Report does not attempt to provide or to recommend definitive management techniques for particular lochs or lochside sites.

1.2 Review of Related Work

The commissioning of this Report constitutes, in itself, a recognition by the Countryside Commission for Scotland of the lack of a coherent body of information about this topic. The first major survey of the recreational use of a large loch and its surroundings was that of Loch Lomond, undertaken by the Countryside Commission for Scotland in 1973. *The Loch Lomond Recreation Report* (Brown, R., and Chapman, V., 1975) illustrated, very clearly, the high level of recreational use of Loch Lomondside, and it drew attention to the nature and extent of the damage to the physical and biological resources around the loch resulting from this use. The study presented here is an extension and development of an unpublished, and essentially exploratory, analysis of environmental damage around Loch Lomond undertaken by the author in 1973, as a contribution to the Commission's recreation survey.

The way in which this more extensive study was undertaken was inevitably influenced by the paucity of relevant data about inland water bodies in general, and about Scottish lochs in particular. Lochs have not been the focus of integrated studies leading to a body of widely applicable theories. The geologist and physical geographer have been concerned largely with their origin, the ecologist with the components, the functioning of and, increasingly, the biological productivity of the freshwater ecosystem.

Despite the significance of the loch as a characteristic element of the landscape, studies of Scottish lochs are relatively few. One of the most ambitious and still an invaluable, if incomplete, source of primary physical data is *The Bathymetrical Survey of Scottish Lochs* undertaken by Murray and Pullar in the first decade of this century. More recent and detailed work has tended to concentrate on lochs where biological research facilities are available, as at Loch Lomond, or which have been selected for special study as in the case of the International Biological Productivity Programme on Loch Leven (Fife Region). There is, however, a large volume of more general literature concerned with such topics as descriptive topography, game-fishing and wildfowling, in which the loch often figures prominently. Illustrative of these are Tom Weir's books on Scottish lochs and McLaren and Currie (1972) on fishing. Readily available, up-to-date data about the recreational use of freshwater lochs and lochsides is even more meagre. The only recent, detailed publication is the Loch Lomond Report referred to above. *The Amenity Use of Reservoirs in Scotland* compiled in 1971 by the now defunct British Waterworks Association is already out-dated.

1.3 Method of Study

To satisfy the requirements of the remit, it was necessary to undertake basic studies of the physical and biological characteristics of the lochside, of its use as a recreational resource, of its relative vulnerability to the impact of a wide range of types and intensities of recreational activity, and of its suitability for various types and combinations of recreational activities. These aims have placed the emphasis and direction of this Report firmly on the nature of the resource base. Characteristics of the users have been considered only in so far as the type and intensity of recreational activity is concerned.

The methodology of the study was also influenced by the number and distribution of freshwater bodies in Scotland. Some 4 per cent of the area is water space. Given the large number of lochs, selection for both general and detailed study was necessary. This was undertaken initially on the basis of two criteria:

(a) Size: it was decided that only those lochs and reservoirs identifiable on the Ordnance Survey 1:250,000 map series would be considered; that is lochs over 5 ha in surface area. It is estimated that there are about 3,000 of these in Scotland.

(b) Proximity: on the assumption that recreational impact is unlikely to be significant unless a loch or reservoir is easy to reach, it was decided to select for study only those with at least one point of their shore within 100 m of a motorable road. This gave a final population of 760 water bodies, or 25.3 per cent of all those with a surface of 5 ha or more.

1.4 Lochside Data Set

Having decided the basis for the selection of lochs and reservoirs, the study was carried out at two scales.

4

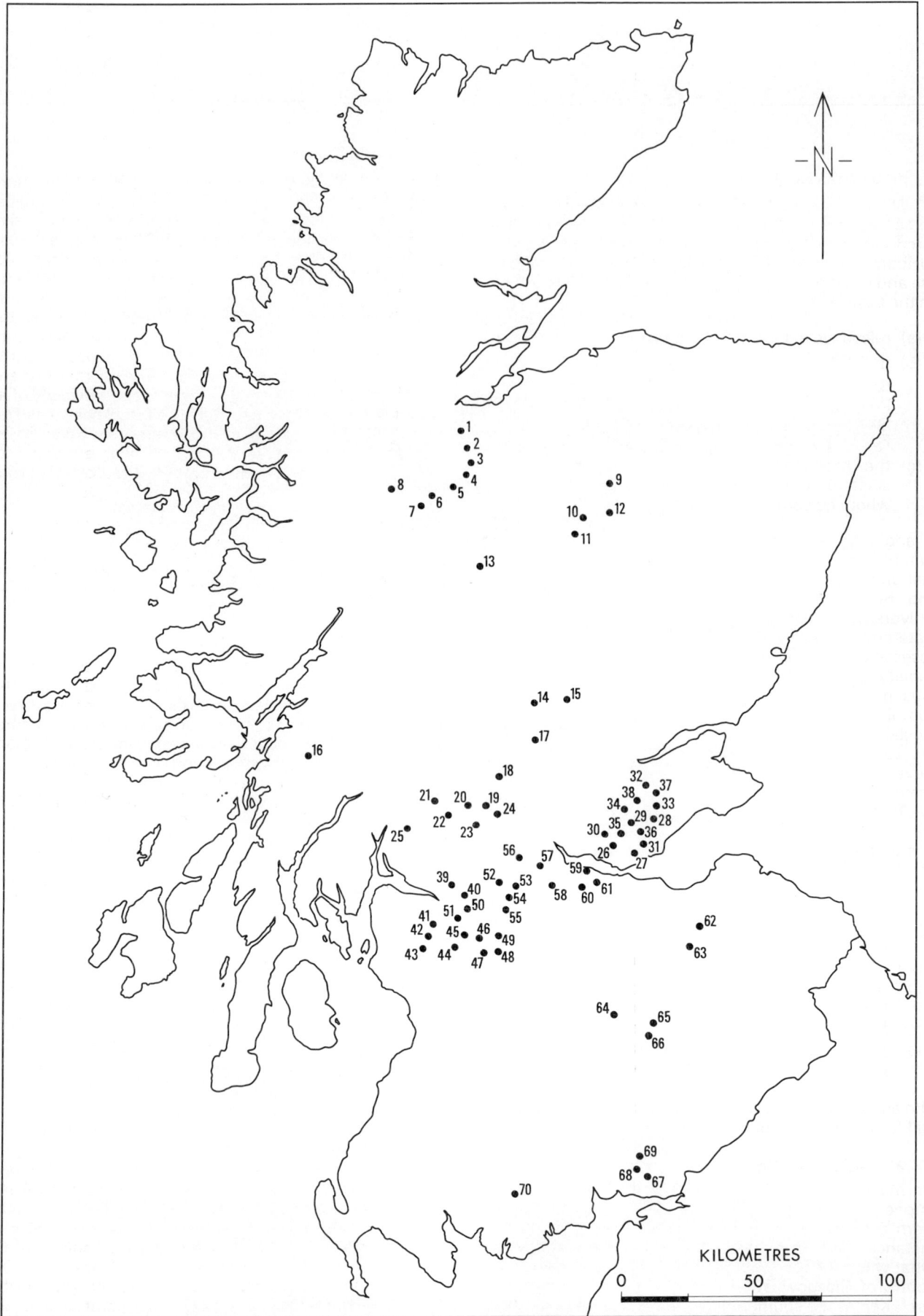

Fig. 2   Location of all lochs and reservoirs from which data were collected

1.4.1   National scale

For this purpose, a data set was collected for the 760 water bodies selected on the basis of size and proximity. The categories of data (covering such aspects as physical attributes, riparian tenure and water rights, recreational use and facilities for each loch and reservoir) were chosen on the basis of:

(a) relevance to the aims of the study;

(b) the type of data available in published form (particularly on Ordnance Survey Maps) or in unpublished but accurate records kept by public bodies such as Water Departments, Hydro-Electric Board, Forestry Commission, Nature Conservancy Council;

(c) the probability of collating data from widely scattered private organisations and individuals for whom no comprehensive register exists.

In addition, the Commission required that the data be suitable for entry to the Tourism and Recreation Information Package (TRIP) data-bank located at the Tourism and Recreation Research Unit (TRRU) of Edinburgh University. A list of the variables, together with a statement about the range of sources used and an assessment of the overall comprehensiveness and reliability of the lochside data are given in Appendix II. It must however be borne in mind that most of this data was compiled in the period July 1974 - December 1975, as the availability of the data set was a necessary preliminary to the selection of lochs for more detailed on-site study. Nevertheless, the data set has been subject to continuous amendment, and addition, as new material became available.

The data set serves several important purposes, not least of which are:

(a) to provide a general statement about the resource-base and recreational use of the more easily accessible lochs;

(b) to contribute to an assessment of the extent to which recreational use and impact are or might become, in the near future, serious problems on Scottish lochsides;

(c) to provide a general framework to which detailed studies of individual lochs could be related.

The analysis of the national scene on the basis of this data will form the subject of the next chapter.

1.4.2   Regional and local scales

On the assumption that intensity of recreational use and impact can generally be related to time and/or distance from a major urban centre, it was decided, in the first instance, to select lochs for detailed on-site studies within what was estimated as a possible day's car-trip from the centre of Glasgow. With Loch Tay as the northern and Loch Ken as the southern limit, the area west of the most direct road-link between Aberfeldy and Dumfries was designated the special study area (see Fig. 1).

Data for the 200 proximal lochs in this area is more comprehensive and reliable than elsewhere in Scotland, the desk compilation being supplemented by an extensive field survey, because it had originally been intended to use this more complete data as the basis for classification of lochs and for the selection of a representative sample for on-site study within the special study area. This idea had to be rejected. Firstly, it proved unrealistic because the data, though comprehensive and reliable, was not complete for all lochs; important variables such as loch-depth and range of water-level had never been recorded in many instances. Also, it was realised that the significance of all the variables recorded had still to be established. Lastly, it was necessary to curtail field transport costs. Hence it was decided to identify a small group of adjacent lochs for detailed study, which would provide a reasonable range of:

(a) size and physical setting;

(b) types of recreational activity;

(c) levels of recreational use.

1.5   Stage One – Special Study Area

On the above basis, the Trossachs Lochs provided the most suitable group within the special study area. This group did not, however, include any very large lochs; a representative number of lowland lochs; nor any Southern Upland lochs. These deficiencies were made good by the inclusion of the following lochs (see Fig. 1):

(a) The Lochwinnoch group, comprising Castle Semple, Barr and Kilbirnie Lochs, south-west of Paisley. These are small lowland lochs with a wide range of water- and shore-based recreational activities. Barr and Castle Semple Lochs were used for the preliminary pilot surveys during which field methods were established.

(b) Loch Ken, to the west of Dumfries, is one of the impounded lochs of the Galloway Hydro-Electric Power Scheme (now within the South of Scotland Electricity Board). It has experienced a recent rapid development particularly of water-based recreational activities following the improvements to the A75 Dumfries to Stranraer trunk road, and an increasing demand for coarse fishing and water-skiing from north-west England.

(c) Loch Lomond, the largest body of fresh water in Britain, is a composite loch, straddling the geological divide between Highland and Lowland Scotland. As such it should be considered in a class of its own as it exhibits a large range of the lochside resource and recreational characteristics found in Scotland as a whole. The relevant findings of the Countryside Commission for Scotland's *Loch Lomond Recreation Report*, together with further site analyses, have been incorporated into this study. Loch Lomond is particularly important in this respect since it is probably the most intensively

6

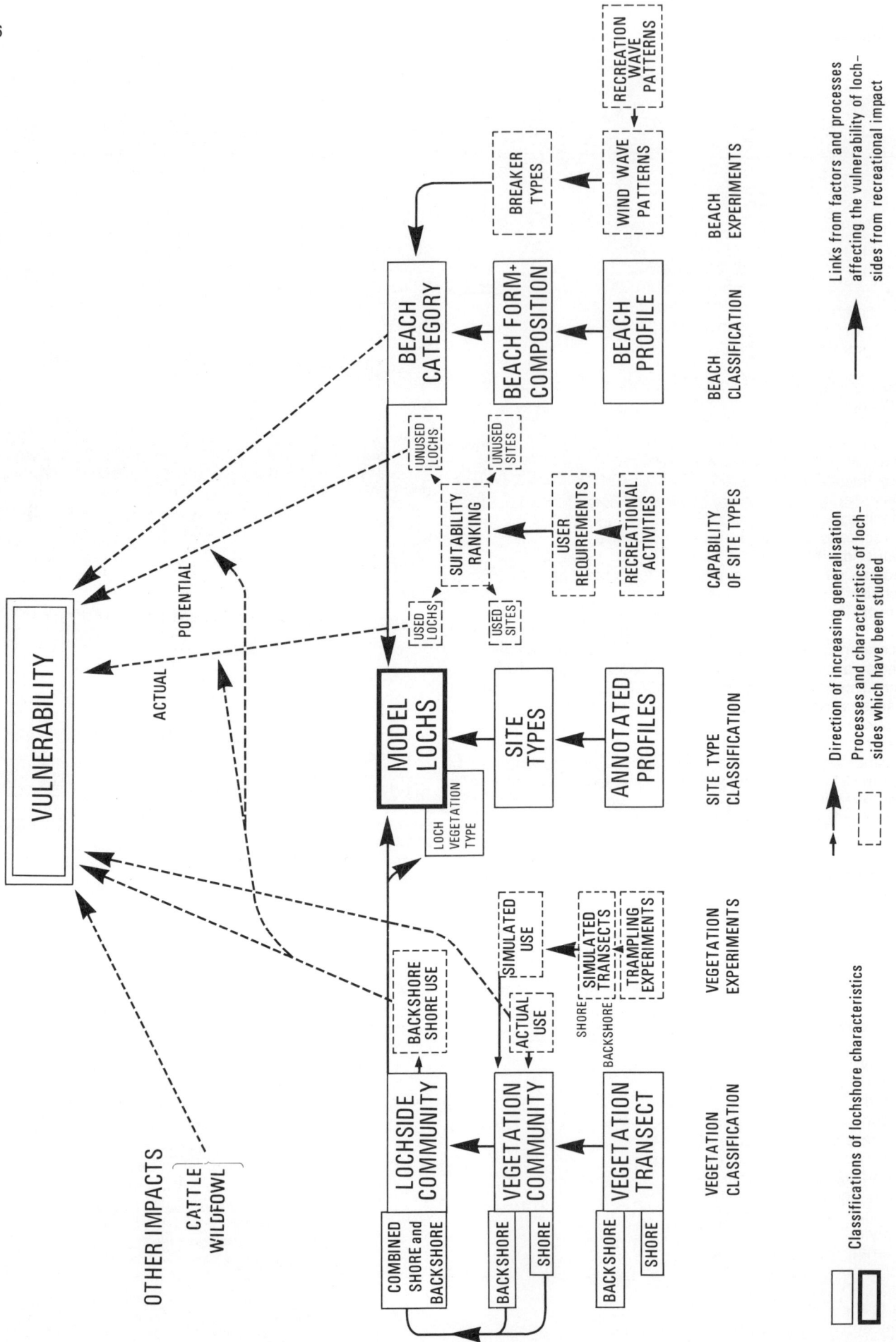

Fig. 3   Lochside recreational impact study: work flow diagram

used loch in Scotland, with a level and extent of lochside impact unequalled elsewhere in the country. As such, it provides an important base-line against which other lochsides can be compared. The main resource and recreation-use characteristics of the thirteen lochs on which initial detailed on-site studies were undertaken are summarised in Table 1.

The detailed field studies of these lochs included the collection of relevant quantitative and qualitative data about the physical form and composition of the lochside; the shore and backshore vegetation; and the shore and beach forms and composition in relation to the range of loch level and the height and direction of loch waves. These studies provided the empirical data for the classification of lochsides on the basis of features relevant to recreational use; for the formulation of hypotheses about possible physical and biological indicators of types and intensities of recreational use; and for the ranking of lochside components, lochside sites and entire lochsides in terms of their vulnerability to recreational impact.

A concurrent survey was carried out of the distribution of formal and informal recreational activities around the lochs, and their associated facilities. Finally, in order to assess varying levels and intensities of recreational impact, two user-surveys were undertaken on all the major informal (and some formal) sites around the Trossachs Lochs in July/August 1975 and around Loch Ken in the same months in 1976. In both seasons, survey employed a simple questionnaire to establish numbers of visitors, their mode of travel, the duration of their stay and the group-activities pursued on each site; surveys were undertaken simultaneously on all sites between 11.00 and 17.00 hours on four peak days (three Sundays and a holiday Monday) in July and August. The questionnaire used is included in Appendix IV. The results of these user surveys will be considered in Chapter 5.

1.6 Stage Two – Testing Hypotheses Outside the Special Study Area

The second stage of the work at the regional and local scale was devoted to testing the results of the initial surveys on other lochs within and outwith the boundaries of the special study area. The object of this second stage was threefold: firstly, to assess the extent to which the methods and hypotheses initially formulated in the special study area were applicable to a wider segment of the whole population of Scottish lochs, as originally identified; secondly, to ascertain whether there were either lochside resources or recreational features of significance which had not been covered in the special study area; and lastly to give a more representative cover of small (less than 55 ha) water bodies, and of Highland lochs, than was possible in the special study area.

As in the first-stage survey in the special study area, the selection of lochs for the above purposes was aimed to be as representative of the whole population as possible; to illustrate the full range of resource and use characteristics,

as well as types of area that might be found within Scotland; and to minimise field travel-costs. This economic constraint, together with the variability in cover for each variable in the data set, did not permit the use of statistical classification of lochs or any form of random or stratified sampling. The selection was therefore a pragmatic one based on accumulated familiarity with, and knowledge about, the range of loch types, gained during the course of compilation of the lochside data set. The lochs chosen (see Fig. 2 and Table 2) fall into three groups:

(a) Southern Scotland – Lochmaben Lochs, Talla Reservoir and St Mary's Loch (with the nearby Loch of the Lowes)

(b) Central Scotland small lochs
   (i) south-west of the Clydeside conurbation
   (ii) east of the Clydeside conurbation
   (iii) Fife and Kinross area

(c) Highlands (excluding north-west Highlands and Islands)
   (i) Loch Ness and Whitebridge lochs plus Loch Cluanie
   (ii) Spey Valley lochs and Loch Laggan
   (iii) Lochs Tummel, Rannoch, Earn and Awe

In total, fifty-one lochs and nineteen reservoirs – nearly 10 per cent of the original population – were inspected, and a variable range of recorded and experimental data was collected about the whole, or particular components or site types, of these lochsides.

1.7 Method of Work

The work flow diagram (see Fig. 3) illustrates the detailed stages followed in the organisation of the study of recreational impact, from the collection of empirical and experimental data, and its analysis, through to the application of standardised analyses to the assessment of lochside vulnerability. As the diagram shows, for the purpose of study the lochside was sub-divided into its major components: vegetation (cultivated or otherwise); site-types based on a combination of vegetation form and slope facets; and beach form and composition. The raw data collected has been assembled and classified, at three levels of increasing generalisation and scale from component type through lochside-type to loch-type dependent on characteristic combinations of types of components (vegetation, site-types and beach types). At each level of generalisation, the condition of the recreationally used component or combination of components has been compared with that of unused components. In the case of vegetation, simple trampling experiments have allowed simulated vegetation stands or communities, produced by known levels of impact, to be compared with those produced by actual recreational use. In the case of similar types of beaches, the form and arrangement of the material on used and unused sites has also been compared. Finally, the suitability of both site-types and loch-types to support

Table 1

Characteristics of lochs in special study area (see Fig. 1)

	Size (ha)	Depth mean (m)	max	Mean range water level (m)	Special recreational features	Other uses
Loch Lomond (1)	7,123	37.0	190.0	1.22	Camp and caravan sites: boating/ sailing/loch cruises	Water supply
Trossachs Lochs						
Chon	106	9.1	22.9	c0.91	Forestry Commission picnic sites	—
Ard	235	13.4	32.6	c0.91	Sailing club	—
Achray	79	11.3	29.7	c1.2	Water-ski club Forest scenic drive	—
Venachar (I)	394	12.5	33.8	3.4	Sailing club	River Teith compensation
Lubnaig	216	6.4	44.5	1.2	Forest cabins	—
*Drunkie (I)	48	11.8	29.6	5.9	Forest scenic drive	Compensation reservoir; draw-down from Loch Venachar
*Rusky	20	7.0	14.0	Not known	Private fishing club	Water supply
Menteith	264	6.0	23.5	c0.61	Inchmahome Island with Priory: Menteith Fisheries Ltd	—
Lochwinnoch Lochs						
*Castle Semple	78	1.3	1.5	2.1	Water-Park Marina and Sailing club, Rowing club, Canoeing	Part of Muirsheil Regional Park
*Barr	114	1.8	3.1	1.6-2.2	Bird-watching RSPB reserve	—
*Kilbirnie	79	3.0	9.1	Not known	Water-skiing	—
Loch Ken (I)	464	6.4	15.2	1.8	Water-skiing: Caravan site	Hydro-electric power

(I) = impounded reservoir.
* Lochs not included in 1975 or 1976 user surveys.

Table 2

Lochs and reservoirs from which data was collected in the field (see Fig. 2)

Loch (L) or Reservoir (R)	Grid ref.	Area (ha)	Loch (L) or Reservoir (R)	Grid ref.	Area (ha)
Highland Lochs			35. L. Fitty	NT 122914	58
1. L. Ness	NH 500230	5641	36. L. Gelly	NT 201935	60
2. L. Ashie	NH 630350	134	37. Balla R.	NO 225050	78
3. L. Duntelchaig	NH 620310	281	38. L. Leven	NO 146015	1376
4. L. Ruthven	NH 620276	149	39. Possil L.	NS 585701	5
5. L. Mhor	NH 542201	44	40. Hogganfield L.	NS 642674	17
6. L. Knockie	NH 455135	92	41. Castle Semple L.	NS 365590	78
7. L. Tarff	NH 425100	69	42. Barr L.	NS 353575	114
8. L. Cluanie	NH 150095	214	43. Kilbirnie L.	NS 330545	79
9. L. Garten	NH 973180	80	44. L. Libo	NS 434557	10
10. L. Alvie	NH 865095	57	45. Harelaw Dam	NS 475537	33
11. L. Insh	NH 830045	170	46. Black L.	NS 497513	59
12. L. Morlich	NH 965095	122	47. Bennan R.	NS 523504	69
13. L. Laggan	NN 490873	699	48. Lochgoin R.	NS 537477	52
14. L. Rannoch	NN 600580	1902	49. High Dam	NS 559509	6
15. L. Tummel	NN 815595	621	50. Ryat Linn R.	NS 521573	7
16. L. Awe	NN 000150	3884	51. Glanderston Dam	NS 499562	8
17. L. Earn	NN 640236	913	52. Johnston L.	NS 697688	11
18. L. Lubnaig	NN 575137	216	53. Queens L.	NS 717687	8
			54. Woodend L.	NS 705667	24
Trossachs Lochs			55. Lochend L.	NS 705663	11
19. L. Venachar	NN 575055	394	56. Banton L.	NS 739786	25
20. L. Achray	NN 515064	79	57. Fannyside L.	NS 800735	26
21. L. Chon	NN 420053	106	58. Black L.	NS 860700	44
22. L. Ard	NN 465019	235	59. Linlithgow L.	NT 002776	42
23. Lake of Menteith	NN 578006	264	60. Lochcote R.	NS 977737	18
24. L. Rusky	NN 615035	20	61. Beecraigs R.	NT 010744	8
25. L. Lomond	NS 375975	7125	62. Gladhouse R.	NT 300535	152
			63. Portmore L.	NT 261502	41
Lowland Lochs					
26. Craigluscar R. (1)	NT 063309	8	**Southern Upland Lochs**		
27. Cullaloe R.	NT 88875	9	64. Talla R.	NT 120214	110
28. Arnot R.	NO 206024	10	65. St Mary's L.	NT 250228	244
29. Roscobie R.	NT 094933	11	66. Loch of the Lowes	NT 237197	40
30. Craigluscar R. (2)	NT 069908	12	67. Castle L.	NY 088815	78
31. Stenhouse R.	NT 211877	14	68. Kirk L.	NY 079823	15
32. Harperleas R.	NO 213054	16	69. Mill L.	NY 077833	12
33. Holl R.	NO 225035	18	70. L. Ken	NX 690698	464
34. L. Glow	NT 088958	51			

particular recreational activities, or combinations of such activities, has been assessed on both recreationally used and unused examples.

Other factors which may directly or indirectly affect the actual or potential vulnerability of the lochside to recreational impact had also to be considered. The two most important are loch-wave patterns and breaker types. There is a close relationship between these and the nature and stability of the lochshore. In addition, recreational modification of the lochside may affect its susceptibility to wave and breaker action and set in train a process of greatly accelerated erosion. This led to an attempt to assess the inter-relationship between the loch wave and the vulnerability of the lochside to recreational impact. Also, activities other than those associated with recreation, particularly grazing and trampling by cattle and wildfowl can, locally, cause considerable damage. However, in comparison to wave action, their effect tends to be local and usually highly concentrated. Both of these types of impact can combine to exacerbate those due initially to recreation or they can, conversely, act as triggers to subsequent recreational impact.

The purpose of the analytical framework summarised in Figure 3 is twofold:

(a) to provide a basis for readily applied or interpreted, standardised descriptions and classifications which are essential to any comparative resource evaluation;

(b) to allow the assessment of the suitability and vulnerability, not of particular lochsides, but of types of lochside components, of loch-sites and of entire lochsides to recreational impact. The difference in this respect is between the case study of a particular loch, and the elucidation of general characteristics and concepts about lochsides which are universally applicable and which can be used to assess the actual or potential vulnerability of a particular loch or group of lochs, as the case may be.

In the achievement of these aims, it is hoped that the Report will provide data and methods which will aid those involved in the planning and/or management of lochs and lochsides for recreational uses. The findings of the study will be dealt with under a number of main headings. The general resource and use characteristics of all proximal Scottish lochs are reviewed in the first section using information obtained from the lochside data set. The nature of the lochside as a recreation site is then discussed with particular reference to the seashore. This is followed by a more detailed consideration of the lochside resource characteristics; with particular reference to the main beach and vegetation types. The next chapter deals with lochside recreational activities and their associated facilities. The main theme of these chapters is the way in which the resource and user characteristics affect lochside impacts. Factors affecting such impacts and their type and location is the subject of the next chapter. The vulnerability of the lochside to these impacts is then discussed with particular reference to the lochside resource. Finally, a review is given of matters relating to the management of lochsides in Scotland.

The object of this chapter is to analyse, on the basis of the data collected, the resource characteristics of the proximal or accessible Scottish loch (as defined in 2.1.5) and the extent to which the lochside is presently used for recreational purposes. This will demonstrate how far the lochs which have been studied in detail are representative of the known national range, and set the problem of recreational impact on the lochside in a national context.

2.1 Nature of the Resource

In both absolute and relative terms, Scotland is richly endowed with water space. It has been estimated that of the 5,505 inland water bodies over 4 ha in Britain (excluding rivers and canals), approximately three-quarters (3,791) are in Scotland; they account for approximately 1.0 per cent of the total surface area in Scotland, and for 0.05 per cent in England and Wales. The recreational value of these water bodies and their surrounding areas is largely dependent on distance either from the major centres of population and/or from motorable roads. Of some 3,000 inland water bodies in mainland Scotland (including the Inner Hebrides but excluding the Outer Hebrides, Orkney and Shetland), it has already been noted that only 760 are proximal, that being within 100 m of a motorable road. At present then, only about a third of the actual resource is readily available for outdoor recreational activities.

2.1.1 Size

In the established image of the Scottish landscape and the Highland Loch, the number and importance of small water bodies tend to be overlooked. Seventy-five per cent of all proximal lochs are less than 55 ha in surface area (see Fig. 4 and Table 3) and the median size class is 10-14.99 ha.

As might be expected, the frequency distribution of loch length (median 1,000 m) and shore length (median 3,000 m) shows a high correlation (0.800) with loch area. There are less than twenty lochs over 1,000 ha, and these include many whose renown and impact on the landscape is a function of their size (see Table 4).

2.1.2 Nature and distribution

There is a marked but by no means perfect relationship between loch size and form on the one hand and distribution on the other. Nearly half of the 569 small lochs (less than 55 ha) are in central Scotland (see Figs. 4 and 5) in close proximity to the bulk of the country's population. Lochs over 55 ha in this area are relatively few and sufficiently well known to merit listing (see Table 5). In the tabulation, these lochs are listed in relation to the 'comfort line' for living of 183 m or 600 ft, above which altitude, the summer season is shorter and cooler; there is increased wind force, lack of shelter, and a higher prevalence of low cloud and hill-fog.

In terms of altitude and regional relief then, many of the small lochs of Central Scotland are upland in character. Deterioration of climate with altitude and the marginality of climate for many types of development are not only extremely relevant to the recreational potential of water

Table 3
Size distribution of proximal lochs by quartiles

First quartile	0 – 9.99 ha
Second quartile	10 – 14.99 ha (median)
Third quartile	15 – 54.99 ha
Fourth quartile	over 55 ha

Table 4
Dimensions of proximal lochs over 1,000 ha in area

Loch	Area (ha)	Loch length (km)
L. Lomond	7123	36
L. Ness	5641	40
L. Awe	3884	36
L. Tay	2648	23
L. Shin	2554	27
L. Shiel	1959	27.5
L. Rannoch	1902	15
L. Ericht	1867	17
L. Arkaig	1554	19
L. Lochy	1551	18
L. Morar	1426	15
L. Leven	1376	6
L. Katrine	1282	14
L. Loyne	1036	12

Table 5
Proximal lochs and reservoirs in Central Scotland over 55 ha in size above or below the 600' 'comfort line'

Loch (L) or Reservoir (R) Above the 'comfort line' (area in ha)		Loch (L) or Reservoir (R) Below the 'comfort line' (area in ha)	
Balla R.	78	L. Rescobie	60
Gladhouse R.	152	Castle Semple L.	78
Thriepmuir R.	87	Barr L.	114
Harperrig R.	91	Kilbirnie L.	79
Loch Coulter R.	64	Beecraigs R.	96
Carron Valley R.	373	Lake of Menteith	264
L. Thom	132	Loch of Clunie	54
Gryfe R.	63	Loch of the Lowes	88
		L. Leven	1376
		L. Gelly	60
		L. Fitty	58
		Gartmorn Dam	57

12

bodies in Central Scotland, but assume greater significance here than in other parts of the country. About 40 per cent of all proximal lochs in Central Scotland are situated above the 'comfort line'. The majority are upland plateau or deep valley lochs, surrounded by uncultivated moorland, characteristic of the hill ranges of the Sidlaws, Campsies, Ochils, Pentlands and the Renfrewshire Heights. A high proportion of these water bodies are impounded public water-supply reservoirs. Below the 'comfort line' the small loch is lowland in setting and in character. It is frequently round or eliptical in shape, usually comparatively shallow and extremely varied in age, origin and both past and present use. Surrounding land use ranges from urban and/or industrial through sub-urban to open cultivated land; many are set in areas of derelict or near-derelict land.

The remaining small lochs are more widely scattered throughout northern and southern Scotland (see Fig. 6), with notable concentrations in the north-eastern lowlands, the extreme north-west, and the south-west. Although their distribution has been shown (for comparability) in relation to the 'comfort line', this does not have the same significance in relation to regional and local

relief as in Central Scotland. In both the north and south of Scotland the lowland coastal area, particularly in the north-west, can be more exposed than the higher inland, and hence sheltered, valleys where higher maximum daytime temperatures in summer can compensate for the shorter summer season. Also, those lochs situated in the wide eastern valleys or straths such as that of the River Spey are, at the local scale, more lowland in size, depth and surrounding relief than those of the north-west.

In the latter area deep, steep-sided, glacially scoured rock-basins set in bare moorland, much less than 180 m in altitude, are more like upland or some highland, than lowland lochs. Two areas stand out by reason of their paucity of lochs – the Eastern Grampians and the Borders.

2.1.3 Lochshore length and shape
As important for the recreational value of the lochside as the loch area, is the length and form of the shoreline together with the width, gradient and composition of the lochshore. As would be expected, there is a close correlation between length of shore line and surface area of a loch. Shore length determines the potential lochside space available; its value, however, is enhanced by its

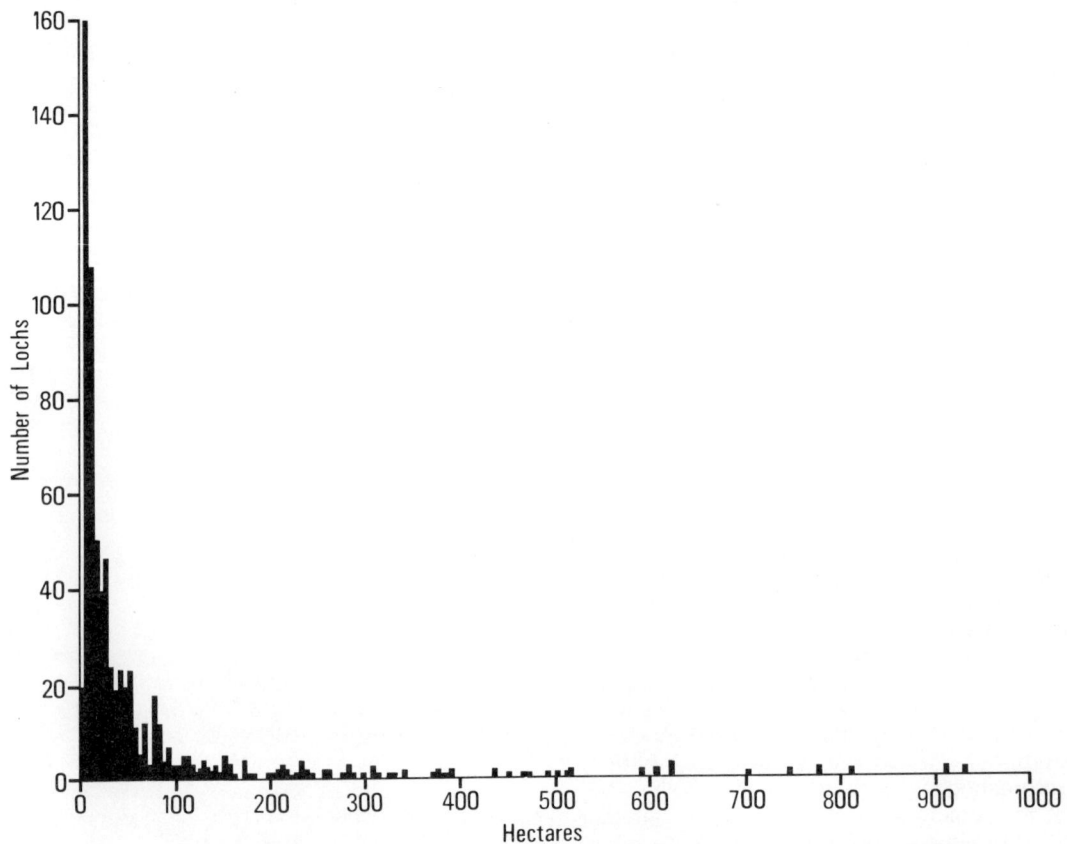

Fig. 4   Size distribution of proximal freshwater lochs and reservoirs less than 1,000 ha

degree of irregularity. An irregular or highly indented shoreline is likely to give greater possibilities for shelter and enjoyment of sunshine, as well as for a diversity of scenery, and sites for recreational activities, than the smoother and more regular shoreline.

The problem of describing and measuring the degree of shoreline irregularity is extremely difficult, since it is not just a function of increasing size of loch but also of variations in basic loch shape. Here the relationship between expected and actual shoreline has been used to compute an irregularity index (see Glossary) which allows the proximal lochs to be ranked according to degree or extent of shoreline irregularity; an index of over 1.0 indicates an increasingly and relatively smoother shoreline, or less than 1.0 an increasingly and relatively irregular shoreline. The graphical expression of this spectrum is given in Figure 7 and this has been used to define the following class boundaries (see Table 6).

Table 6

Frequency distribution of lochs according to degree of irregularity of shoreline (see Fig. 7)

Index of irregularity	Number of lochs (total 732)	Percentage	Degree of irregularity
0.130-0.999	82	11.2	High
1.000-1.049	25	3.4	Moderately high
1.050-1.199	35	4.8	Moderately low
1.200-1.299	61	8.3	
1.300-1.399	72	9.8	
1.400-1.499	89	12.1	
1.500-1.599	100	13.6	
1.600-1.699	83	11.3	
1.700-1.799	65	8.8	Low
1.800-1.899	42	5.7	
1.900-1.999	25	3.4	
2.000-2.999	45	6.1	Very low
3.800	8	1.0	

While an index of irregularity provides a basis for a comparative description of shorelines, it should be borne in mind that it is based on the difference between the actual and expected irregularity of lochs of given length, breadth and circumference, and while it reflects amount of irregularity it does not take the amplitude of the irregularities into account. Its significance in relation to the existing recreational use of lochsides will be examined later. It is interesting to note that a relatively small proportion of the proximal lochs have very irregular or very smooth shorelines; moderate irregularity would appear to be the norm. This distribution needs to be

treated with some caution, however, given the practical problems of measuring the shorelength of small lochs from Ordnance Survey maps.

2.1.4 Water-level fluctuation

The form of a lochshore is dependent on the fluctuation of the loch-level in conjunction with the loch floor gradient. Data on water levels are unfortunately relatively sparse. Reliable records, other than those held by the Hydro-Electric Boards and the Regional Water Departments, are few, and for only 151 (18.9 per cent) of the proximal lochs are there records available for the mean range of water-levels. In fact, this total is less than the number of lochs and reservoirs recorded in the data set as impounded – 253 (nearly a third of the proximal lochs). The accuracy of this figure and the value of the categorisation into impounded and natural lochs must be treated with caution. It is not possible to identify with certainty from existing maps whether a loch is impounded or not. Also, many impounded lochs and reservoirs no longer serve the purpose for which they were designed; hence their levels are no longer subjected to the same controls as in the past and records are therefore not available. The combined total of proximal lochs and reservoirs used for hydro-electric generation and public water supply is 198 (26 per cent). Some of the lochs so used are not impounded and records of water-levels are therefore not available. The frequency distribution of the mean range of water-levels for lochs for which there are reliable records is given in Table 7.

Table 7

Frequency distribution of ranges of water-level on 151 reservoirs for which records are available

Mean range water-level	Number of lochs	
Over 40 m	1	(Loch Loyne 40.30 m)
21 – 40 m	1	(Loch Cluanie 29.00 m)
11 – 20 m	10	
6 – 10 m	25	
1 – 5 m	78	
Less than 1	36	
Total	151	

While these records are all from controlled lochs and reservoirs, they may well encompass the extreme range for all Scottish lochs. The range of water-level fluctuation in these cases can be regulated, within given supply limits. In some, particularly the very large hydro-electric power and water-storage reservoirs (see Table 8), the range of water-level is much greater than would occur under natural conditions as at Loch Loyne and Loch Cluanie; in others the installation of dams and other forms of barrage have served to reduce formerly much higher seasonal ranges of water level, as on Loch Lomond. Indeed, in some areas the maximum draw-down or extraction is determined by statutory regulations, which require that there is sufficient compensation water to maintain stream-flow downstream and salmon migration upstream. The draw-down is minimised as far as practicable at some impoundments in the interests of amenity or for the safety of water.

Table 8
Reservoirs with mean water-level ranges of 10-20 m

Reservoir	Size (ha)	Mean range water-level
1. L. Sloy	80	17.10
2. L. Lednock	154	15.24
3. Clatteringshaws Loch	390	12.19
4. Daer R.	205	12.19
5. L. Doon	814	12.19
6. Glenfinglas R.	139	12.15
7. Cruachan R.*	38	12.20
8. Muirhead R.	41	12.04
9. L. Lyon	513	10.70
10. Lochan Shira	142	10.90

R. = public water supply, the remainder are used for hydro-electric power generation; *pump-storage loch.

2.1.5 Proximity to classified (A or B) roads

In addition to the intrinsic physical attributes of the loch and its surroundings, recreational use is influenced by the proximity of the loch to a classified (A or B) road. Of the 759 lochs proximal to motorable roads, 195 (25.69 per cent) are within 100 m or less, 40 per cent within 1 km, and 82 per cent within 6 km of an A or B road (see Table 9).

More important from the point of view of recreational impact are the relative and absolute amounts of shore proximal to these types of roads. The frequency distribution of lochs whose shorelines are within given distances of a classified road is given in Table 10.

Absolute length gives a measure of recreational potential; the percentage shorelength indicates the potential impact on a given loch. The first twenty-five lochs in each of the two categories are identified in Table 11, and their distributions are shown in Figures 8 and 9. Nearly half the listed lochs are common to both lists.

While a lochside may be in fairly close proximity to trunk roads and/or to major population centres, the ultimate factor determining its recreational use will be whether it is accessible from the road. On the one hand, this may be difficult or virtually impossible because of physical restrictions to movements – constraints which will be discussed in more detail later in the Report. On the other hand, access to and use of the loch and the lochside may be subject to less tangible legal restrictions associated with land tenure and water rights.

2.2 Legal Constraints to Access

The lochside is subject to two bodies of law; those pertaining to ownership of the land around the loch and the solum under the water, and those pertaining to the use of the water and rights to game-fish (salmon and trout), and wildfowl. Access to the waterside is by the explicit or implicit agreement of the current land-holder or holders, except in cases where there are long-standing, and often historic, rights-of-way. Public right of access to a loch surface is most frequently found at former fording or launching points. In the case of more than one holder, tenure of the sub-loch solum extends to mid-loch except where the rights of a public body using the water include the solum as well. Where a motorable road comes within 100 m of a lochshore, the land area between the road and loch is often relinquished to public use by the owners because it is so limited in space and/or use.

Under long-established riparian law, land-holders are entitled to use the water which impinges on their property provided they do not seriously impair its quantity or quality beyond their legal boundaries. Game-fishing and wildfowling rights have, since mediaeval times, been vested in the hands of riparian landowners, who may, and often do, retain or lease them quite independently of the riparian land-tenure system.

Table 9
Number and percentage of lochs at given distance classes from a classified (A or B) road

Distance from A or B road	Number of lochs	Percentage of lochs
100 m or less	196	25.7
101 m – 999 m	112	14.7
1.0 km – 1.999 km	125	16.4
2.0 km – 2.999 km	75	9.8
3.0 km – 3.999 km	50	6.5
4.0 km – 4.999 km	44	5.9
5.0 km – 5.999 km	20	2.6
6.0 km – 6.999 km	14	1.8
7.0 km – 7.999 km	11	1.4
8.0 km – 8.999 km	13	1.7
9.0 km – 9.999 km	7	0.9
10.0 km – 14.999 km	32	4.2
15.0 km – 19.999 km	14	1.8
20.0 km and over	21	2.9

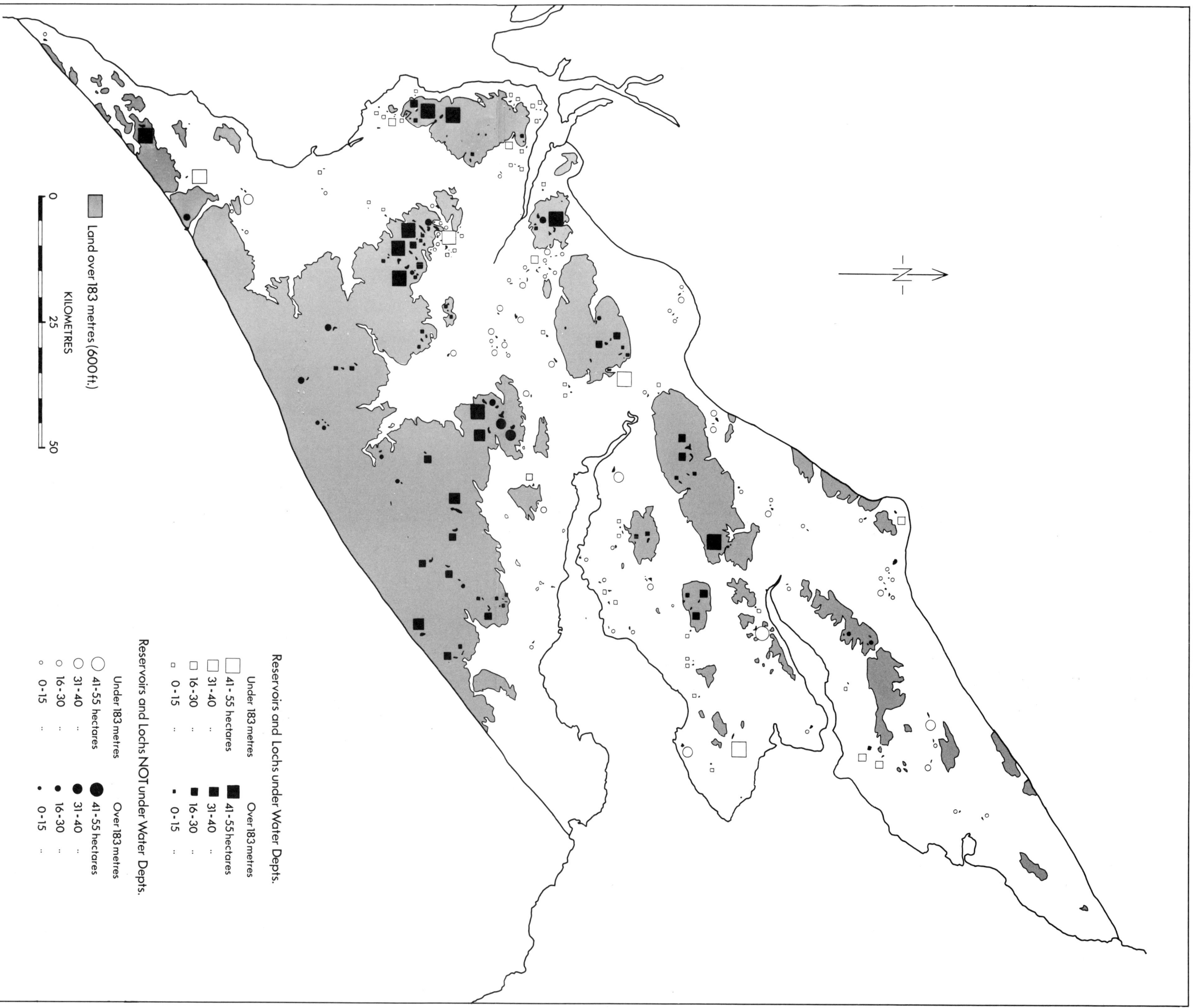

Fig. 5 Distribution of proximal lochs and reservoirs less than 55 ha in Central Scotland in relation to the 183 m contour line

Land over 183 metres (600 ft.)

KILOMETRES
0   25   50

Reservoirs and Lochs under Water Depts.

	Under 183 metres	Over 183 metres
41 - 55 hectares		
31 - 40		
16 - 30		
0 - 15		

Reservoirs and Lochs NOT under Water Depts.

	Under 183 metres	Over 183 metres
41-55 hectares		
31-40		
16-30		
0-15		

Land over 183m (600 ft)

-N-

Refer to map of
Central Scotland.

KILOMETRES

0   50   100

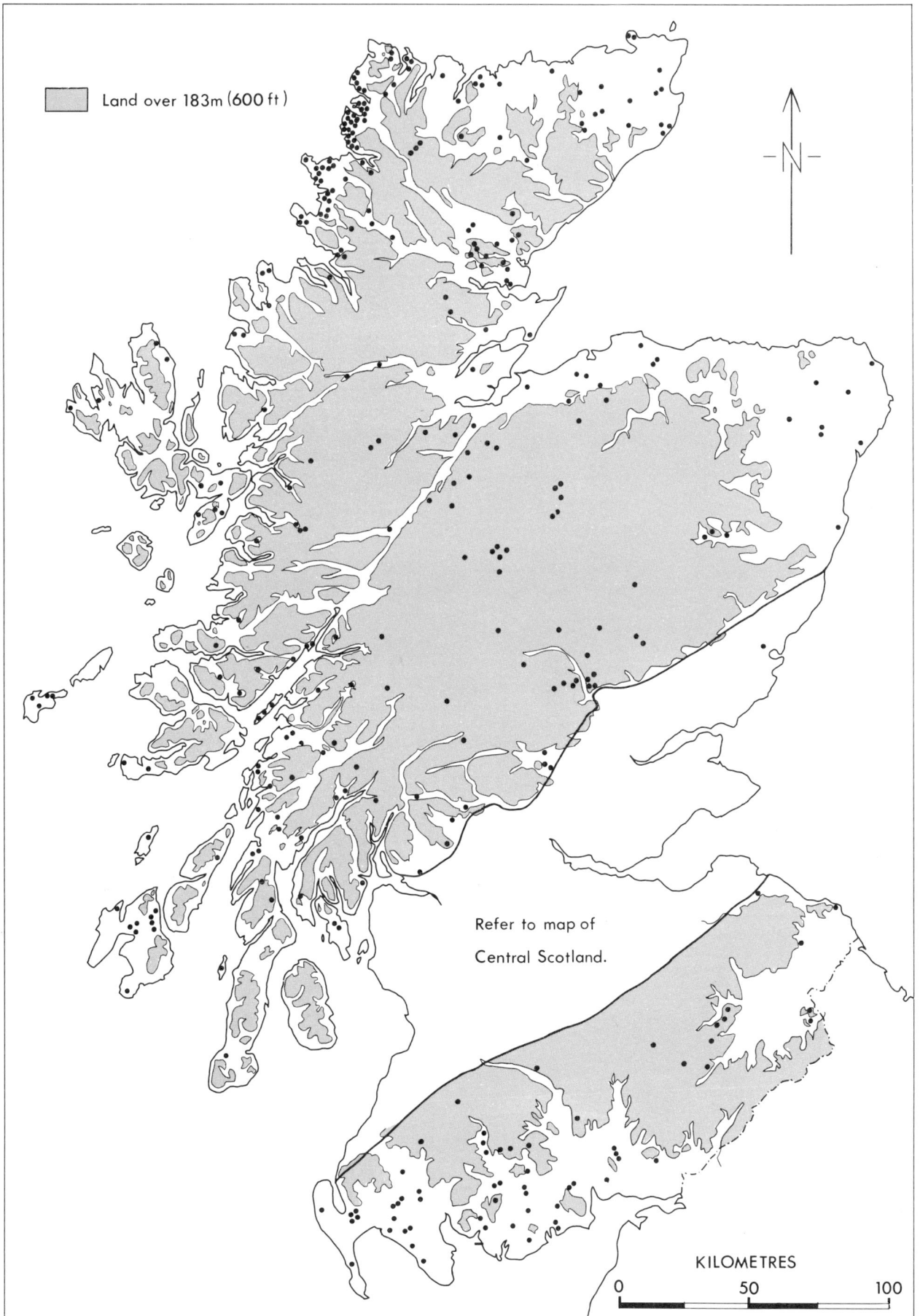

Fig. 6   Distribution of proximal lochs and reservoirs less than 55 ha in the north and south of Scotland

Table 10

Frequency distribution of lochs according to absolute shorelength and percentage shorelength within 100 m of a classified (A or B) road

Absolute shorelength within 100 m of A or B road	Number of lochs		Percentage shorelength within 100 m of A or B road	Number of lochs	
20 km and over	5	First ranking	over 66	8	First ranking
10 – 19 km	10		50-65	18	
5 – 9 km	7		25-49	45	
1 – 4 km	34		10-24	46	
less than 1 km	58		less than 10	24	

2.2.1 Water undertakings

This relatively simple legal system was complicated, particularly in the late eighteenth and nineteenth centuries, by increasing demand for water for urban and industrial use at locations often some distance removed from the source of supply. Existing lochs were impounded and their levels raised; new reservoirs were created. Development, however, was piecemeal and was undertaken particularly for industrial supply by private individuals or corporations, as well as local authorities such as Burgh Councils for public water supplies. In both cases, many of the reservoirs so created were small. Arrangements regarding the tenure of surrounding land and sub-water solum, together with the associated fishing and shooting rights, were negotiated with the landowner, but there was no uniformity of procedure at either the national or local level. Initial concern with water purity, at a time when treatment was minimal, resulted in the acquisition not only of the area occupied by the water but part of the surrounding catchment area, especially for the larger reservoirs designed to supply main cities and towns. This was usually sufficiently large to control run-off and prevent contamination, particularly by domestic animals, from the surrounding agricultural land. The most common method of control and protection of the quality of the water supply at this time was to establish coniferous plantations around part or the whole of the reservoir. In other cases only a narrow peripheral zone of agreed but variable width was either bought or leased and surrounded by a high iron fence, and at many of the nineteenth century reservoirs the water-works included a residential headquarters of substantial proportions.

The subsequent history of the organisation of the water supply industry on a national and local level has been chequered. The industry was finally nationalised under the terms of the 1946 Water (Scotland) Act when responsibility for public water supply was vested with local government authorities until 1967, when Water Boards, independent of local authorities, were set up. Less than ten years later, in 1975, reorganisation of local government resulted in the regionalisation of the water supply industry under the present Regional Council Water Departments (see Fig. 10). These new departments then have

Table 11

First-ranking lochs according to percentage shorelength and absolute shorelength from a proximal (A or B) road; lochs common to both lists are in italics

Over 50 per cent shorelength within 100 m A or B road	Over 5 km shorelength within 100 m A or B road
1. *L. Oich*	1. *L. Ness*
2. *L. Rannoch*	2. *L. Awe*
3. Linlithgow L.	3. L. Lomond
4. L. of Clunie	4. L. Shiel
5. *L. Ness*	5. L. Shin
6. *L. Earn*	6. *L. Arkaig*
7. *L. Eck*	7. L. Lochy
8. L. of Butterstone	8. *L. Rannoch*
9. *L. Awe*	9. L. Ken
10. Glenboig L.	10. L. Katrine
11. Balgavies L.	11. *L. Assynt*
12. Lochan na Glanhaidh	12. *L. Earn*
13. Castle L.	13. L. Luichart
14. Earlstoun L.	14. *L. Naver*
15. *L. Arkaig*	15. *L. Eck*
16. L. Ordain	16. L. Glascarnoch
17. L. Fada West	17. *L. Oich*
18. *L. Assynt*	18. L. Loyal
19. *L. Eilt*	19. L. Lubnaig
20. L. Beannacharan	20. L. Venachar
21. L. Dochfour	21. *L. Eilt*
22. L. Loirston	22. L. Ard
23. *L. Naver*	
24. Clar L.	
25. L. n'Bargh Ghanmhich	

Loch Bà
(Rannoch Moor)
0.4

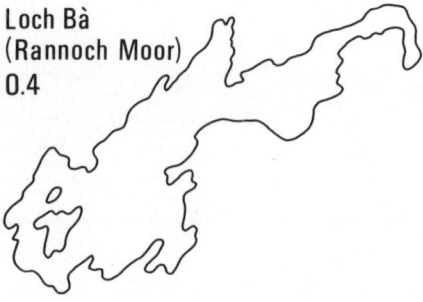

Loch Kernsay
0.5

Clatteringshaws
Loch
0.6

Harelaw Dam
0.7

Loch of Strathbeg
0.8

Loch Hempriggs
1.02

Milton Loch
1.15

Loch na
Beinne Baine
1.25

Loch Ronald
1.36

Garsfad Loch
1.45

Loch Ra
1.57

Loch Mayberry
1.66

Loch a'Bhadaidh
1.76

Loch of Butterstone
1.86

Loch Garten
2.0

Loch Glow
2.1

Loch Pattack
2.5

Loch Glenbuck
3.0

Loch Cluanie
3.8

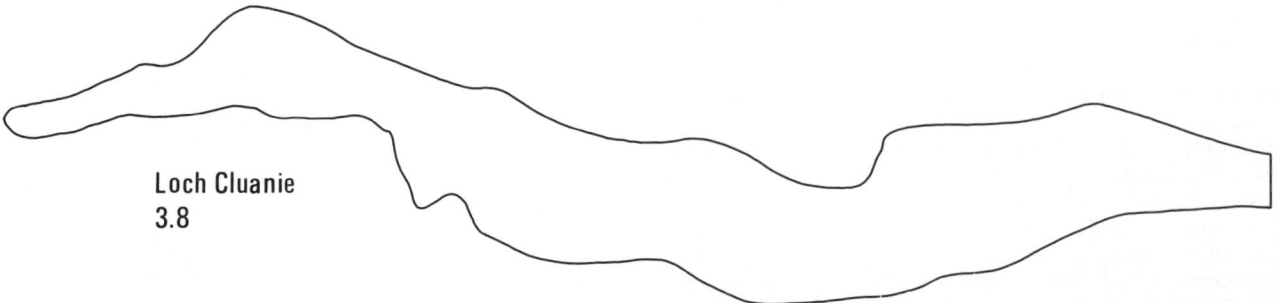

0   1   2 kms

Fig. 7   Range of shoreline irregularity

inherited a system of land-tenure and water-rights of varying degrees of complexity and so far attempts at rationalisation have not substantially altered or simplified matters.

2.2.2 Hydro-electric power

The development of hydro-electric power during the nineteen-thirties in Galloway and in the Highlands necessitated a further modification of existing riparian water-rights in those areas where storage reservoirs either raised existing loch levels, or created new water-bodies, and where water was diverted within a catchment, or from one catchment to another in order to provide as high a head or volume of water as possible at the power station. The South of Scotland Electricity Board and the North of Scotland Hydro-Electricity Board also acquired authority to control water-flow and levels of the storage reservoirs and associated supply lochs, as well as rights to the sub-water solum and to a narrow riparian zone of varying width dependent on the nature of the lochside and its off-shore gradient. This zone provides an essential flood easement which is either bought or leased from the landowner. The fishing rights on the loch either pass to the Boards or are retained by the previous riparian owners. In the case of the hydro-electric power reservoirs, legal problems are often more complex than in water-supply reservoirs because of the greater size of the former and the greater likelihood of multiple public and private riparian ownership.

2.3 Riparian Tenure and Water Rights

Data about riparian ownership, fishing rights and water use and/or control were compiled from a wide range of sources for the proximal lochs. From the fifty-six possible combinations given by a cross tabulation of the following seven types of holder of riparian tenure and of water rights, there are eighteen categories that cover the 759 proximal lochs, as set out in Table 12.

Private individual – single owner (I_s)
Private individual – multiple ownerships (I_m)
Private organisation– single owner (O_s)
Private organisation– multiple ownership (O_m)
Public body – single owner (P_s)
Public body – multiple ownership (P_m)
Complex ownership– all other types (C)

Riparian tenure is dominated by the private sector, which accounts for just over two-thirds of all lochs, compared with only 13 per cent entirely surrounded by public land. In contrast, the proportion of lochs with private water rights is slightly less (58 per cent) with public rights greater (30 per cent). Over half the lochs have both their riparian tenure and water rights in private hands compared with 13 per cent under public organisations. Other than those characterised by single or multiple private riparian tenure and water rights, the most important categories are: (3) in which all are Water Department Reservoirs or North of Scotland Hydro-Electric Board lochs or reservoirs; category (7) is dominated by Water Department, other local

Table 12

Categories of riparian land tenure and water rights for proximal lochs and reservoirs

Land tenure		Water rights or user	Number of lochs	Per cent total recorded
(1)	I_s	I_s	290	38.3
(2)	I_s	O_s	7	0.9
(3)	I_s	P_s	43	5.7
(4)	I_m	C	1	0.1
(5)	I_m	I_s	3	0.4
(6)	I_m	I_m	136	17.9
(7)	I_m	P_s	54	7.0
(8)	I_m	C	5	0.7
(9)	O_s	O_s	13	1.7
(10)	P_s	P_s	99	13.1
(11)	P_m	P_m	2	0.3
(12)	P_m	C	1	0.1
(13)	C	I_s	15	0.2
(14)	C	I_m	12	1.6
(15)	C	O_s	4	0.5
(16)	C	P_s	29	3.8
(17)	C	P_m	1	0.1
(18)	C	C	44	5.8

authority reservoirs and a few lochs used for hydro-electric power generation; (10) the third last category in the public sector is very mixed, the three most important bodies being Water Departments, the Forestry Commission and District Councils. Category (9) contains only thirteen lochs which belong to a wide variety of private organisations including the Royal Society for the Protection of Birds, the Scottish Wildlife Trust, Angling Clubs, the British Aluminium Company, and the Roman Catholic Church. Category (18) is the most complex, in so far as riparian tenure and water-rights are under the jurisdiction of completely different ownerships, one or both of which may involve multiple public and/or private ownership.

The number of proximal reservoirs in each of the nine main categories in each of the Regions is shown in Table 13. The widest range of combinations can be found in Strathclyde, Fife and Lothian Regions which have high levels of urban and industrial development.

2.4 Recreational Policies of Water Departments

The type and amount of recreational use of the loch or reservoir will largely be dependent on the particular policies of the private or public bodies who control the riparian and water-rights. In this respect, functional constraints on the use of Water Department Reservoirs (and lochs) are stricter and more consistently applied than

Table 13

Types of riparian land tenure and water (fishing) and sporting rights for proximal Water Department reservoirs by Regions in Scotland

Region	Total number of reservoirs	T:W P:P	T:W C:C	T:W P:C	T:W C:P	T:W P:P+C	T:W C+P:C	T:W C:C+P	T:W C+P:P	T:W C+P:C+P
Strathclyde	(86)	13	17	7	—	—	16	—	23	10
Strathclyde/ Lothian (shared)	(1)	1	—	—	—	—	—	—	—	—
Central	(12)	—	—	—	—	—	6	—	—	6
CSWDB	(2)	1	—	1	—	—	—	—	—	—
Tayside	(8)	2	5	1	—	—	—	—	—	—
Fife	(22)	8	2	3	1	6	1	1	—	—
Lothian	(22)	2	—	4	1	—	8	—	4	3
Lothian/ Borders (shared)	(5)	2	1	—	—	—	1	—	1	—
Borders	(2)	2	—	—	—	—	—	—	—	—
Dumfries and Galloway	(13)	6	—	4	—	—	3	—	—	—
Highland	(6)	5	—	—	1	—	—	—	—	—

P = private; C = Regional Council; T = land tenure; W = water rights; CSWDB = Central Scotland Water Development Board

in any other of the tenure/water-right categories. Recreational policy must take into consideration the need, firstly, to supply water of a high quality and hence protect it from possible sources of pollution; secondly, to have regard for the safety of the public using these reservoirs under the terms of the Reservoir Safety Provision Act (1930), and lastly, to avoid conflicts which might arise from the use of water and riparian land for various recreational purposes.

2.4.1 Factors determining type and level of recreational use

Since the late nineteen-thirties, a series of Acts have been passed to either strengthen or increase the powers of the Water Departments to protect their interests and to allow them to regulate the way in which their waters should or should not be used for recreation. Three principal factors determine the type and level of recreational use to which a Water Department reservoir may be put:

(a) Type of reservoir in terms of the intended use of its waters. In this respect there are three types of reservoirs:

(i) Impounded or direct supply reservoirs hold water draining from their catchments, which is transported, normally by pipeline, and used for direct supply after treatment to ensure that it is of accepted standards of purity and clarity.

(ii) Compensation or river regulating reservoirs (which may or may not be impounded) store water drawn directly off their surrounding catchments, which may then be fed into direct supply reservoirs or into the river to regulate its flow. The water, with very few exceptions, receives no treatment. Recreation is more acceptable on compensation reservoirs since it does not conflict with the maintenance of predetermined water-quality standards.

(iii) Storage reservoirs have no catchment of their own but are fed from another compensation or impounded reservoir.

Table 14 shows the number of these different types of reservoirs in each of the Regions. The proportion of compensation reservoirs is very small: it is significant only in the Lothian Region where nearly 50 per cent of all reservoirs are of this type. This is an historical legacy from the period when paper making was a more widely dispersed, water-located industry and the existing compensation reservoirs were created to maintain supplies to local mills.

(b) Water treatment. The extent to which treatment, to ensure high quality in terms of biological purity and colour, affects the acceptance of recreation at the reservoir varies, since much depends on the size of the water body and its ability to cope with given levels of pollution. The greater the size of a reservoir and the more frequent its turn-over, the greater the amount of pollutants that can be accepted before appreciable changes in water quality occur.

Table 14

Types of proximal Water Department reservoirs by Region

Region	I	Type of reservoir I/C	C	S	Natural loch	Total
Strathclyde	77	3	1	5	—	86
Strathclyde/ Lothian (shared)	1	—	—	—	—	1
Central	12	—	2	—	—	14
CSWDB	2	—	—	—	—	2
Tayside	8	—	—	—	—	8
Fife	20	1	1	—	—	22
Lothian	15	—	7	—	—	22
Lothian/ Borders (shared)	6	—	—	—	—	6
Borders	2	—	—	—	—	2
Dumfries and Galloway	11	—	—	1	1	13
Highland	3	—	—	3	—	6

I = impounded; C = compensation; S = storage;
CSWDB = Central Scotland Water Development Board

(c) Types of recreational activities. Those involving particularly high risks of immersion or of pollution tend to be less acceptable than those where these risks are not so great.

Most of the Water Department Reservoirs in Scotland are of pre-World War II origin and few large ones have been created since then, in comparison to England and Wales. The high cost of satisfying the requirements which would make direct water supply and recreational use compatible and, until recently, the comparatively low demand (in comparison with England and Wales) for recreational water-space at reservoirs, have maintained recreational development of public supply reservoirs at a limited and fairly low level. The policy with respect to the large new reservoirs, proposed or under construction, is to create systems capable of accepting multiple recreational and water-supply use.

The types of recreational activities and facilities on proximal Water Department Reservoirs in Scotland are given in Table 15. In comparison to fishing, which takes place on about 80 per cent of these water bodies, other recreational activities are limited. The most frequently recorded activities are bird-watching and casual visiting (including walking) pursued by relatively few people and involving little direct contact with the water. Thus

Table 15

Type of recreational activities and facilities associated with proximal Water Department reservoirs in Scotland

	Total number Proximal reservoirs	Private	Fishing club	WD	WD/P	BW	CV	SK	CA	R	S	CP	L	T	PS	CS	H
Strathclyde	(86)	25	28	17	2	1	8	1	1	1	1	7	6	8	—	—	—
													(scenic drive 2. Drunkie-Trossachs)				
Strathclyde/ Lothian (shared)	(1)	—	—	1	—	—	—	—	—	—	—		—				
Tayside	(8)	2	1	5	—	5	—	—	—	—	—	7	6	—	1	2	—
Central	(14)	5	5	2	—	1	—	—	—	—	1	—					
Fife	(22)	5	9	4	—	2	—	—	—	—	—	11	7	6	—	—	—
Lothian	(22)	9	8	2	—	10	8	—	1	—	1	2	—	1	1	—	2
													(+ caravan-type changing room)				
Lothian/Borders	(5)	2	1	2	—	1	2	—	—	—	—	1	—	—	—	—	—
Borders	(2)	—	2	—	—	—	—	—	—	—	—	—	—	—	—	—	—
Dumfries and Galloway	(13)	6	4	2	—	2	—	—	—	—	—	11	7	6	—	—	—
Highland	(6)	—	—	—	—	—	—	—	—	—	—	3	3	—	1	1	—

Fishing: WD = Water Department; P = private. Recreational activities: BW = bird watching; CV = casual visitors and walkers;
SK = water skiing; CA = canoeing; R = rowing; S = sailing. Recreational facilities: CP = car park; L = litter collection; T = toilets;
PS = picnic site; CS = camp site; H = fishing or boat hut.

Lochs with over 50%
shore near A/B road

KILOMETRES

0   50   100

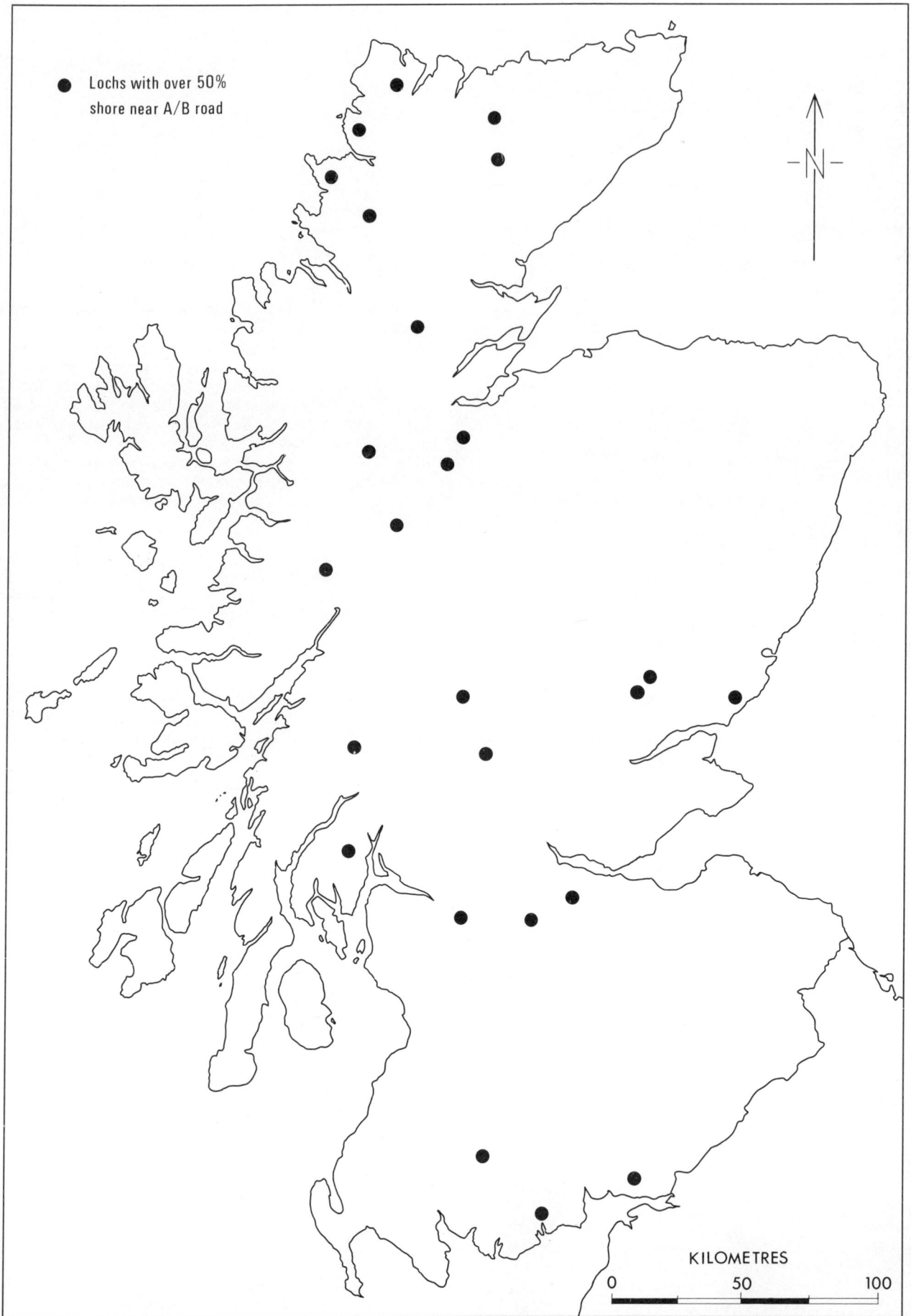

Fig. 8   Distribution of lochs with over 50 per cent proximal to a classified (A or B) road

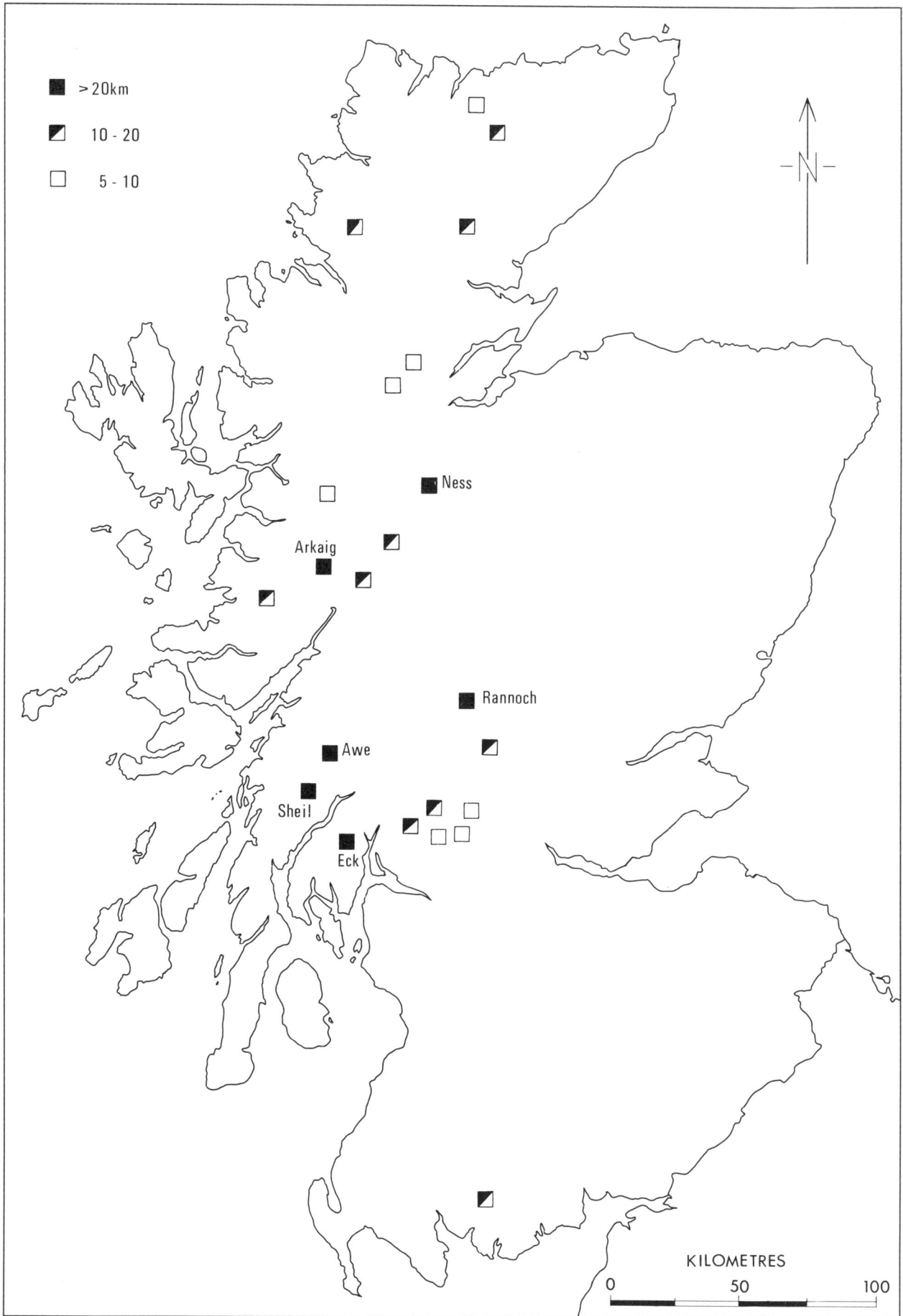

Fig. 9 Distribution of lochs with over 5 km of shore proximal to a classified road

24

Fig. 10 Regional organisation of the water supply industry in Scotland

recreational facilities are limited, but some provision exists for the transient motorist. Among the older reservoirs, some of the smaller ones which are no longer used for public supply, or those which it is proposed to relinquish for this purpose, are being reorganised primarily for recreational use.

2.5 Extent of Recreational Use and Nature of Activities

When the data available on the present recreational use of proximal Scottish lochs and reservoirs are analysed (see Table 16) the very low level of use is striking – 80 per cent of all proximal lochs have one or no recorded activity. The proportion of lochs with two or more activities form a decreasing proportion of the remainder, until the threshold of six activities is reached, when an increase occurs.

The distribution of lochs with two or more activities (see Fig. 11) reveals three clusters:

(a) A south-west 'crescent' which follows the lowland /upland junction across Dumfries and Galloway. Lochs in the mid 3-6 activity range, and with 30-50 per cent of their activities water-based, predominate.

(b) A Central Scotland group, extending from Renfrew District to Fife, and including the line of hill-foot reservoirs along the northern edge of the Southern Uplands, is characterised by a mixture of the 2-4 activity lochs with a significant number devoted primarily to water-based activities.

(c) A south-west to central Highland group, extending from Loch Awe and Loch Eck in the south-west to Loch Ness in the north, is an elongated S-shaped zone of lochs; these include the Trossachs Lochs, Lochs Rannoch and Tummel, and the Speyside lochs, with a number of lochs in the 5 to over 6 activity range.

Figure 12 shows the location of the lochs for which four or more activities have been recorded, showing marked concentration in the Dumfries and Galloway area and the south-west Highlands. There is, however, apparently no relationship between number of activities and size of loch — the latter ranges from Loch Ness to Kirk and Castle Lochs (Lochmaben, near Dumfries) which are among the smallest lochs in Scotland, and correlation with shore-length is not high (0.39). Figure 13 illustrates the relationship, particularly in the south-west and in the Highlands, of the main loch patterns to the line of the major routeways: the A75 (Dumfries to Stranraer) in the south-west and the A80 and A9 in the north. There is, however, no significant correlation between the lochs with the largest number of activities and minimum distance from an A or B road. The reasons for their present relative importance must lie with the scenic and other tourist attractions of areas such as Loch Lomond, the Trossachs, Speyside, the Great Glen, and with a variety of factors particular to one or a few lochs such as location in relation to former rail-links or, more recently, individual

Table 16

Number and percentage proximal lochs with given number of recreational activities

Number of recorded recreational activities	Number of proximal lochs	Per cent total proximal lochs
0	213	28.0
1	396	52.2
2	73	9.6
3	34	4.5
4	9	1.2
5	8	1.1
6 and over	26	3.4
Total	759	100

enterprise in establishing a given loch as a base for a particular recreational activity or group of activities.

Of all the lochside recreational activities recorded, fishing is the most widespread. Some type of fishing has been recorded for 68 per cent of all the proximal lochs, and this may well be an underestimate as it was not possible to obtain a comprehensive cover of all privately owned lochs. The presence of other water-based activities and facilities, such as are shown in Figures 14, 15 and 16, are much more restricted. However, except for the special study area, it was not possible to obtain reliable data about boating activities not associated with a club or marina.

Also, the reliability of data about the use of lochs for casual boating could not be assessed with any degree of accuracy on the basis of a desk study alone.

The extent and level of use of the proximal lochside for land-based activities is even more difficult to analyse, as detailed studies of Loch Lomond, the Trossachs and Loch Ken have shown that a high proportion of people who visit these areas engage in unorganised (informal) and often passive pursuits associated with pleasure driving. Figures 17 and 18 show those lochsides with facilities (other than bed and breakfast accommodation) for overnight stays or with facilities such as educational and information centres and lochsites of high conservation interest (Table 17), which tend to act as foci for visitors.

2.6 Facilities and Level of Recreational Use

Number of activities alone is an inadequate measure of level of recreational activity or degree of recreational development of the lochside, and data on numbers of people and time spent in the given range of recreational activities around proximal lochs was not available. Instead, it was decided to use the number of facilities

Table 17

Proximal lochs with designated sites of conservation interest

	Number of lochs	Per cent total proximal lochs
Nature reserve	11	1.4
Site of special scientific interest	77	10.1
Royal Society for the Protection of Birds Reserve	3	0.4
Scottish Wildlife Trust Reserve	7	0.9
National Trust for Scotland	5	0.7
Total	103	13.5

recorded for those lochs on or around which there were four or more activities, as a measure of the degree of recreational development. The numbers of the following facilities around each loch were summed, without any weighting: toilets, camping and/or caravan sites, marinas, clubhouses, education and/or information centres, hotels, youth hostels, car parks, picnic tables, log cabins, shops and/or cafes, and berthing facilities; jetties and piers could not be used since data about their numbers were among the least reliable for lochs which had not been surveyed. Out of the first 49 lochs with the greatest number of activities, only 10 obtained a score of 9 points or more, and the next highest was 7 points (3 lochs). In terms of available facilities, Table 18 shows the range of recrea-

tional developments within the top ten, and with the exception of Lochs Eck and Ken, the other seven include those lochs which are probably most widely known within and outwith Scotland. Their variation in size and shorelength, however, is considerable. Thus, the relationship of number of facilities to surface area, or preferably shorelength, provides a more realistic index of intensity or degree of lochside recreational development. Ranking on the basis of facilities to area and of facilities to shorelength is given in Table 19 (see also Fig. 19).

Table 19

Ranking of first ten proximal Scottish lochs on the basis of the ratio of the number of facilities to loch area, and of the ratio of the number of facilities to shorelength

	Loch	Ratio of facilities to loch area (ha)	Loch	Ratio of facilities to shorelength (km)
1.	Morlich	7.1	Morlich	0.3
2.	Earn	29.40	Earn	0.7
3.	Eck	43.50	Lomond	1.3
4.	Ken	51.5	Tay	2.3
5.	Tummel	62.10	Eck	3.2
6.	Tay	120.30	Rannoch	3.4
7.	Rannoch	126.00	Tummel	3.2
8.	Lomond	136.8		
9.	Ness	282.05	Ness ⎱	4.5
10.	Awe	353.09	Ken ⎰	
			Awe	8.9

Table 18

Ten first-ranking proximal lochs in Scotland according to number of selected lochside facilities

Loch	Point scale	Area ha	Shorelength km	T	Camping site	Caravan site	Marina	Club house	Educ/Info centre	Hotel	YH	CP	Picnic site	Log cabins	Shops cafes	Berthing facilities
1. Lomond	(83)	7125	66	8	6	7	1	1	1	7	3	24	2	—	8	15
2. Earn	(27)	913	22	1	—	2	3	3	1	4	—	4	1	—	2	6
3. Tay	(21)	2648	51	1	1	4	—	—	2	5	—	2	—	—	4	2
4. Ness	(15)	5641	90.2	—	4	1	—	—	2	1	—	2	—	1	1	3
5. Morlich	(17)	121	5.5	1	1	1	—	1	3	—	1	3	3	—	1	2
6. Rannoch	(13)	1902	51	1	1	2	—	—	2	2	—	1	1	—	1	2
7. Awe	(10)	3884	981	—	—	—	—	—	—	6	—	1	—	—	—	3
8. Tummel	(10)	621	35	1	1	—	—	1	—	1	—	2	—	—	—	4
9. Eck	(10)	435	32	—	1	2	—	—	—	2	—	2	2	—	—	1
10. Ken	(9)	464	44	1	2	2	—	1	—	—	—	1	—	—	—	2

T = toilet, CP = car park, YH = youth hostel.   Based on data set and supplemented (apart from Loch Eck) by data collected in the field.

c

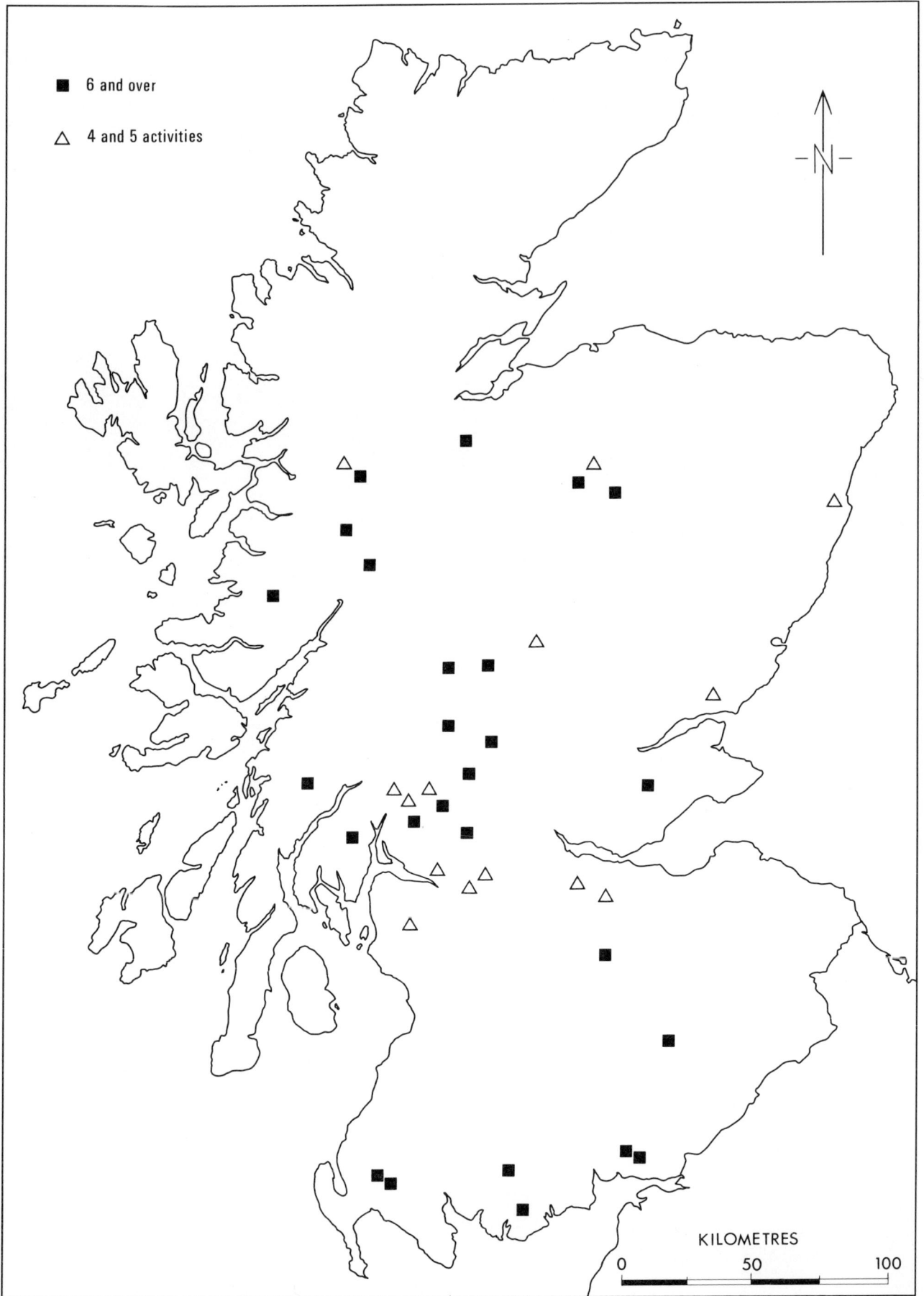

Fig. 12   Proximal lochs for which four or more recreational activities have been recorded

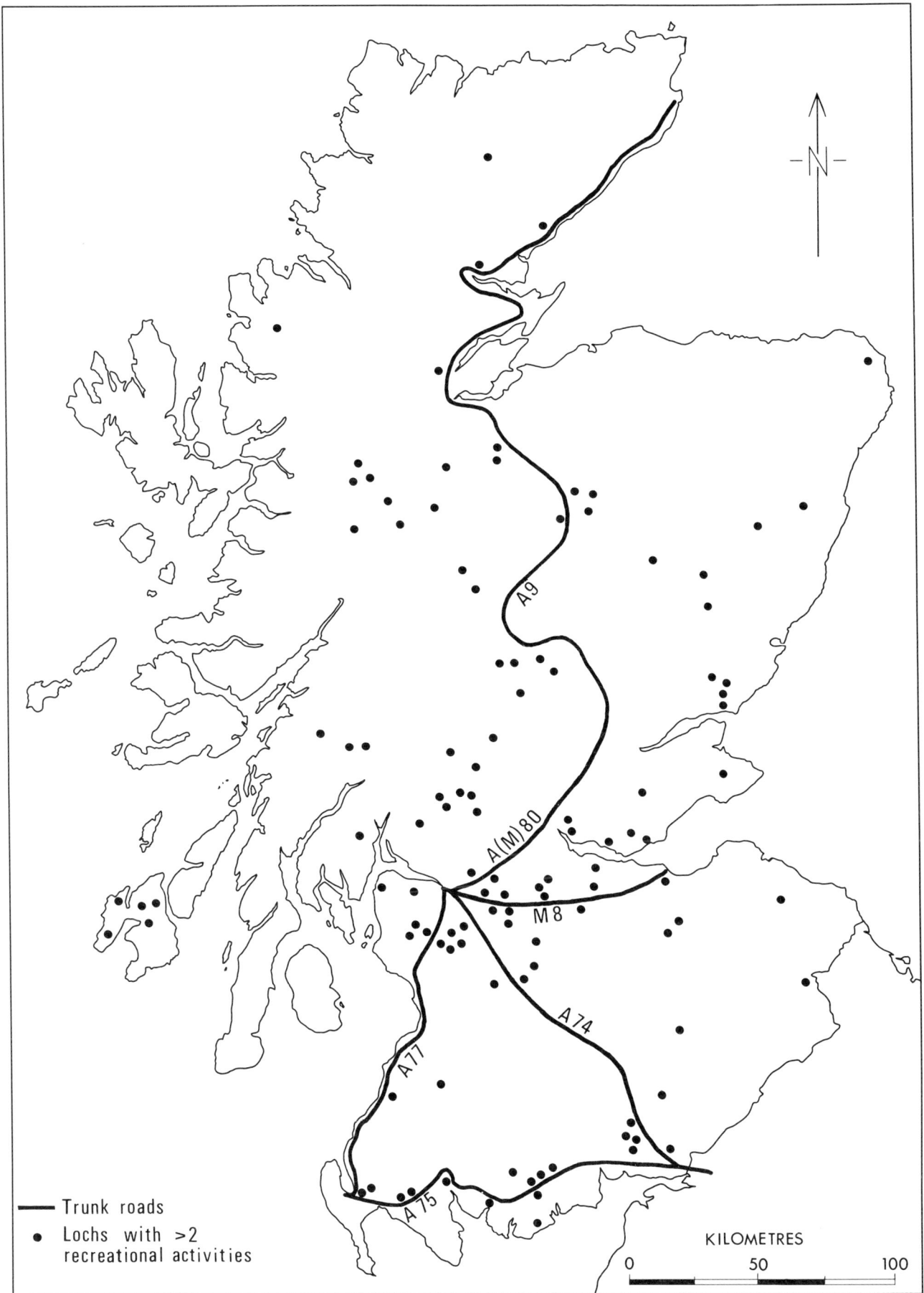

Fig. 13 Major trunk roads in relation to lochs with more than two recreational activities

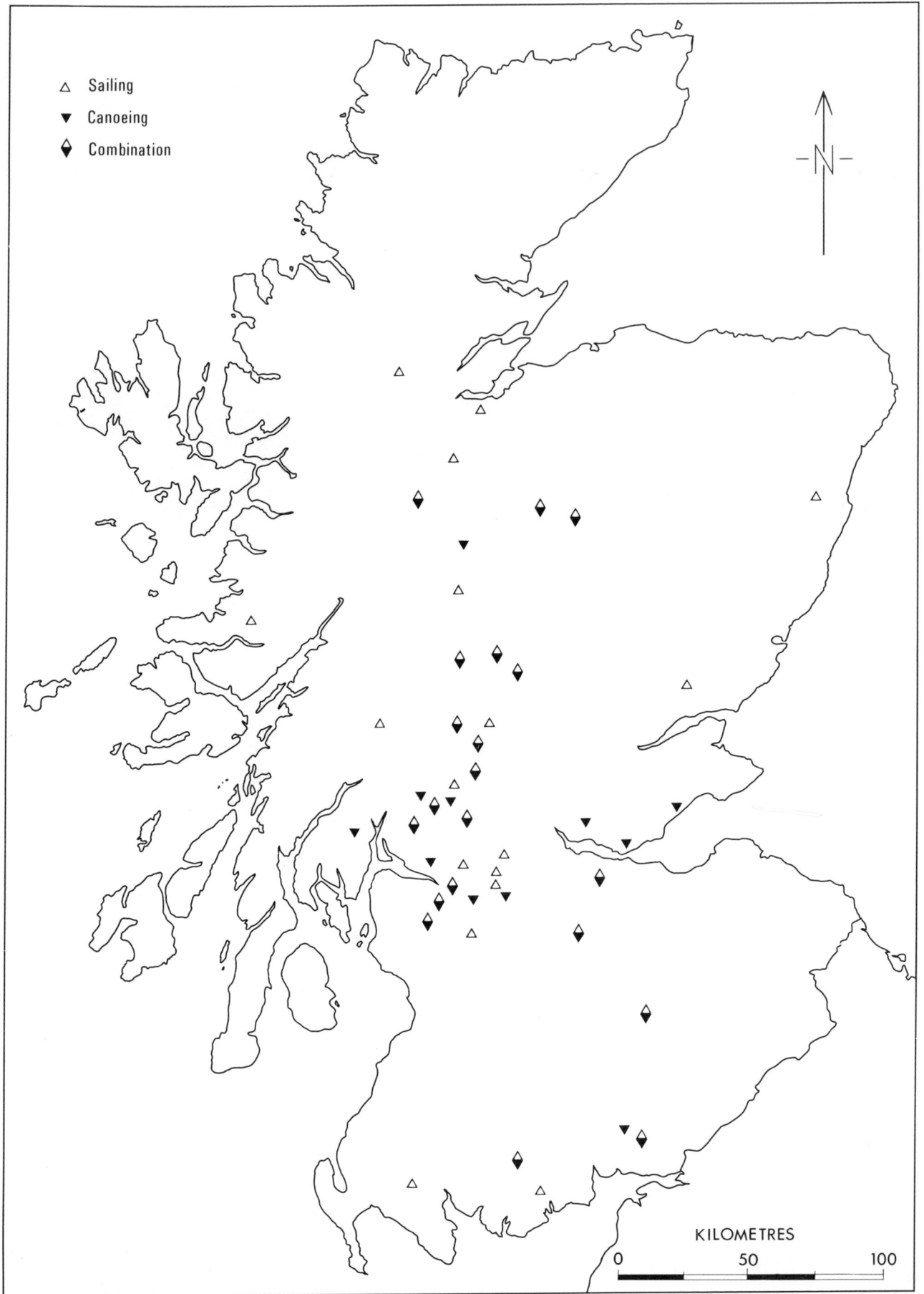

Fig. 14   Proximal lochs with sailing and canoeing

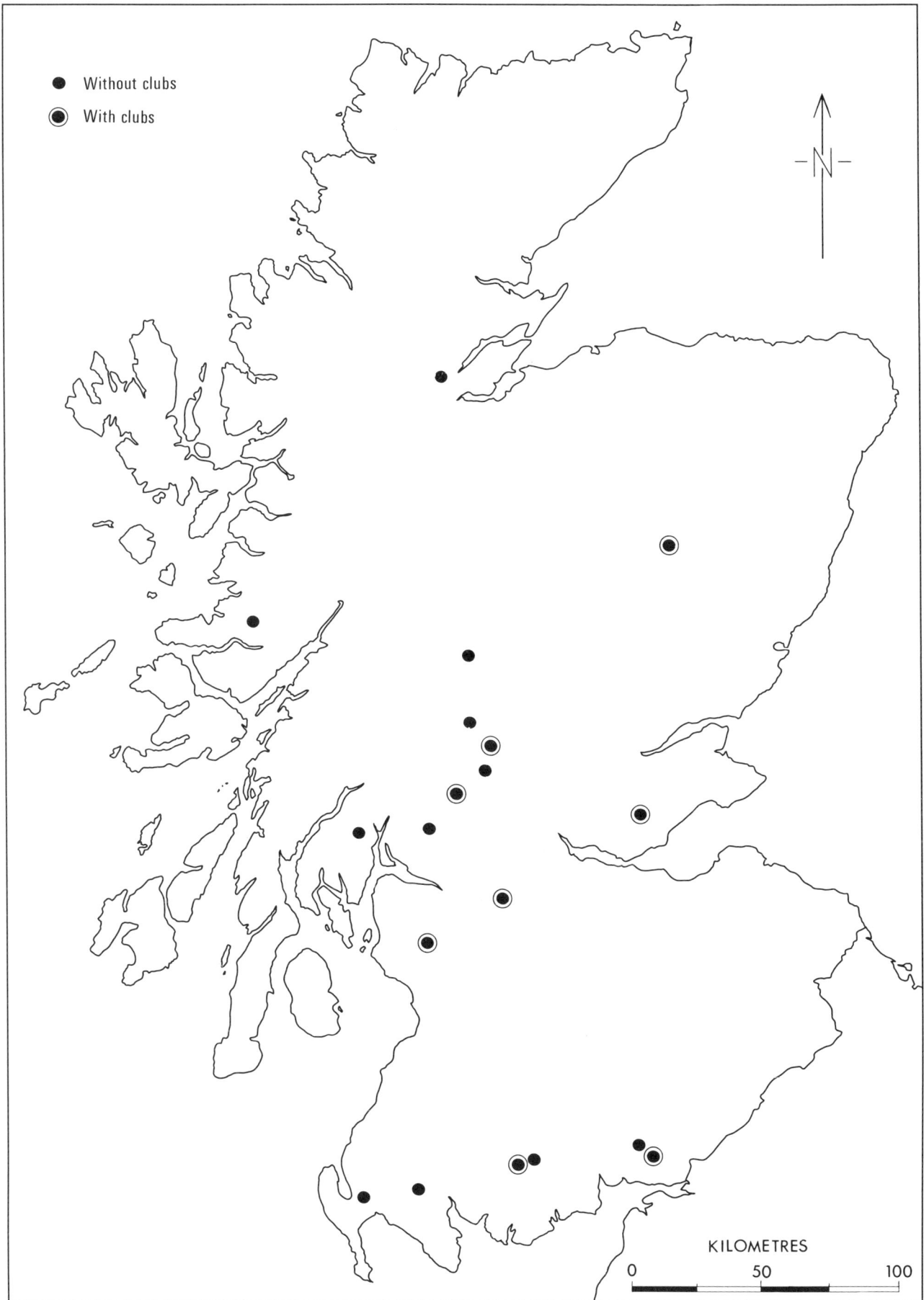

Fig. 15 Proximal lochs with water-skiing

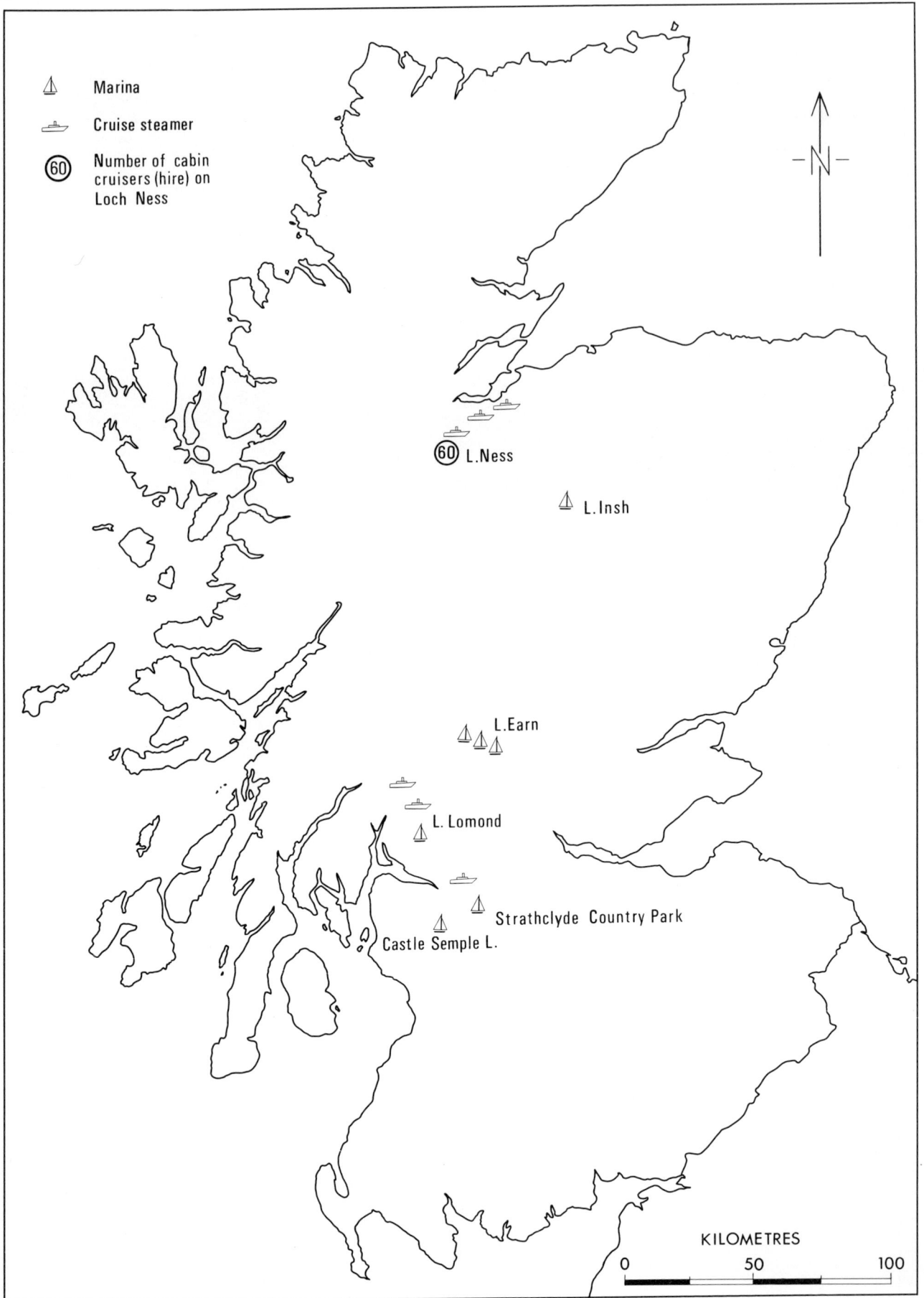

Fig. 16   Proximal lochs with marinas and cruise steamers

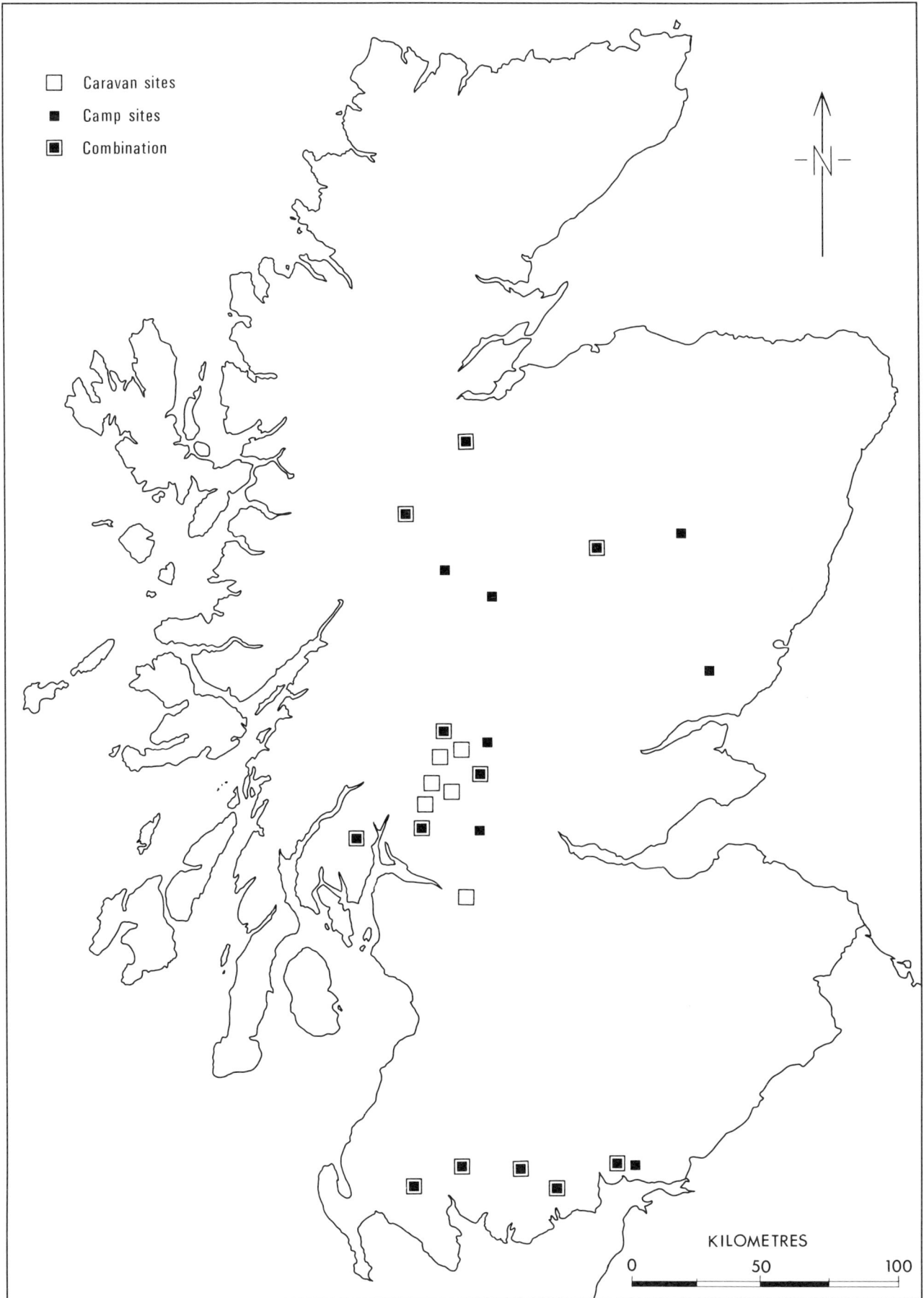

Fig. 17   Proximal lochs with caravan and camp sites

34

H Hotel
▲ Youth hostel
◐ Residential } centre
   Educational }
○ Information centre

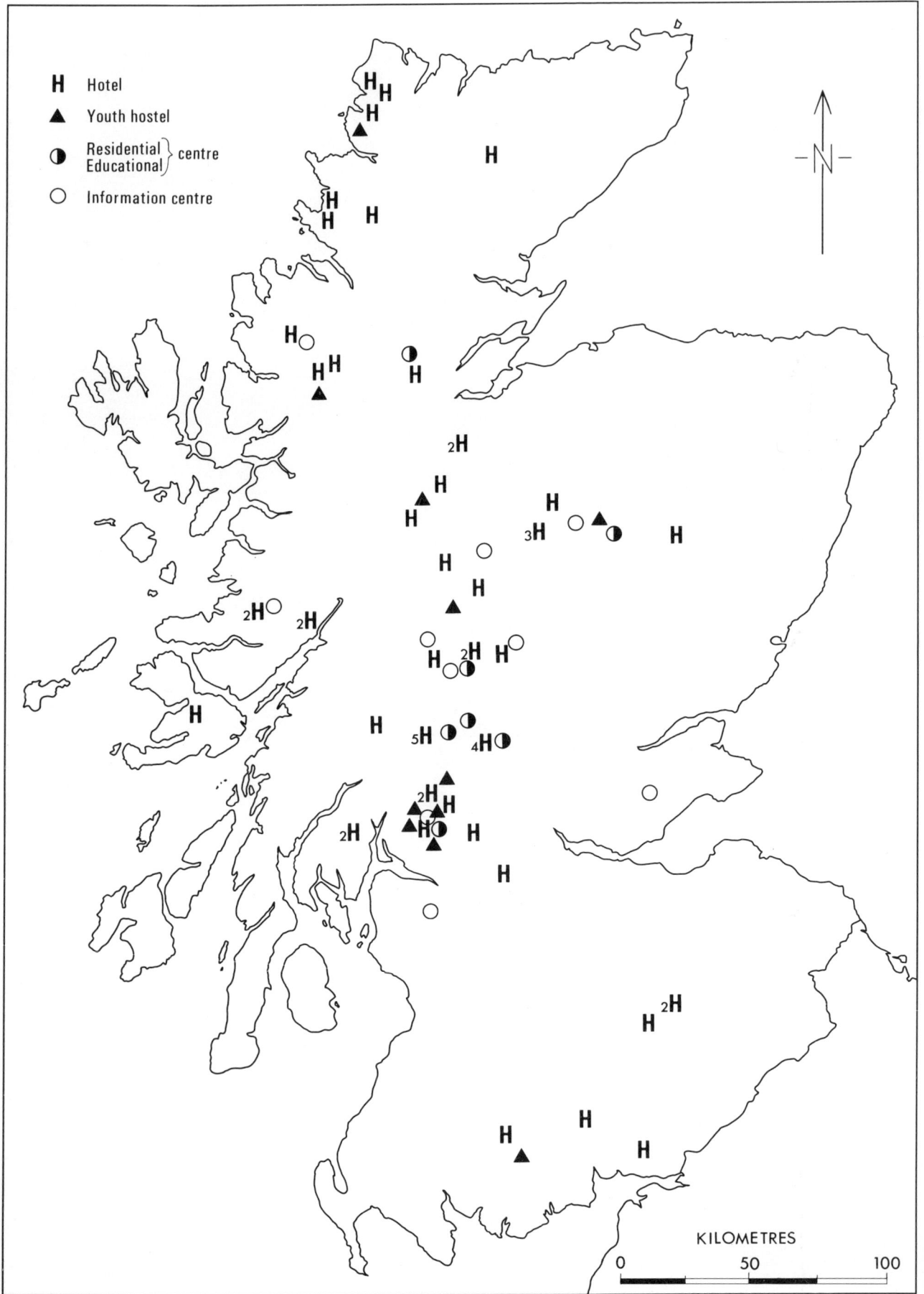

Fig. 18   Proximal lochs with residential and educational facilities

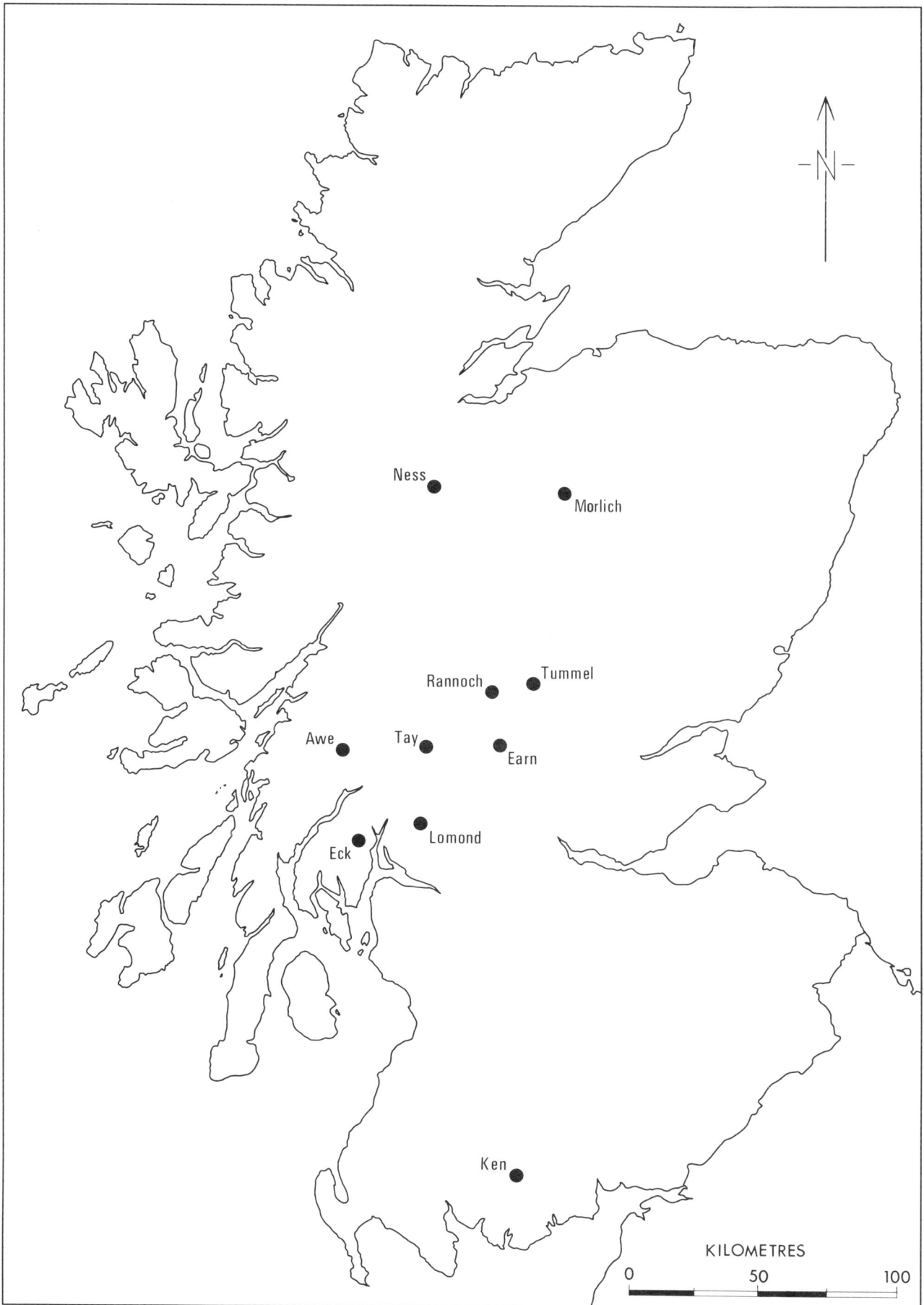

Fig. 19   Ten most highly developed recreational lochs in Scotland

It is interesting to note that, in terms of area and shorelength, Lochs Morlich and Earn are the two with the highest, and Lochs Awe and Ness the lowest, indices of development. The contrasts and comparisons between them give some indication of the factors which have accounted for the distribution of lochside recreational development in Scotland. Loch Earn's essentially private, and originally land-based development, was initiated by railway and hotel building in the nineteenth century; Loch Morlich's publicly developed facilities are a twentieth-century phenomenon to which the construction of the road up to the Cairngorm ski-area has made an incalculable contribution. In both cases, their presently accessible shores and lochsides have reached the maximum or near maximum development possible given their existing patterns of land tenure. Loch Ness and Loch Awe are two of the largest and most scenic lochs in Scotland, and have strong historic associations as well as other tourist attractions. On both, the proportion of the lochside accessible from the main roads which skirt them is small and views of the lochs are relatively limited; in contrast their proximal lochsides and shores are comparatively underdeveloped, partly because they lie off the main trunk roads, and on both, water-based recreational activities are more numerous than land-based ones.

The most significant points to emerge from the analysis of recreational activities are, given the potential resource, the limited scale of development and the extent to which high levels of lochside recreational use are concentrated on such a small number of lochs.

The loch is a characteristic feature of the Scottish landscape, an important scenic and scientific element and a recreational resource of considerable potential. Indeed, in Scotland, surveys have found that lochsides (together with riversides) are second only to beaches in attracting informal recreationists (TRRU, 1976).

Its attraction for recreation is a function of:

(a) a land-water interface, where the very basic attraction which water bodies have for a wide spectrum of people, is combined with a scenic diversity as great, if not greater, than is found around the coast;

(b) the opportunities which lochs afford for water-based recreational activities or those for which the loch may be more suitable than the sea;

(c) the very considerable tourist attraction of the historic, literary and/or cultural associations which have made certain Scottish lochs internationally renowned.

3.1 Comparison of Lochside with the Seashore

3.1.1 The shore

In comparison with coastal sites, the lochside and particularly the lochshore is, for a variety of reasons, generally more spatially restricted (Fig. 20). Fluctuation of water-level in natural lochs is dependent on seasonal variations of rainfall and evapo-transpiration, and relative rates of surface water run-off, and/or percolation of water into the soil within their natural catchment basins. While winter water-levels are, on average, generally higher than summer ones, marked short-term fluctuations, dependent on daily variations in rainfall, can occur at any period of the year. On impounded lochs, water-levels are modified by regulation of in- and out-flow. The range of water-level may, as in water-supply reservoirs during summer droughts, be greatly increased, or it may be decreased to the extent that loch levels are virtually stabilised as in many flood control and hydro-electric power reservoirs. Also, the regulation of level on one loch/reservoir may affect that of an unimpounded loch downstream such as Loch Achray when the Loch Katrine sluices are opened, and Loch Venachar when Loch Drunkie compensates. The shore area available for recreational use can in consequence be considerably extended or drastically reduced as the case may be.

However, while the exposed lochshore is, on the whole, more limited in extent than the seashore, the former can be less variable in width particularly during periods of low summer water-level. Furthermore, the lochshore may remain constant or even increase in width at the period when recreational use is heaviest, unlike the seashore which is subjected to regular diurnal tidal fluctuations throughout the year. Consequently, those who use the lochshore do not have to advance and retreat during the course of a day as on coastal sites.

3.1.2 Waves

Nevertheless, both the loch- and the sea-shore are zones of turbulence and inherent instability is caused by breaking waves. The point at which waves break occurs where wave-height is three-quarters of the water-depth, and where the transition from deep to shallow water causes the crest to travel forward at a faster rate than the base, to steepen and finally to fall over and break. The break point is determined around both loch- and sea-shores by off-shore gradients. However, loch waves are of much smaller dimensions – in height, length, steepness, and frequency – as a result primarily of the limited fetch (uninterrupted extent of water) across Scottish lochs in comparison with the open sea. The longest fetch is that on Loch Ness (33.4 km); Loch Lomond, which has a greater total length, has a maximum fetch of only 9.6 km because obstacles such as promontories and islands interrupt the continuity of the water surface. On most lochs the fetch is much shorter: the mean of proximal lochs studied in Scotland is of the order of 5.7 km. Even on Loch Ness, recent work on deep-water waves suggests that wave amplitude is only about a tenth of that in the open North Atlantic (per. comm. Lanchester Polytechnic). Observations and recording further indicate that wave development starts sooner on a loch than in the open sea; after ten to twenty minutes in the former, usually after at least thirty minutes in the latter situation. However, the length of time during which the minimum wind velocities from one direction are sustained is frequently not sufficient to produce waves on many lochs.

Also, wind speed and direction on lochs are considerably influenced by the local relief of the surrounding land areas; lochs are notoriously susceptible to wind veering and gustiness. Effects of relief are most pronounced in the larger upland and highland lochs situated in deep valleys. These lochs are frequently subjected to wind-funnelling, to cold-air drainage from surrounding steep slopes, and to a diurnal movement of local morning and night-air consequent upon the relative changes of temperature over land and water, particularly during the summer months. On the larger lochs, local relief and atmospheric conditions may produce several wave-trains running in different directions which are, according to their periodicity, capable of cancelling or reinforcing the effect of each other. Indeed, water-surface conditions on a loch can change, in response to varying wind direction and speed, with greater rapidity than at sea. Because of the relatively small scale and the inter-visibility of the shores of Scottish lochs, these hazards tend either to be forgotten or to be not fully appreciated by recreation users or managers.

3.1.3 Sediment movement

A direct consequence of the difference in scale of the loch and marine wave is that around lochs mobile beach material is rarely as well-sorted as around the seashore. The upper size-threshold of the local beach material is smaller; the highly-prized sandy, marine beaches are rare, while those composed of stony-silt and clay are more

MARINE

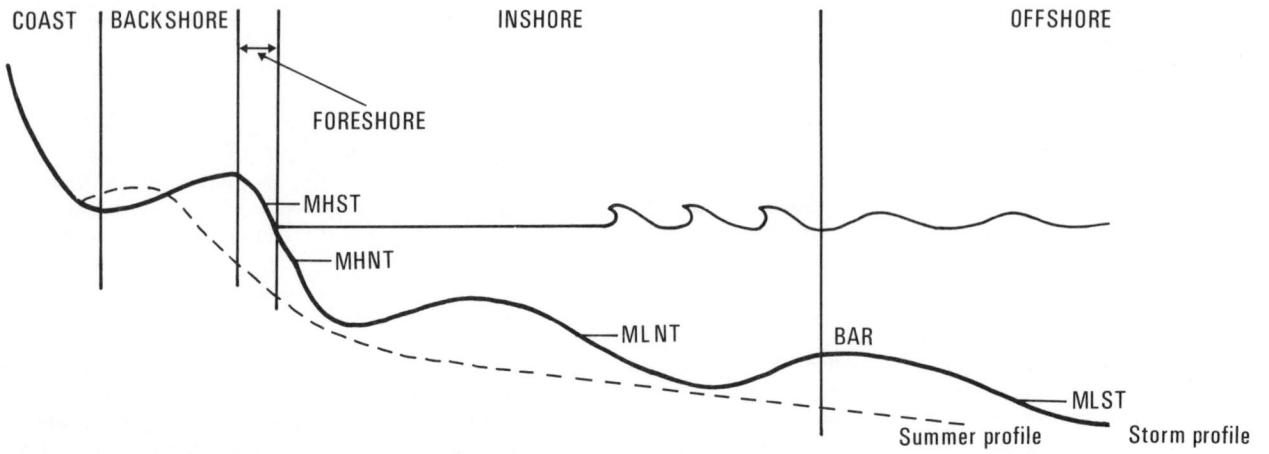

COAST | BACKSHORE | INSHORE | OFFSHORE

FORESHORE

MHST

MHNT

MLNT

BAR

MLST

Summer profile

Storm profile

LOCHSHORE

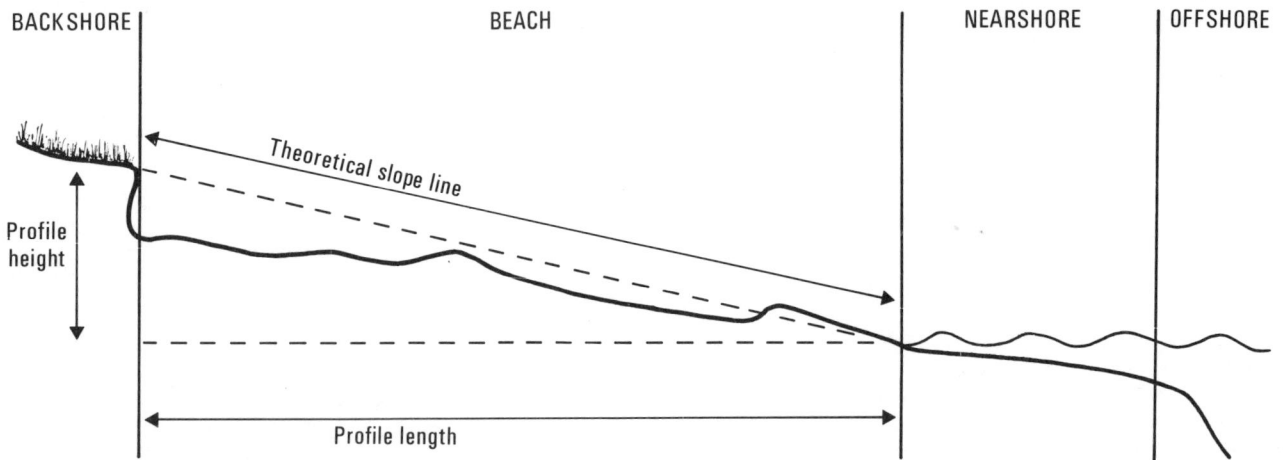

BACKSHORE | BEACH | NEARSHORE | OFFSHORE

Theoretical slope line

Profile height

Profile length

MHST	Mean high spring tide
MHNT	Mean high neap tide
MLST	Mean low spring tide
MLNT	Mean low neap tide

Fig. 20   Marine and lochshore beach profiles

common. Furthermore, the active surface layers of the lochshore beach material are frequently very limited in width and quite shallow in depth. They do not appear to change in distribution or surface form as rapidly as those of coastal beaches. Also, loch beach material is less well-sorted, is less mobile and is a less effective agent of erosion than that on seashores, mainly because the wave-energy generated is much less. On balance, the process of accumulation of mineral and organic matter tends to be more prevalent in lochs. It occurs particularly where rivers and burns enter a loch, and in the smaller and shallower lochs the process can result in a relatively rapid infilling of the basin and the ultimate disappearance of the loch. This process is admirably illustrated by Spence (1964). In peaty highland lochs, however, where there is no significant in-flow, the process is negligible. Small lowland lochs surrounded by agricultural land are probably most likely to be subjected to relatively rapid silting.

3.1.4 Vegetation

Both the marine shore and the lochshore are characterised by a marked parallel zonation of vegetation consequent on variation of water-depth, amount and frequency of submergence and exposure of the beach, and of the texture and mobility of the material of which it is composed. Except in sheltered bays and estuaries, the vegetation of the seashore is a less significant landscape element than around lochshores. It is dominated by the submerged and floating seaweeds which require a relatively stable anchorage in face of heavy marine breakers and can survive only limited periods of exposure. As such, they are associated more with rocky shores than sandy-cobble beaches. On the latter, uprooted plants tend to be swept in-shore and form temporary strand lines of varying heights above sea-level.

In contrast, the most conspicuous vegetation of the lochshore is composed of emergent reed and sedge beds which, in sheltered areas with some depth of fine bottom sediment, occupy a zone from the highest water-level inland to one metre (maximum 1.5 m) water-depth off-shore. Around both the coast and the loch, shore vegetation as such has limited recreational value. More favoured are those shores with little or no conspicuous vegetation and where the movement of people and vehicles is unimpeded by its presence. Lochshore reed and sedge swamps, as with estuarine salt marshes, are relatively unfamiliar habitats to the majority of people. An unstable substratum combined with tall, dense, wet vegetation makes access difficult, and the lochside swamp and marsh repel rather than attract recreational users.

3.1.5 Wildlife habitat

Both the loch and coast, however, provide a distinctive habitat for wildlife, particularly for those species which live and breed on the landward margin and which feed in the water. The numbers and diversity of wildfowl in both cases are great and make an important and distinctive contribution to the recreational value of these sites.

3.1.6 Recreational use

Both the seashore and the lochside have a similar potential to support a wide range of combined land- and water-based recreational activities; few are exclusive to one or the other. Curling and skating must be the only activities peculiar to the loch, but the occurrence of freezing weather conditions, together with shallow water, restrict these sports to relatively few lochs and their pursuit outdoors has declined in face of all the advantages of indoor ice-rinks. Bank- and boat-fishing are, for obvious reasons, more commonly and frequently pursued around the lochside than around the coast. The advantages of a more constant water-level during the course of a day and calmer water conditions for water-skiing, sub-aqua training, sailing, rowing, canoeing and other casual boating activities can be offset by the obvious disadvantages of a loch. Not least of the latter are the limited water-space together with the incompatibility of many water-based activities, as well as the unpredictable and potentially hazardous weather conditions referred to previously. For swimming, the disadvantages of fresh and relatively cold loch-water are marked in comparison to warmer more buoyant sea-water.

3.1.7 Recreational impact

The range and intensity of recreational impacts, as well as the relative susceptibility of lochsides and coasts, are similar in type if not in scale. On both, the main physical impacts come from trampling by people and compaction by vehicles; digging is more common on coastal than on loch beaches, while lighting picnic fires is a more frequent activity around the lochside. The latter combines a nearby source of wood for fuel with a greater degree of local shelter than on the coast. The remains of fires, and charred tree trunks are widespread features of all recreationally used lochsides in Scotland. The lochside, however, is more susceptible to damage because the potential recreation site is spatially more limited than on the coast. Hence recreation activity becomes even more concentrated on fewer and smaller favoured sites. The well-known edge-effect is exaggerated on the lochside where the width of the shore or indeed the whole site may be so narrow as to form an edge in itself.

However, there are two other important factors which make the lochside more vulnerable to deterioration:
  (a) the impact on the lochside is two-directional, from the land lochwards and vice-versa, which tends to increase intensity of use; in contrast impact from the landward side tends to be more important if not dominant on coastal sites;
  (b) given the differences in scale, the lochside is part of a circular closed system, in comparison with the more open ocean system.

Impacts, of the same order, which cause water pollution of one kind or another will be greater on a lochside than a seashore of similar length; in the sea pollutants are dissipated more rapidly. In the loch, the effects are rapidly cumulative at an exponential rate and may be suddenly increased by periods of intense drought and high evapo-transpiration, as occurred during the summers of 1975 and 1976 in Britain.

The object of this chapter is to define and describe the lochside resource characteristics in order to provide a basis on which to assess variations in the levels of recreational use and the vulnerability of the resource to impacts.

The lochside is a zone of varying width which, unlike the seaside, has not, at least in Britain, been formally delimited for any particular purpose. Its definition is difficult, because of a variable water-line on the lochward side and the lack of any consistently recurring boundary, or discontinuity, landward. The definition of the lochside as a recreational site for the present purpose is "the area of land extending from the water's edge to a road, track, or railway line within a distance of 100m; or a major obstruction to movement caused by a change of land form or vegetation cover; or a distance of 100m in the absence of any man-made or natural limitation". As such, it comprises one or a combination of two or three distinctive physical components or habitats:

(a) the shore – the intermittently submerged zone between mean low water-level and the highest water-level;

(b) the back-shore – that part of the lochside immediately above the highest water-level and which is characterised by terrestrial rather than aquatic or semi-aquatic vegetation;

(c) the shore/back-shore discontinuity – the main cliff or bank (for difference see Glossary) of varying height and gradient between the shore and back-shore.

The recreational use of one of these components can rarely be treated in isolation from the others since the use of one can be affected by or, in turn, can affect the use of the others. In order to analyse the recreational use of the lochside and assess the impact of recreation, it is necessary to understand the basic resource characteristics of the site and their relative value for a wide range of recreational activities. Hence, it is important to establish some framework of generalisation or means of categorising the lochside on the basis of resource attributes significant for recreational use. To date, lochs have been classified on the basis of their form, origin, stage of evolution and, most commonly for biological and ecological purposes, according to the nutrient status of their waters. None of these schemes provides a satisfactory basis on which to classify the lochside for the current purpose. Given the number of attributes involved, together with the variability of lochsides, the task is as formidable as it is essential.

One of the aims of this report has been to construct a flexible framework within which the lochside could be described and categorised. The two principal criteria which have been used to identify lochside site-types are those considered most important in terms of recreational user requirements. They are:

(a) land form, with degree of slope as the most important physical attribute for the satisfactory pursuit of the majority of lochside based recreational activities;

(b) vegetation form (and appearance) defined primarily in terms of ease of movement across or through the dominant vegetation type present on the lochside.

These criteria have been used to describe site-types based on assemblages of landform, vegetation and shore features along adjacent profiles around the lochside (see Fig. 21 and Glossary).

4.1 Classification of Lochside Site-types

The classification devised is summarised diagrammatically in Figure 21. The primary features indicative of the site-type are the dominant slope and dominant vegetation-form of the lochside profile. These primary classes are sub-divided by secondary characteristics, namely, shore features.

The dominant slope is the average slope of the lochside profile. Dominant vegetation form is that type of vegetation (grass and forb/shrub/tree) which visually occupies the greatest area; thus although grasses and forbs may cover 80 per cent of the ground, if there is a 30 per cent cover of open woodland above it, the latter is the dominant vegetation form.

The presence of shore features, including rocky shore, wetland fringe, beach, cliff, bank, strip and embankment, forms the secondary part of the classification. These terms are defined in the Glossary, however the word strip requires further explanation. It is used to describe a narrow strip or verge of land of gentle or moderate gradient which parallels and is close to the shoreline. It must be less than 5m in width, and be backed landward by a steeper slope. The strip is a common lochside feature, of particular value for walking and bank fishing when the primary facet behind it is very steep.

In Figure 21 the numbers in the left-hand column refer to the dominant slope category which increases in 5° intervals (see Glossary) from site-type 2 to 6. Site-type 1 is always flat; 7 is undulating, 8 is a low embanked shore; 9 is a high embanked shore; 10 is a rocky promontory; and 11 is a high, steep, rocky shore. The lower case letters indicate the vegetation form of the site-type; a = grass and forb, b = shrub and dense woodland, c = open woodland (see Glossary). From the application of the two primary criteria of slope and vegetation-form to data collected from approximately 60 lochs surveyed in the field on a scale of 1:10,000, twenty-three site-types have been identified. A large number of variations can be described dependent on the presence of shore features. The right-hand column of Figure 21 gives examples of some of the possible combinations of slope, vegetation-form and shore features. These may be further complicated, particularly in the case of wide lochsides, by the presence of one or more subsidiary site-types. The site-type nearest

	BASIC PROFILE	Essential Features		POSSIBLE SHORE FEATURES							EXAMPLE OF PROFILE with Shore Features
		Dom. Slope	Dom. Veg. Form	B Beach	R Rocky Shore	W Wetland Fringe	C Cliff	S Strip	K Bank	E Embank-ment	
1		G	M		✓	✓	✓			✓	E
2a		G	F	✓	✓	✓	✓		✓	✓	B
b		G	S	✓	✓	✓	✓		✓	✓	B C
c		G	T	✓	✓	✓	✓		✓	✓	K E
3a		M	F	✓	✓	✓	✓	✓	✓	✓	C
b		M	S	✓	✓	✓	✓	✓	✓	✓	C B
c		M	T	✓	✓	✓	✓	✓	✓	✓	B C S
4a		S	F	✓	✓	✓	✓	✓		✓	S
b		S	S	✓	✓	✓	✓	✓		✓	B C S E
c		S	T	✓	✓	✓	✓	✓		✓	B
5a		VS	F	✓	✓	✓	✓	✓		✓	B
b		VS	S	✓	✓	✓	✓	✓		✓	B S
c		VS	T	✓	✓	✓	✓	✓		✓	B C
6a		ES	F	✓	✓	✓	✓	✓		✓	S
b		ES	S	✓	✓	✓	✓	✓		✓	R S
c		ES	T	✓	✓	✓	✓	✓		✓	—
7a		M	F	✓	✓	✓	✓	✓	✓	✓	B C S
b		M	S	✓	✓	✓	✓	✓	✓	✓	C S
c		M	T	✓	✓	✓	✓	✓	✓	✓	E
8	H<2m X<20m	>M	F S T	✓	✓		✓	✓			B C S
9	5m>H>2m X<20m	>M	F S T	✓	✓		✓	✓			B
10		G–S			✓	✓	✓	✓		✓	B
11		S–VS		✓	✓						B

Slope : **G** = Gentle (0-5°), **M** = Moderate (6-10°), **S** = Steep (11-15°), **VS** = Very Steep (16-20°), **ES** = Extremely Steep (OVER 21°)

Vegetation Form : **M** = Wetland, **F** = Grass and Forbs, **S** = Shrub and Dense Woodland, **T** = Open Woodland

Fig. 21   Classification of lochside site-types

the shore is termed the primary site-type because it is likely to be the most important for recreation. If a change in dominance of one or both of the primary criteria (i.e., the slope and vegetation-form categories) occurs in a direction perpendicular to the shoreline, a subsidiary site-type is defined. It is classified in the same way as the primary site-type, without the shore features (see Fig. 22). Conversely, where a change occurs in a direction parallel to the lochshore (i.e., where there is a change in the lochside profile) a new site-type is defined. Detailed instructions for the delimitation and mapping of site-types is given in Appendix V.

It was decided that the most concise and practical method of describing the site-type categories identified was in diagrammatic form as illustrated in Figure 21. These are presented as visual models, with which a particular lochside site-type can be compared. The problems and methods of assessing the suitability of these site-types for a variety of recreational activities, and their vulnerability to recreational impact will be discussed in the succeeding chapters.

4.2 Classification of Lochside Types

The recreational value of the lochside site and its vulnerability to impact cannot be assessed in isolation from the particular lochside of which it may form the whole, a dominant, or a minor part. This highlighted a need, which emerged early in the project, to find some method of generalisation on the scale of the lochside type for the purpose of:

(a) providing a systematic and standardised, descriptive framework that could be applied to an entire lochside;

(b) assessing impact generally on the whole lochside.

The final classification arrived at is summarised in Table 20 in which lochsides are grouped according to the percentage of shoreline occupied by the dominant site-type. The following five categories are recognised:

1. 81-100 per cent of shoreline occupied by dominant site-type
2. 61-80 per cent of shoreline occupied by dominant site-type
3. 41-60 per cent of shoreline occupied by dominant site-type
4. 21-40 per cent of shoreline occupied by dominant site-type
5. 0-20 per cent of shoreline occupied by dominant site-type

This refers to the site-type as defined by the primary features of slope and vegetation form only, therefore the emphasis is on the land rather than shore features. This gives an indication of the comparative variety of site-types around a loch; those with over 80 per cent of shoreline occupied by a dominant site-type have little variety,

whereas those with less than 20 per cent occupied by the dominant site-type must have a lochside composed of a great number of site-types. Having established this framework, it is possible to detect further variation in terms of other measured parameters.

The scheme, however, does not allow for the wide size range of lochs surveyed, from a minimum of 4 ha to a maximum of over 7,000 ha. A small loch such as Loch Cote Reservoir (18 ha) with 21-40 per cent of the shoreline occupied by site-type 2a is not directly comparable with a much larger loch, such as Loch Ken (464 ha), with a similar proportion of its shoreline occupied by the same site-type; the actual lengths are 0.7 km and 12.5 km respectively. For this reason, it was considered necessary to group lochs further on the basis of their size, using frequency of loch area on the basis of the data set (see Fig. 4) for all proximal lochs in Scotland. The four categories selected are:

(a) 0 - 25 ha – small lochs;
(b) 26 - 100 ha – medium-sized lochs;
(c) 101 - 400 ha – large lochs;
(d) over 400 ha – very large lochs.

For any purpose, then, only lochs within defined size-categories can or should be compared.

While the site-type classification has been based on dominant land-form and vegetation-form characteristics, there are two features whose influence on actual use is greater than their relative extent might suggest. The first is the physical nature of the shore and, more particularly, the presence or absence of a beach. The second is the composition as well as the form of the shore vegetation and the particular combination of shore and back-shore vegetation associated with a site-type.

4.3 Classification of Lochshores and Beaches

On the basis of width, gradient, and surface-material four general types of lochshore can be distinguished:

(a) cliffed shore of varying height and composition, from a low peat-bank to a high rock-face, where there is a water-line but no shore in the accepted sense of the term;

(b) rock shore of varying width and slope, composed of rock outcropping in situ;

(c) vegetation bound and/or covered shores;

(d) beaches which, in this context, are types of shores in so far as they are defined as "the zone of accumulated (and unconsolidated) inorganic material (mud – boulders) extending from the summer (or lowest sustained) water-level to the highest point reached by storm wave" (Steigler, 1976). Their surfaces are normally either bare or carry only a sparse, open vegetation cover. The lochshore beach is particularly attractive for recreational use, especially where it provides a bare and open, usually

44

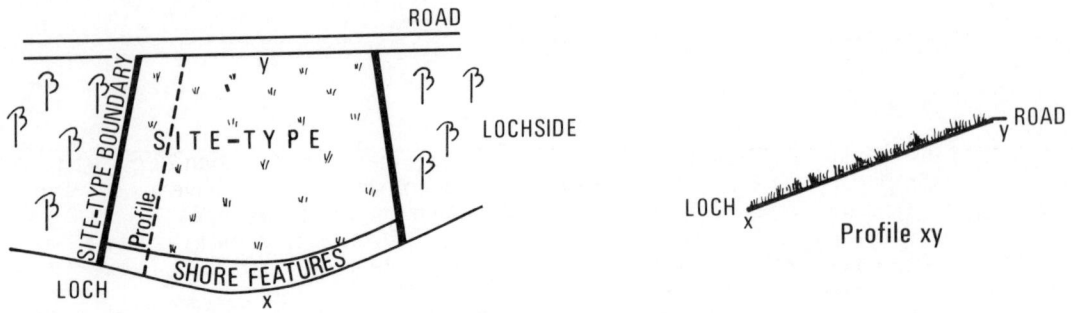

(a) Plan and profile of lochside showing SITE-TYPE BOUNDARIES defined by a change in vegetation form.

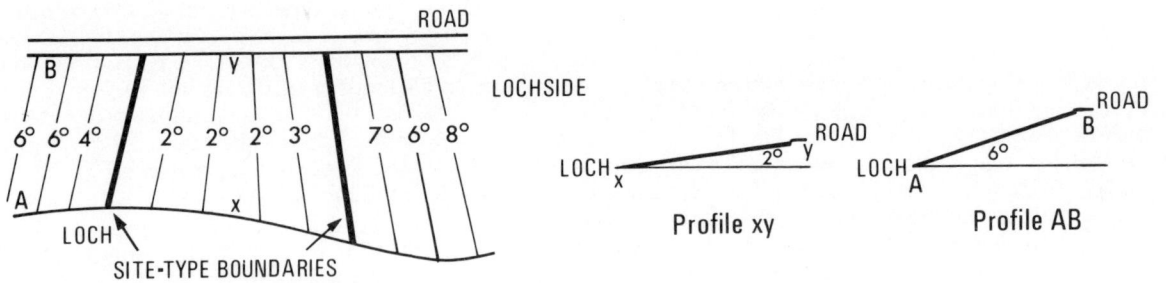

(b) Plan and profiles of lochside showing SITE-TYPE BOUNDARIES defined by changes in slope categories.

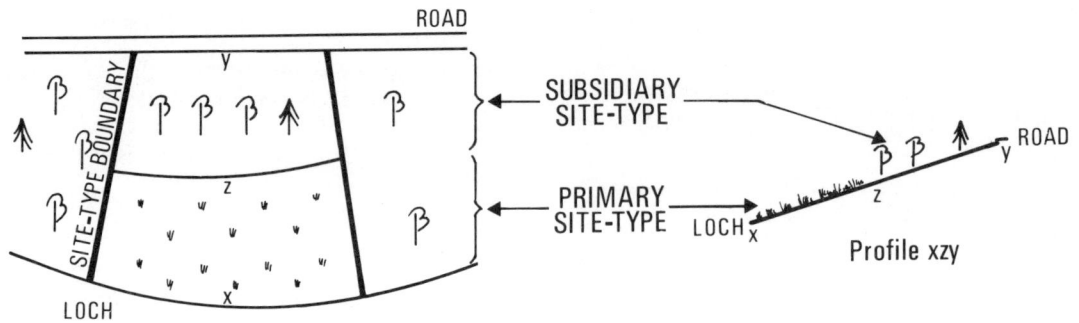

(c) Plan and profile of lochside showing PRIMARY AND SUBSIDIARY SITE-TYPES defined by a change in vegetation form.

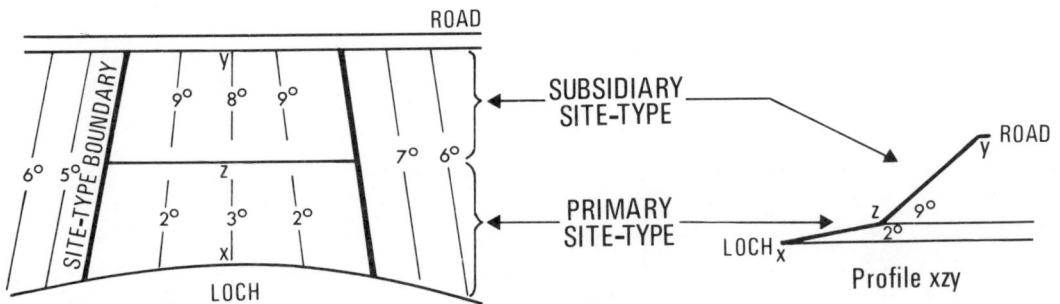

(d) Plan and profile of lochside showing PRIMARY AND SUBSIDIARY SITE-TYPES defined by a change in slope category.

Fig. 22 Boundaries and characteristics of primary and subsidiary site-types

Table 20

Loch-type description

Loch	Wet fringe	Beach	Cliff	Bank	Strip	Rock	No. divisions	content/% shoreline
Lochend L.	—	—	35	—	—	—	1	
Roscobie R.	3	—	4	82	—	—	3	—2a
L. Garten	16	84	25	14	—	—	2	—2c 81-100
Lochgoin R.	—	1	91	—	—	—	4	—2a
Ryat Linn R.	—	56	38	14	—	—	4	
Mill L.	3	59	—	70	—	—	3	—2a 61-80
L. Tarff	—	100	100	—	66	—	0	—4a
Linlithgow L.	—	13	6	23	2	—	4	
Glenboig L.	—	—	—	—	—	—	—	
Kirk L.	22	—	—	—	—	—	2	
Balla R.	31	—	12	—	—	—	3	
Holl R.	—	81	81	—	—	—	—	—2a
L. Fitty	31	—	22	24	—	—	4	
L. Gelly	34	5	—	23	—	—	3	
Cullaloe R.	—	—	—	—	—	32	2	
Beecraigs R.	69	—	—	—	—	—	—	—2c 41-60
L. Morlich	—	100	12	48	—	—	1	—2a
Harperleas R.	—	—	47	—	35	—	—	
Woodend L.	44	—	—	20	5	—	3	
Glanderston Dam	—	29	73	—	—	8	4	
Bennan R.	13	—	71	—	—	1	6	
Harelaw Dam	8	5	17	1	—	—	4	—3a
Craigluscar R. (1)	47	—	—	43	—	—	1	
Craigluscar R. (2)	—	—	—	—	—	—	1	
Stenhouse R.	—	—	49	34	85	—	1	—4a
L. Ness	—	79	2	—	7	2	7	—8
Lochcote R.	1	35	11	1	—	—	3	
L. Insh	12	61	60	10	9	—	2	—2a
St Mary's L.	—	94	5	28	—	—	8	
L. Ken	32	40	36	9	16	—	24	
Johnston L.	40	—	8	24	—	—	1	—2b
L. Glow	9	—	89	22	—	—	1	—2a 21-40
L. Duntelchaig	—	74	69	30	—	15	5	
Arnot R.	4	—	11	—	—	—	1	—3a
L. Rannoch	—	68	5	—	—	—	3	—3c
Black L.	—	29	74	—	47	—	1	—4a
Talla R.	—	83	37	—	—	—	4	—5a
L. of the Lowes	—	97	11	35	1	—	—	—8
Castle L.	60	9	7	2	—	—	3	—1
L. Tummel	4	11	20	—	—	—	3	—5c 0-20
L. Ruthven	7	65	29	15	15	1	—	—5b

well-drained and hence dry site beside water. Its recreational value on the lochside, however, is all the greater because of its relatively small size and limited development compared to that of the coastal beach. However, its value and also its vulnerability to recreation impacts are dependent to a certain extent on the form of the beach and the size of the material of which it is composed.

4.3.1 Beach material

Beach material consists of mineral particles derived from the material of which the lochside is composed or from deposition by inflowing streams and rivers. The relatively weak wave-action on most Scottish lochs is such that little beach material is derived from sources beyond its present location. Longshore drift, so characteristic of coasts, only occurs to a very limited extent on the largest lochs. However, redistribution of bottom sediments appears to take place in the shallower lochs such as Lochs Leven and Kilbirnie.

The two most obvious attributes of beach material are:

(a) the size-composition (texture) of the material as determined by the relative proportions of the particle-size classes (such as clay, silt, sand, gravel, cobbles, boulders);

(b) the variations in texture across the surface of the beach.

The majority of loch beaches examined were composed of fairly coarse material; those formed of gravel or mixed sand and gravel predominate, with the remainder composed of large cobbles and boulders. Beaches of fine sandy material (other than the machair and other coastal lagoon lochs) are the exception; those on Lochs Laggan and Morlich, therefore, have a rarity value, which is enhanced by their width and lateral extent. Finally, the material of which lochshore beaches is composed, in comparison to coastal sites, is poorly sorted and shows no consistent variation in size-composition along or across their surfaces.

4.3.2 Beach profile

The beach profile or elevation is the cross-sectional form. The gradient and form of this profile is a function of wave-action alone, or of wave-action modified by recreational use. Lochshore beach profiles are characterised by a series of more or less parallel ridges at successively lower levels from the backshore to the water's edge. Each ridge corresponds to a temporary halt in a gradually falling loch-level and is separated from the adjacent one by a smooth slope or step produced when the loch-level was falling. During autumn when the loch rises again, these ridges are destroyed; the beach material is transported landwards and is re-worked again in the succeeding summer.

Standardisation of description for comparative purposes is difficult because of the generally low gradients of the beaches and the small scale of the ridges formed. A right-angled triangle constructed from actual field measurements of profile length and gradient has been used here to provide the basis for the calculation of other descriptive parameters. (see Fig. 23).

These include:

(a) overall profile slope or gradient – the profile height divided by profile length;

(b) ridge-frequency – the number of ridge-crests along the profile divided by the profile length;

(c) profile-type – described in terms of the position of the actual profile to the theoretical slope line, namely, convex, concave and combined (see Fig. 23);

(d) relative relief, or range of relief between the highest and lowest points on the ground-surface; the mean relative relief of a beach profile can be calculated by dividing the area above and/or below the theoretical slope line by the length of the theoretical slope-line. (For the theoretical basis of this index see Bruun, 1954).

4.3.3 Beach types

On the basis of plan, or horizontal form, three types of beach can be distinguished:

(a) Line beach – a more or less unbroken stretch of beach material which may vary from a few centimetres to several metres in width; the widest loch beaches in Scotland are in this category (over 80 m on Loch Morlich). On some lochs, a suite of parallel line beaches may occur, each representing a former loch-level. The line beach is probably the most common type of beach found around Scottish lochs and certainly predominates in terms of the length of shore-line it occupies. For instance, on a random sample of twenty lochs, fourteen had shores of which over 60 per cent were composed of line beach, and none had more than 10 per cent occupied by the two succeeding types of arc and fan beach. Line beaches can be composed of any type of material from mud, or very occasionally sand, to extremely large boulders with a corresponding variation in width, gradient and ridge-frequency.

(b) Arc beach – generally widest at its centre, and marked at either end by a more or less abrupt change of shore alignment away from the loch. It is found, characteristically, at the head of a bay or inlet, and it may completely fringe a wide bay, grading imperceptibly into a line beach. Some narrow inlets, however, may only have an arc beach right at their heads or along one arm only of their shore, and in very constricted inlets arc beaches may be poorly developed because their situation shelters them from most wave-action. They often occur in an angle formed by a natural promontory, or a pier which forms an arc with an adjacent natural feature such as at Rowardennan on Loch Lomond. To a greater extent than the others, arc beaches tend to be formed of finer material such as gravel and mixed gravel/sand.

(c) Fan beach – an area of beach material generally widest at its centre which is marked at either end by a more

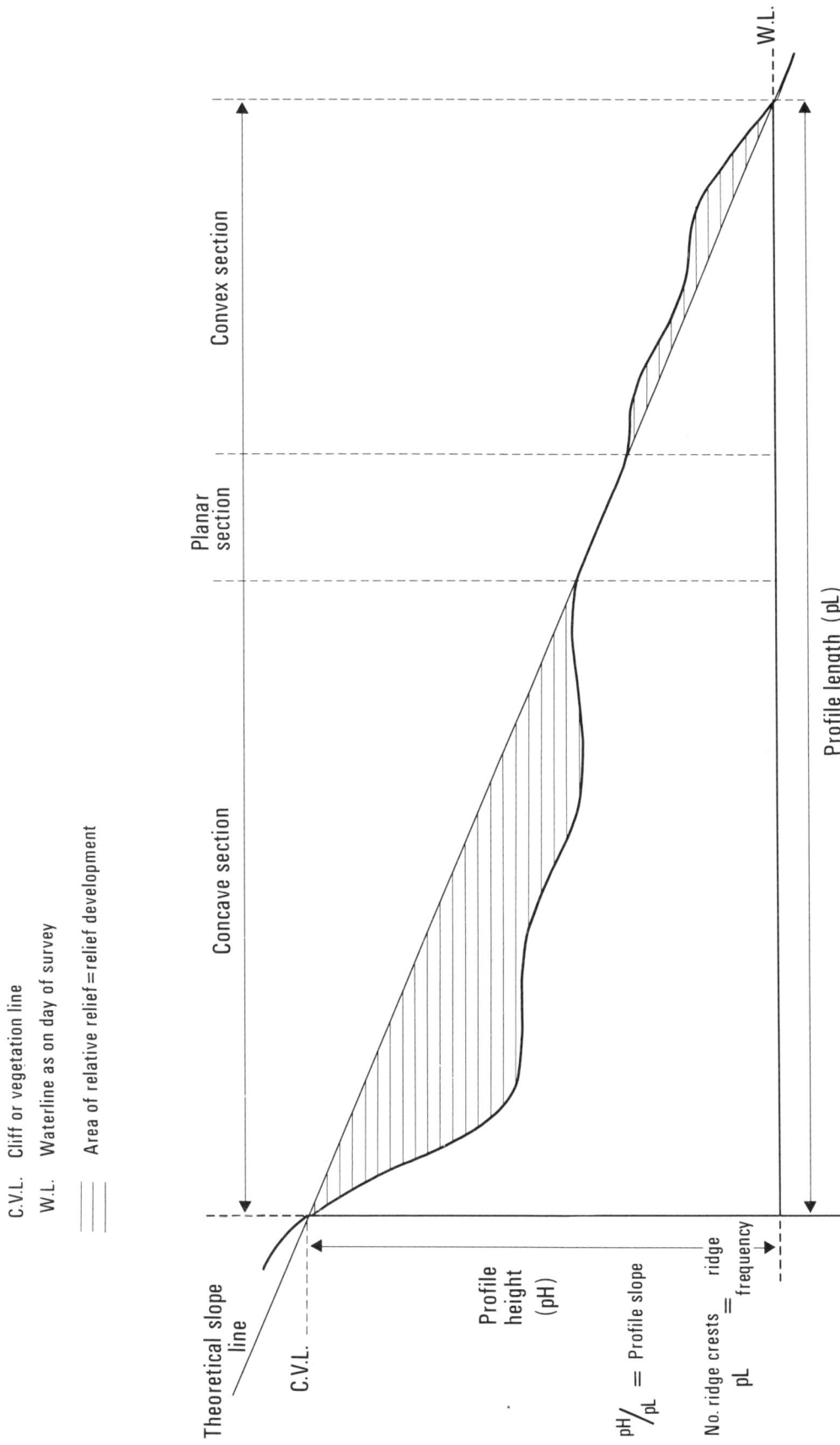

C.V.L. Cliff or vegetation line

W.L. Waterline as on day of survey

⦀⦀⦀ Area of relative relief = relief development

Theoretical slope line

C.V.L.

Profile height (pH)

$\dfrac{pH}{pL}$ = Profile slope

$\dfrac{\text{No. ridge crests}}{pL}$ = $\dfrac{\text{ridge}}{\text{frequency}}$

W.L.

Concave section

Planar section

Convex section

Profile length (pL)

Fig. 23 Lochshore beach profile parameters

Table 21

Lochshore beaches: classes of ridge-frequency, slope and relief-development

Ridge frequency		*Slope*		*Relief development*	
None	0-0.1	Flat	0°- 5°	*Smooth	0-1.0
Low	0.1-0.3	Gentle	5°-10°	Poor	1.0-1.5
Medium	0.3-0.5	Moderate	10°-15°	Moderate	1.5-2.0
High	0.5-1.0	Steep	15°-25°	High	2.0-3.0

*Smooth beaches show the closest approximation to the theoretical slope-line discussed earlier.

or less abrupt change of shore alignment towards the loch. Fan beaches are closely associated with stream or river deltas, some of which are still active, but many of which are now abandoned or have become choked with debris for all or part of the year. A fan beach may fringe the entire delta or develop along only one side. They also form around the tip of small promontories or at the end of a loch which narrows sharply towards its outflow.

Both fan and line beaches tend to be composed of coarser material with a greater number of more pronounced and persistent ridges than an arc beach. In addition, each of these beach categories exhibit variations in terms of their overall slope, ridge-frequency and relief development. For descriptive purposes, four classes for each of the three parameters are used (see Table 21). The three beach categories can thus be further characterised:

(a) line beaches – a flat or gentle slope with any number of ridges but with, at best, a poorly developed overall relief;

(b) arc beaches – a gentle, moderate, or steep slope with, at best, a low ridge-frequency but with a smooth to highly developed overall relief;

(c) fan beaches – a flat to moderate slope, usually with a medium to high ridge frequency and, at best, a moderately developed overall relief.

During the course of field surveys, variations within these beach-types have been identified by comparing their profile parameters as indicated in Table 22. It should be noted that all these beach-types are mutually exclusive except for types: 1(e)/3(c) gentle slope – high ridge frequency; smooth relief development; and 1(b)/3(b) gentle slope – low ridge frequency; smooth relief development.

4.4 Classification of Lochside Vegetation

As already indicated, vegetation-form (or, appearance) is a primary attribute of the lochside site. Its recreational value is related to the varied and changing colours and textures it imparts to the landscape; to the wildlife it

Table 22

Lochshore beach-types

Beach type		*Variations of beach-types*
1. Line beaches:	(a)	Flat – no ridges; smooth relief development
	(b)	Gentle slope – low ridge frequency; smooth relief development
	(c)	Gentle slope – low ridge frequency; poorly developed relief
	(d)	Gentle slope – medium ridge frequency; smooth relief development
	(e)	Gentle slope – high ridge frequency; smooth relief development
2. Arc beaches:	(a)	Gentle slope – no ridges; smooth relief development
	(b)	Gentle slope – low ridge frequency; highly developed relief
	(c)	Moderate slope – low ridge frequency; poorly developed relief
	(d)	Steep slope – low ridge frequency; smooth relief development
3. Fan beaches:	(a)	Flat – high ridge frequency; smooth relief development
	(b)	Gentle slope – low ridge frequency; smooth relief development
	(c)	Gentle slope – high ridge frequency; smooth relief development
	(d)	Moderate slope – low ridge frequency; smooth relief development
	(e)	Moderate slope – medium ridge frequency; smooth relief development
	(f)	Moderate slope – high ridge frequency; moderately developed relief

supports; to its scientific interest; and to the cover and shelter it provides for outdoor recreationists. Lochside vegetation, however, may have a positive or negative recreational value; it may attract or deter people from a particular site, dependent on the form and composition of the vegetation communities and on the preferences of the recreationist.

The lochside spans a transitional zone between a predominantly aquatic habitat and a predominantly terrestrial habitat. The lochshore combines the two and is characterised by a distinctive type of wetland vegetation composed of assemblages of plants rooted in the surface material, but tolerant of varying degrees of inundation. This swamp wetland has been defined by Spence (1964) as:

"Vegetation which exists on permanently, or seasonally, submerged substrata dominated typically by total or partial hydrophytes with linear emergent leaves (emergent aquatics). The summer water table in such a piece of ground may vary from more than 100 cm above, to a few centimetres below, the soil surface. Swamp vegetation is restricted as a term to vegetation on areas adjoining open water, water which may be present in winter only."

In so far as it abuts on open water, this type of wetland is particularly easily accessible by boat. Also, the lochshore wetland vegetation is of particular relevance, because it may, dependent on the type present, dampen wind-generated loch-waves, promote the accumulation and stabilisation of loch-floor sediments (either mineral or organic) and provide a habitat for a variety of wildfowl.

In the following sections, the characteristics of the lochshore and backshore vegetation, and their recreational significance, will be considered separately, before analysing the shore and backshore combinations most frequently found around the Scottish lochsides.

4.5 Lochshore Vegetation

Lochshore vegetation is characterised by three types of vegetation which may occur singly or in combination around part or the whole of many freshwater lochs or reservoirs in Scotland. The most prominent and conspicuous of these are discussed below. Plant nomenclature follows Clapham, Tutin and Warburg (1962). Details of the lochs on which vegetation transects were recorded are given in Appendix III.

4.5.1   Emergent reed and sedge beds

Emergent reed and sedge beds are composed of tall grass-like plants with, particularly in the case of the reeds, large feathery or spike-like flower-heads much prized for decorative purposes. While reeds are still cut in parts of Britain for a variety of uses, such as thatching, protective covering in horticulture, animal litter, the present demand is small. No reed cutting for commercial or domestic purposes was encountered on Scottish lochshores.

The principal species of which these beds are composed are relatively few in number (see Tables 23 and 24). The reedmace (*Typha* spp.), reed (*Phragmites communis*), reed canary-grass (*Phalaris arundinacea*) and reed sweet-grass (*Glyceria maxima*), bottle sedge (*Carex rostrata*), water horsetail (*Equisetum fluviatile*) and the yellow iris (*Iris pseudacorus*) are the most conspicuous and most commonly occurring. While they are rooted in a permanently wet or saturated substratum, usually composed of organic mud, they can tolerate fluctuating water-levels, which in summer can vary from 50 cms below the soil surface up to one metre's water-depth. Indeed, water-depth is the principal factor limiting the growth of a particular species.

Most of the tall grass-like reeds occur in dense beds all or part of which grow in some depth of water. Reedmace is commonly found in water over 30 cm so that, although the flower heads are attractive, they are difficult to reach from the shore. Apart from occasional breakage of stems by boating, they do not appear to suffer greatly from recreational impacts. The reed and reed sweet-grass usually grow on sites above the summer water-level. Reed canary-grass is more characteristic of backshore margins and often occurs in a zone behind the others. While it sometimes forms almost pure stands, it also occurs mixed with a variety of other species.

Other reed or reed-like communities are composed of the common spike-rush (*Eleocharis palustris*), the branched bur-reed (*Sparganium erectum*), the bulrush (*Scirpus lacustris*) and the water horsetail. Of these the common spike-rush is widespread. It usually occurs on a firm substratum, and is frequently found on relatively exposed sandy-stony promontories separating more sheltered bays in which other emergents are dominant. It may occur in dense stands without any other associated species, or in a more open community often with shoreweed (*Littorella uniflora*) growing beneath it. It is also found mixed with other species such as pineapple weed (*Matricaria matricarioides*), knotgrass (*Polygonum aviculare*) and annual poa (*Poa annua*) (many of which are characteristic invasion weeds) particularly on sites used for recreation. Almost as widespread is the water horsetail, though this species rarely forms a dense community. It is characteristic, like the common spike-rush with which it is often associated, of stony/sandy shores where it grows with shoreweed. None of the remaining species is found in large amounts, but all have either a marked visual impact, like the yellow flag, or are of scientific interest because of their relative scarcity. When in flower, the yellow flag usually forms vivid patches in beds of reedmace or reed canary-grass. An exception worth noting is Queen's Loch, Glenboig (NS 717687), a small urban loch, where the yellow flag forms an almost continuous peripheral fringe and, despite a high intensity of bank- and boat-fishing, shows little sign of impact. The branched bur-reed occurs in patches in nutrient-rich water usually in sheltered sites. The bulrush although conspicuous is not a common lochshore species; it is characteristic of deep water and frequently forms a discontinuous lochward fringe round reed beds; on the sites investigated it showed little signs

50

Table 23

Lochshore vegetation communities showing frequency of diagnostic plant species
(for detailed species lists see Appendix III)

Communities	Diagnostic species		Frequency (percentage)
Emergent Communities			
(a) Reed beds			
1. Common spike-rush/shoreweed	(*Eleocharis palustris/*		15-65
	Littorella uniflora)		55
2a. Dense common spike-rush	(*Eleocharis palustris*)		65
2b. Mixed common spike-rush	(*Eleocharis palustris*)		65
		Associates	35
3. Reedmace	(*Typha* species)		80
4a. Reed – dense or open in water	(*Phragmites communis*)		40
4b. Reed – transitional to backshore	(*Phragmites communis*)		25
		Other species	75
5. Reed sweet-grass	(*Glyceria maxima*)		75
6a. Reed canary-grass dominant	(*Phalaris arundinacea*)		60
		Associates	45
6b. Reed canary-grass open or mixed	(*Phalaris arundinacea*)		60
		Bare ground	10
7. Branched bur-reed	(*Sparganium erectum*)		50
8. Bulrush	(*Scirpus lacustris*)		25-50
9. Water horsetail	(*Equisetum fluviatile*)		50
10. Yellow flag	(*Iris pseudacorus*)		80
(b) Sedge beds			
11a. Bottle sedge-dense	(*Carex rostrata*)		85
11b. Bottle sedge-mixed	(*Carex rostrata*)		50-80
11c. Bottle sedge-open, mixed	(*Carex rostrata*)		50
12. Bladder sedge	(*Carex vesicaria*)		60
13. Slender sedge	(*Carex lasiocarpa*)		50
Floating-leaved Communities			
14. Water-lilies	(*Nymphaea alba, Nuphar lutea*)		90
15. Bogbean	(*Menyanthes trifoliata*)		50
16. Water bistort	(*Polygonum amphibium*)		50
17. Broad-leaved pondweed	(*Potomageton natans*)		40-60
18. Floating bur-reed	(*Sparganium angustifolium*)		20-40
Submerged/Low-shore Communities			
19. Fontinalis moss	(*Fontinalis* species)	Occasional individual	
20. Water-milfoil	(*Myriophyllum* species)	Scattered individuals	
21. Shoreweed/water lobelia	(*Littorella uniflora/*		35-75
	Lobelia dortmanna)		5-30
22. Dense shoreweed	(*Littorella uniflora*)		75
23. Discontinuous shoreweed	(*Littorella uniflora*)		37-75
		Associates other than water lobelia or bulbous rush }	25
24a. Open shoreweed ± bulbous rush	(*Littorella uniflora*)		30
	(*Juncus bulbosus*)		30
		Bare ground	45-75
24b. Sparse shoreweed ± bulbous rush	(*Littorella uniflora*)		30
	(*Juncus bulbosus*)		30
		Bare ground	75
25. Northern sedge	(*Carex aquatilis*)		40
		Bottle sedge	40
Sparsely vegetated stony shores			
26a. Common sedge-open	(*Carex nigra*)		10-70
		Bare ground	15-50
26b. Common sedge-sparse	(*Carex nigra*)		30
27. Jointed rush-open	(*Juncus articulatus*)		20
		Bare ground	25
28. Stony shores with weeds	Various species	Variable	

Table 24

Preferred habitat conditions for principal lochshore communities

Vegetation	Summer water depth	Water movement	Base status	Substratum
Emergent Reeds and Sedges				
1. Common spike-rush/shoreweed (*Eleocharis palustris/Littorella uniflora*)	5-50 cm	Exposed sites	Poor-rich	Sandy/stony
2. Common spike-rush (*Eleocharis palustris*)	Above water 20 cm	Variable	Poor-rich	Sandy
3. Reedmace (*Typha* spp)	0-60 cm	Non-turbulent	Mod-rich	Sandy-organic; variable
4. Reed (*Phragmites communis*)	Above water 100 cm	Non-turbulent	Poor-rich	According to associates
5. Reed sweet-grass (*Glyceria maxima*)	Above water 30 cm	Non-turbulent	Mod-rich	Brown/black mud
6. Reed canary-grass (*Phalaris arundinacea*)	Above water	—	Poor-rich	Firm sandy/clay
7. Branched bur-reed (*Sparganium erectum*)	Water-level 50 cm	Very sheltered	Mod	Sandy mud
8. Bulrush (*Scirpus lacustris*)	30-150 cm	Relatively exposed	Poor-rich	Firm silty/mud
9. Water horsetail (*Equisetum fluviatile*)	20-150 cm	Exposed sites	Poor-rich	Sandy/stony or organic mud
10. Yellow flag (*Iris pseudacorus*)	Above water 30 cm	Very sheltered	Rich	Brown/black mud
11. Bottle sedge (*Carex rostrata*)	Above water 50 cm	Non-turbulent	Poor-rich	Sandy/mud/peat
12. Bladder sedge (*Carex vesicaria*)	Above water 10 cm	Non-turbulent	Poor-mod	Loam
13. Slender sedge (*Carex lasiocarpa*)	Above water	—	Poor	Peat
14. Northern sedge (*Carex aquatilis*)	Above water 10 cm	Non-turbulent	Poor-rich	Peaty mud
15. Common sedge (*Carex nigra*)	Above water	—	Poor-rich	Variable – open on stony ground; close cover on damp loam
16. Jointed rush (*Juncus articulatus*)	Above water	—	Poor-rich	Variable – open on stony ground; more dense on damp loam
Floating-leaved				
17. Water-lilies (*Nymphaea alba, Nuphar lutea*)	25 cm	Non-turbulent	Poor-rich	Soft brown/black mud
18. Bogbean (*Menyanthes trifoliata*)	20 cm	Non-turbulent	Poor-rich	Soft brown/black mud
19. Water bistort (*Polygonium amphibium*)	20-200 cm (occasionally above water)	Non-turbulent	Poor-rich	Sand/brown mud
20. Broad-leaved pondweed (*Potomageton natans*)	10-250 cm	Non-turbulent	Poor-rich	Sand/brown mud
21. Floating bur-reed (*Sparganium angustifolium*)	50 cm	Non-turbulent	Poor	Brown mud/peat
Submerged/Low-shore				
22. *Fontinalis* moss	20 cm	Slow moving	Poor-rich	Black mud, often beneath stones
23. Water-milfoil (*Myriophyllum* spp)	20 cm	Non-turbulent	Poor-rich	Sandy
24. Shoreweed/water lobelia (*Littorella uniflora/Lobelia dortmanna*)	5 cm	Non-turbulent	Poor	Sandy/stony
25. Dense shoreweed (*Littorella uniflora*)	Above water 3 m	Non-turbulent	Poor-mod	Sandy
26-27. Shoreweed/bulbous rush (*Littorella uniflora/Juncus bulbosus*)	Above water	Non-turbulent	Poor-mod	Sandy/stony

Habitat conditions

of recreational impact.

The most widespread emergent sedge is the bottle sedge, which occurs on almost all the lochs studied. This is hardly surprising since it can tolerate a variety of sites and a wide range of water quality. When undisturbed, it forms dense beds which become more open as the water deepens; towards the backshore it becomes increasingly mixed with other plants. However, there is sufficient evidence to suggest that the more open stands of bottle sedge may be closely associated with recreational use.

4.5.2 Floating vegetation

Floating vegetation is a characteristic feature of the more sheltered waters of small lochs, or the bays and inlets of the larger water bodies. It is composed of plants rooted in a substratum of usually soft, relatively unconsolidated sand or organic mud, some or all of whose leaves float on or just below the water surface. Their flowers always occur on or project above the water surface. They are off-shore rather than shore plants, usually occurring in water at least 30 cm deep in summer and extending into water depths of about 2 metres.

This type of vegetation occurs on the lochward side of the emergent reed and sedge beds and is almost always composed of one main species. The most conspicuous, and visually attractive, are the white (*Nymphaea alba*) and yellow (*Nuphar lutea*) water-lilies; the latter usually occurring in the less oligotrophic lochs. Others include the pale, pink-flowered bogbean (*Menyanthes trifoliata*) and the deeper pink water bistort (*Polygonum amphibium*). All cover considerable areas of water, have attractive flowers not easily accessible to the shore-based recreationist and all are effective wave-dampeners. In addition, broad-leaved pondweed (*Potomageton natans*) occurs, often in small lochs; and the floating bur-reed (*Sparganium angustifolium*) forms patches of floating leaves in the more acid water of peaty lochs, being most characteristic of small reservoirs in moorland areas.

4.5.3 Submerged/low-shore vegetation

Submerged/low-shore vegetation is composed of low-growing plants the commonest of which are Fontinalis moss, water-milfoil (*Myriophyllum* species), water lobelia (*Lobelia dortmanna*), shoreweed and the bulbous rush (*Juncus bulbosus*). All are subject to varying periods of submergence. The first three occur as scattered plants growing on or between stones or on a stony/sandy loch floor; they are hardly ever exposed by low water-levels and none can tolerate complete drying out. In contrast, the shoreweed and bulbous rush can both withstand fairly long periods of exposure and desiccation. They dominate what can be described as low shore-vegetation, which is characteristic of exposed lochshores composed of varying proportions of sand, gravel and small stones. Given a sandy substratum, with some organic content, shoreweed can form a dense turf-like sward fairly easily disrupted by recreational impact. However, it also grows on stony sites, often in conjunction with the bulbous rush, where the vegetation cover is naturally sparser and more open;

under these circumstances the plants gain some protection from trampling from the larger stones between which they grow.

4.5.4 Sparsely vegetated stony shores

Other types of shore are those on which there is very little, if any, vegetation cover. They are characteristically composed of sand, gravel, cobbles or boulders whose mobility and/or size makes the establishment of a continuous mat difficult; under these conditions the sites are occupied by a sparse weedy cover.

4.6 Backshore Vegetation

Given the widespread distribution of lochs, together with their varying location and habitats, it is hardly surprising that a very wide range of vegetation types (cultivated or uncultivated) can be found around the backshore of Scottish lochsides. Some thirty-four vegetation types were identified from those studied in the field. These have been grouped into seven major categories (see Table 25) mainly on the basis of overall appearance. The visual characteristics are particularly important in either attracting recreational users to or repelling them from the lochside.

4.6.1 Short grass and forbs

The most attractive vegetation type for the recreationist of those summarised in Table 25 is short grass and forbs (type 2), which is very commonly found on that part of the backshore immediately adjacent to the shore; type 2a accounts for about a third of all recorded sites. The remainder, used more occasionally for recreation, have a fairly typical meadow-grass mixture of bent (*Agrostis* species), Yorkshire fog (*Holcus lanatus*) and sweet vernal-grass (*Anthoxanthum odoratum*) on lowland sites; and of bent/fescue pastures (*Festuca rubra* and *Festuca ovina*) on better hill land. A frequent feature is a narrow fringe of the soft rush (*Juncus effusus*) at the loch edge. Under wetter, normally less well-drained site conditions, what looks like a short-grass sward is often found to be dominated by the common sedge (*Carex nigra*) (type 2c); Spence (1964) suggests that the dominance of this sedge can be stimulated by grazing.

4.6.2 Long grass and forbs

Long grass and forbs (type 3) frequently form a narrow strip between the road or path and the lochshore. On a moderately steep slope, it is characteristically composed of the tall grasses (type 3a) and, sometimes, reed canary-grass which extends up-slope from reed beds at the back of the shore. On more gently sloping, less well-drained sites, tufted hair-grass (*Deschampsia caespitosa*) (type 3b) and its associates are common, as is vegetation dominated by the jointed rush (*Juncus articulatus*) (type 3c). The only other widespread vegetation type dominated by forbs rather than grasses is that composed of meadow-sweet (*Filipendula ulmaria*) (type 3d); this usually occurs

Table 25

Major categories of backshore vegetation (for detailed species lists see Appendix III)

Backshore vegetation type	Vegetation community
1. Wetland	a. Cotton-grass bog (*Eriophorum* species)

2. Short grass and forbs (over 10 cm)	a. Short grassland and meadow
	   (i) grazed
	   (ii) used for recreation
	b. Short grassland with rush fringe
	c. Short grassland with common sedge (*Carex nigra*) and jointed rush (*Juncus articulatus*)

3. Long grass and forbs (over 10 cm)	a. Weedy grass with oat-grass (*Arrhenatherum elatius*) and cocksfoot (*Dactylis glomerata*) b. Tufted hair-grass (*Deschampsia caespitosa*) c. Jointed rush (*Juncus articulatus*) with other grasses d. Meadow-sweet (*Filipendula ulmaria*)
4. Moor grass	a. Purple moor-grass (*Molinia caerulea*)
5. Low shrub (less than 1 m)	a. Heather (*Calluna vulgaris*) b. Heather/purple moor-grass c. Bog-myrtle (*Myrica gale*)/purple moor-grass d. Bog myrtle e. Bracken (*Pteridium aquilinum*)
6. Shrub (1-3 m)	a. Gorse (*Ulex europaeus*)/broom (*Cytisus scoparius*) b. Alder/willow (*Alnus glutinosa*/*Salix* species)

7. Woodland (trees more than 3 m)	a. Deciduous (i) undisturbed
	             (ii) used for recreation
	b. Coniferous (i) undisturbed
	            (ii) used for recreation

to the landward side of beds of reed canary-grass, and provides for visitors a visually attractive, sweet-smelling site. Occasionally, patches of Michaelmas daisies (*Aster* species) occur on similar sites while the rose bay willow herb (*Epilobium angustifolium*) is common, particularly on disturbed sites such as embankments.

4.6.3. Woodland

An open woodland community (type 7), particularly when a loch can be seen through it, is second in attractiveness only to short-grass vegetation. However, woodland of any kind rarely extends right down to the lochshore; there is generally a fringe, albeit very narrow, of low shrub, grassland, or common sedge community at the lochshore margin.

4.6.4 Other vegetation types

Moor grass (type 4) and low shrub (type 5) vary in terms of recreational value; the drier grass heaths and heather communities being more attractive to recreationists than the damper and longer purple moor-grass (*Molinia caerulea*), bog myrtle (*Myrica gale*) and bracken (*Pteridium aquilinum*) stands. Shrub communities are, in general, unattractive because of their dense, sometimes prickly plants (type 6a), often damp ground surface (type 6b) and, usually, relative impenetrability. Wetland peat communities (type 1) only rarely occur adjacent to the lochshore; there is usually a zone of varying width of grass or low shrub between it and the shore. Some peaty upland lochs such as Loch Goin (NS 537477) and a few lowland ones such as Johnston Loch (NS 697688) are bordered by cotton-grass bog.

4.7 Types of Lochside Vegetation

Shore and backshore sites are frequently quite clearly separated one from the other by a cliff which may vary in height from one site to another from only 25 centimetres up to 3 metres. This results in a marked discontinuity between the two main lochside components in terms of substratum, slope and drainage which is reflected in differences in vegetation type. They are ecologically unrelated and, given the diversity of both shore, and more particularly backshore sites, the possible combinations of associated vegetation types would appear to be infinite. However, from the point of view of recreational use, the separation of the continuous backshore and the shore is unrealistic. The vegetation of one may influence the use or non-use of the other, while the use of one may affect the nature and level of use of the other.

In order to assess the extent to which recreationally used shores and backshores coincided, and to identify any combinations of shore/backshore vegetation which might be particularly vulnerable to recreational use, an attempt has been made to categorise the principal types of lochside vegetation. Twelve possible combinations of lochshore and backshore communities were identified (see Table 26) of which three (1, 4 and 5) occur most frequently. In the majority of cases, there is only one type of shore vegetation, and only in the emergent reed/sedge beds does more than one community commonly occur

Table 26

Lochside vegetation types

Lochshore vegetation type			Number of transects recorded
*	1.	Bare sandy/stony shore, with or without weeds	(147)
	2.	Sparse floating-leaved vegetation	(5)
F	3.	Sparse submerged vegetation	(15)
*	4.	Submerged/low-shore vegetation	(86)
F*	5a.	Emergent vegetation (firm ground)	(154)
	5b.	Emergent vegetation (soft ground)	
	6.	Emergent – submerged vegetation	(5)
F	7.	Sparse submerged vegetation – emergent vegetation	(9)
*	8.	Low shore vegetation – emergent vegetation	(30)
W	9.	Floating vegetation – emergent vegetation	(24)
W	10.	Floating vegetation – low-shore vegetation	(1)
W	11.	Floating low-shore – emergent vegetation	(1)
W	12.	Floating – emergent – low-shore vegetation	(2)

* = those most attractive to the land-based recreationist
W = those avoided by water-based recreationists
F = those particularly attractive to bank-fishermen

between the backshore and the water's edge. Table 27 shows the relationships between lochshore and backshore vegetation for lochside types 1-5; the backshore community is that immediately adjacent to the shore, so that where a narrow strip of grassland occurs between the shore and a wooded area the former rather than the latter is taken as the backshore type. For lochside types 6-12 (Table 26), detailed data about backshore vegetation is not given, first because the backshore is adjacent to a lochshore type already indicated on the Table, and second, with the exception of 8 and 9, recorded sites in these cases are fewer and their diversity is such as to make sound and meaningful generalisations difficult. Table 27 also shows the percentage of shore and backshore sites used for recreation, and the percentage of backshore sites showing changes through recreational use (see also Table 54).

In the majority of cases, the backshore vegetation types associated with each lochshore type are surprisingly limited; one or two account for a large proportion of the combinations shown in Table 27. Grassland accounts for a significant proportion of backshores around lowland lochs, as do grass moorland and low shrub around upland and highland lochs. The same is true for backshore deciduous and coniferous woodland (or plantation), although their incidence tends to be low because of the intervening grassland zone previously noted. Of the lochside vegetation types identified in Table 39, the most attractive from the recreational point of view are undoubtedly those with a bare shore and short grass backshore (type 1) and those with a low-shore and backshore of either grassland, well-drained open woodland or heath vegetation (type 4). The latter are particularly attractive in highland lochs where the water lobelia with its emergent blue flowers is a characteristic component of the shore vegetation. All these shores are gently sloping and are separated from the backshore by a low cliff varying in height from a few centimetres upwards.

As a result of the tall, often wet plants and frequently soft ground surface, emergent lochshore vegetation (type 5) tends to repel rather than attract high levels of recreational use. There are two exceptions: common spike-rush with shoreweed which tends to be associated with sites comparable to those on which low-shore vegetation is found, and reed canary-grass which often grows on a firm loamy soil and is rarely submerged (see Table 24). This community is frequently found adjacent to fairly deep water, and has a particular attraction for bank fishermen.

Of the other lochside types of vegetation identified, the two most important are: type 8 which is characteristic of firm, sandy shores usually with a cliff separating the emergent reed/sedge beds from a backshore with either grassland or open woodland; and type 9 which is a characteristic sequence on small lochs and in sheltered bays. Although it exhibits considerable variation, on 80 per cent of the sites examined the emergent community was composed of common spike-rush or bottle sedge with short grass or deciduous woodland the characteristic

Table 27

Relationship between lochshore and backshore vegetation for Types 1-5 (see Table 26), the percentage of sites used for recreation and the percentage of backshore showing change through recreation

Number of sites	Shore vegetation at water margin	% Sites used for recreation	Intervening vegetation (where present)	% Sites where this occurs	Backshore vegetation	% Sites where present	% Backshore used for recreation	% Backshore showing change through recreation
Type 1.	**Bare sandy/stony shore, with or without weeds**							
39	Sandy, bare	94.8			−*Short grassland	64.8	100 ⎱	79
					− Moor grass	2.7	100	
					− Low shrub	8.1	66 ⎬ 94.8	
					− Woodland decid.	24.3	89	89
					− Woodland conif.	5.4	100 ⎰	100
2	Rocky, bare	100			−*Woodland conif.	100	100	100
83	Stony, bare	80.7			−*Short grassland	55.5	75 ⎱	67.4
					− Long grassland	4.9	—	—
					− Moor grass	9.8	25	
					− Low shrub	4.9	100 ⎬ 65.1	50
					− Shrub	6.2	—	
					− Woodland decid.	16.1	92.3 ⎰	92.3
27	Stony, weedy	29.6	Carex nigra/ Juncus articulatus	14.8	−*Short grassland	44.4	8.3 ⎱	8.3
					− Long grass	37.0	— ⎬ 22.2	—
					− Woodland decid.	14.8	100 ⎰	50
Type 2.	**Sparse floating-leaved vegetation**							
5	Sparganium angustifolium	—			− Short grassland	20	—	—
					−*Moor grass	40	—	—
					−*Low shrub	40	—	—
Type 3.	**Sparse submerged vegetation**							
15	Fontinalis spp	—			−*Woodland decid.	100	80	80
6	Myriophyllum spp	—	Littorella uniflora sparse/open/ discontinuous	100	− Moor grass	50	50 ⎱	—
					− Low shrub	50	50 ⎰ 50	—
Type 4.	**Submerged low-shore vegetation**							
5	Littorella uniflora dense	—			− Short grassland	40	—	—
					− Moor grass	20	—	—
					− Woodland decid.	40	—	—
25	Littorella uniflora discontinuous	48.0	Carex nigra/ Juncus articulatus	16	− Short grassland	22	57.1 ⎱	14.3
					−*Long grass	38	—	—
					− Moor grass	16	— ⎬ 25.0	
					− Low shrub	12	— ⎰	—
13	Littorella uniflora /Lobelia dortmanna	—			− Moor grass	46.2	50 ⎱	
					−*Low shrub	53.8	14 ⎰ 30.8	14.0
43	Littorella uniflora open/sparse	83.7			− Short grassland	18.6	87.5 ⎱	87.5
					−*Long grass	32.5	64.2	28.5
					− Moor grass	23.2	60.0 ⎬ 65.1	10.0
					− Low shrub	9.3	75.0	50.0
					− Woodland decid.	16.2	14.2 ⎰	—

Table 27 (continued)

Relationship between lochshore and backshore vegetation for Types 1-5 (see Table 26), the percentage of sites used for recreation and the percentage of backshore showing change through recreation

Number of sites	Shore vegetation at water margin	% Sites used for recreation	Intervening vegetation (where present)	% Sites where this occurs	Backshore vegetation	% Sites where present	% Backshore used for recreation		% Backshore showing change through recreation
Type 5. Emergent vegetation									
16	Eleocharis palustris/ Littorella uniflora	75.0	Carex rostrata/ Carex vesicaria	18.8	– Short grassland	25.0	25.0		25.0
			Carex nigra/ Juncus articulatus	18.8	– Long grass	6.3	—		—
					– Moor grass	25.0	100	62.5	25.0
					–*Low shrub	37.5	66.6		66.6
					– Woodland decid.	6.3	100		100
30	Eleocharis palustris dense/mixed	30.0	Carex rostrata or Phalaris arundinacea	40.0	–*Short grassland	70.3	43.7		12.5
					– Long grass	6.6	50.0	36.6	—
					– Moor grass	3.3	—		
					– Woodland decid.	20.0	32.2		32.5
6	Typha spp	33.3	Carex rostrata or Phalaris arundinacea	100	– Long grass	50.0	100	83.3	100
					– Woodland decid.	50.0	66.6		66.6
15	Phragmites communis open/dense	53.3	Carex rostrata/ Carex vescaria	20.0	– Short grassland	20.0	100		33.3
			Phalaris arundinacea/ Glyceria maxima	20.0	–*Long grass	33.2	80.0	62.5	20.0
				6.6	– Low shrub	26.6	25.0		25.0
					– Woodland decid.	20.0	66.6		66.6
6	Phragmites communis transitional	50.0			– Moor grass	25.0	—	50.0	—
					–*Low shrub	75.0	66.6		66.6
12	Glyceria maxima	50.0	Phalaris arundinacea	66.6	– Short grassland	16.6	100		100
					–*Long grass	75.0	33.3	50.0	33.3
					– Woodland decid.	8.4	100		100
19	Phalaris arundinacea dominant	36.8 36.8	Carex vesicaria	5.2	– Short grass	5.2	100		100
					–*Long grass	57.8	40	36.8	10
					– Shrub	10.5	—		—
					– Woodland decid.	26.3	40		40
13	Phalaris arundinacea open/mixed	100 100			– Short grassland	23.1	100		100
					–*Long grass	46.2	100	69.2	66.6
					– Shrub	23.1	—		—
					– Woodland decid.	7.6	100		100
3	Sparganium erectum	33.3	Phalaris arundinacea	66.6	– Short grassland	100	100		33.3
4	Scirpus lacustris	25.0	Phragmites communis	75.0	– Short grassland	25	—		—
					– Moor grass	25	—	50.0	—
					–*Low shrub	50	100		—
4	Equisetum fluviatile	—	Iris pseudacorus	75.0	–*Long grass	75.0	100	75.0	—
					– Dam	25.0	—		—
44	Carex rostrata dense/mixed open	65.9	Carex vesicaria Phalaris arundinacea Eleocharis palustris	27.2 13.6 2.2	– Short grassland	20.5	22.2		22.2
					–*Long grass	50.0	27.2	36.3	4.5
					– Moor grass	13.6	50.0		—
					– Woodland decid.	15.9	28.5		14.3

*Indicates most common backshore type associated with each type of shore vegetation.
Dam = In 25 per cent of cases the Equisetum fluviatile is backed by a stone faced dam.

communities.

Of the remaining lochside combinations, type 3 includes two categories. One is on stony sites where fairly deep water occurs immediately off-shore and is, in all cases, associated with open deciduous woodland or scattered trees on the backshore. The other also has a stony shore with permanently submerged vegetation and a variable zone of lowshore vegetation; it is particularly characteristic of Highland lochs and is usually combined with a heath backshore vegetation. Finally, type 2, sparse floating-leaved vegetation with heath backshore adjacent to deep water, is relatively rare and is characteristic principally of peaty lochs, and many upland reservoirs with no shore and little or no emergent vegetation. Of the remainder of lochside types recorded, variation is too great and they are not sufficiently common to allow any useful categorisation. Lochside vegetation communities of relatively uncommon occurrence, which are worthy of conservation, are listed in Appendix III.

4.8 Loch Type Vegetation

Vegetation must also be considered in the context of the entire lochside. The type and extent of the vegetation types, described in the previous sections of this chapter, vary greatly from one loch to another dependent on the number and complexity of the variable environmental factors involved. Of these, the nutrient status of the water (as well as the surrounding land) is one of the most important factors influencing the type and amount of organic material (both plant and animal) it will support (see Table 28). Because of this, water nutrient-status is still the most commonly used basis for the classification of water-bodies whether for ecological purposes or for assessing water quality for a variety of domestic and industrial uses. No analyses of water quality were undertaken during the course of this survey. However, Spence (1964, 1967) has shown that the largest part of all Scottish lochsides abut on water of low nutrient-status (oligotrophic), and only a few, mostly small (less than 25 ha) lochs are nutrient rich (eutrophic).

Table 28

Preferred water nutrient-status of plants characteristic of Scottish lochs

Rich water	Moderately rich water	Poor water	Very tolerant
Duckweed (Lemna species)	Canadian pondweed (Elodea canadensis)	Floating bur-reed (Sparganium angustifolium)	Broad-leaved pondweed (Potomageton natans)
Water bistort (Polygonum amphibium)	Water-crowfoot (Ranunculus aquatilis)	Bulbous rush (Juncus bulbosus)	[1] Yellow water-lily (Nuphar lutea)
Water plantain (Alisma plantago-aquatica)	Branched bur-reed (Sparganium erectum)	Water lobelia (Lobelia dortmanna)	White water-lily (Nymphaea alba)
Reedmace (Typha species)	Reed canary-grass (Phalaris arundinacea)	Bogbean (Menyanthes trifoliata)	Water milfoil (Myriophyllum alterniflorum)
Reed sweet grass (Glyceria maxima)			Shoreweed (Littorella uniflora)
Yellow flag (Iris pseudacorus)			Common spike rush (Eleocharis palustris)
			Water horsetail (Equisetum fluviatile)
			[2] Bottle sedge (Carex rostrata)
			Reed (Phragmites communis)
			Bulrush (Scirpus lacustris)

Rich = over 60 ppm Ca CO3; moderate rich = 16-60 ppm; poor = less than 16 ppm.
[1] Slightly less tolerant of poor water than the white water-lily;
[2] Often absent from rich waters as a result of competition from other species.
Sources: Spence (1964, 1967), Sneddon (1972), British Trust for Conservation Volunteers (1976) and field observation.

KEY for Figs. 24, 25 and 26

BACKSHORE VEGETATION

	Wetland
	Grassland / Farmland
	Moorland grass
	Low shrub
	Damp willow / Alder shrub
	Deciduous woodland
	Coniferous woodland

SHORE VEGETATION

	Reed canary-grass	(Phalaris arundicea)
	Reed sweet-grass	(Glyceria maxima)
	Reed	(Phragmites communis)
	Yellow flag	(Iris pseudacorus)
T T T	Reedmace	(Typha species)
	Bottle sedge	(Carex rostrata)
E	Common spike-rush	(Eleocharis palustris)
Sc	Bulrush	(Scirpus lacustris)
Sp	Branched bur-reed	(Sparganium erectum)
	Shoreweed / Water lobelia	(Littorella uniflora / Lobelia dortmanna)
	Water-lilies	(Nymphaea alba / Nuphar lutea)
▲ ▲	Bogbean	(Menyanthes trifoliata)
o o o	Water bistort	(Polygonum amphibium)

SHORE VIRTUALLY BARE OF VEGETATION

Beach or stony shore, occasional weeds

Sites with no shore are blank

Small lowland loch < 25ha
eg. Johnstone Loch
- 11 ha

Large lowland loch > 25ha
eg. Loch Gelly
- 60ha

Northern lowland loch
eg. Loch Garten
- 80ha

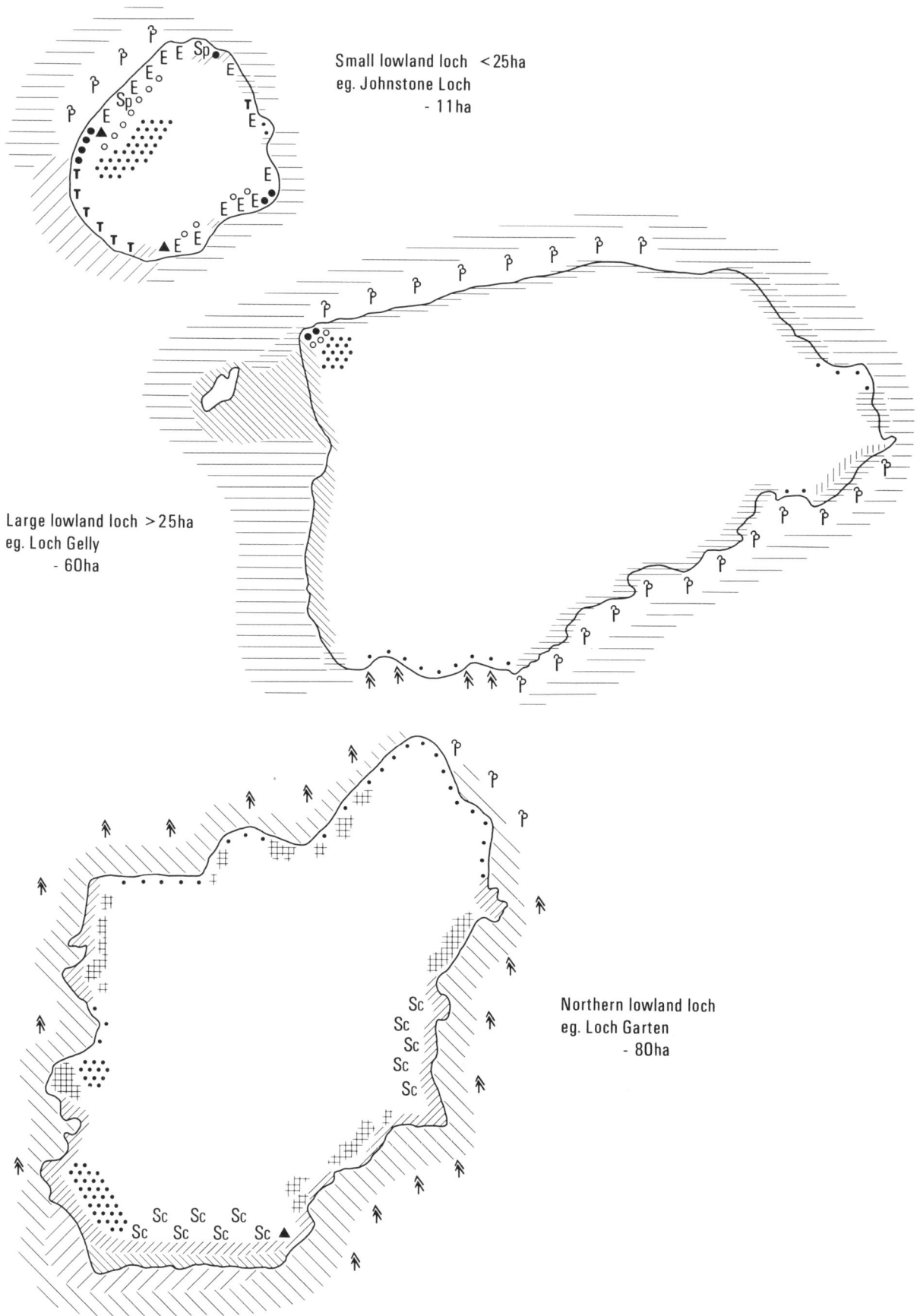

Fig. 24   Vegetation of loch types: central/southern lowland lochs, small and large

On the basis of the data collected during this survey, an attempt has been made to categorise lochs on the basis of their vegetational characteristics (particularly the composition and peripheral distribution of their shore vegetation) in order to assess variations in levels of recreational use and vulnerability to impact in relation to the size, shape, and the landform and land use setting of Scottish lochs.

The lochs examined fall into the following categories:

4.8.1   Central/southern lowland lochs

(a) Small (less than 25 ha) (Fig. 24)

Situated below 100 m O.D., these lochs are generally shallow with muddy bottoms and shores. Wave-action is limited and many, particularly those less than 10 ha, have an almost unbroken fringe of emergent reed/sedge beds. Low-shore vegetation is absent, though Canadian pondweed (*Elodea canadensis*) is a common submerged species. Absence of water turbulence allows mats of floating-leaved vegetation to become established and often cover a considerable proportion of the total water surface; water-crowfoot (*Ranunculus aquatilis*) frequently occurs around the margins.

The most striking characteristic of these small lochs is their great variety of species and communities. Apart from the main species, colourful plants such as the water plantain, trifid bur-marigold (*Bidens tripartita*) and cowbane (*Cicuta virosa*) occur within or at the edges of emergent beds; woody nightshade (*Solanum dulcamara*) is not uncommon on their drier margins. These species contribute considerably to the landscape and scientific value of the lochs, while being self-protecting to a certain degree as the tall, wet emergents do not attract land-based recreational activities.

Usually surrounded by agricultural land or species-rich, damp, peaty ground, the central southern lowland lochs would appear to be among the most nutrient-rich of the Scottish lochs. The algal scum observed on some such as Johnston Loch (NS 697688) may be an indication of incipient eutrophication. Several of the plant species, such as reedmace, reed sweet-grass, yellow flag and the branched bur-reed, which occur commonly in and round these lochs but rarely elsewhere are characteristic of nutrient-rich water. It is also significant that the nutrient-demanding types of vegetation are concentrated at the outflow where sediment tends to accumulate and nutrients levels are perhaps higher than elsewhere.

(b) Large (over 25 ha) (Fig, 24)

Also below 100 m O.D., the larger lowland lochs tend to have less complete emergent fringes, though Castle Loch (Lochmaben NY 088815) and Barr Loch (NS 353575) are notable exceptions, and tend to display a lesser variety of species. The main emergent beds are formed of one of the following dominants – reed, reed sweet-grass and reed canary-grass. There is evidence to suggest that when the reed is damaged the more vigorous reed sweet-grass can spread rapidly and replace it (see Chapter 6). Floating-leaved vegetation is confined to more sheltered bays, with water-lilies being the most common constituents.

During field surveys in 1976, the reed sweet-grass was seen to be yellowing prematurely in some lochs; similar observations were made by Hamilton (*per. comm.*) on the Forth and Clyde Canal at Bishopbriggs. Hamilton has suggested that this could be a sign of incipient eutrophication.

4.8.2.   Northern lowland lochs

Northern lowland lochs (Fig. 24) are lochs which occur at altitudes, particularly in the Highlands, from 250-350 m O.D., but have lowland characteristics in terms of the surrounding landforms and the variety of emergent species which are found on their shores. The latter are generally firm and stony, often with bare sandy/stony beaches, and with peaty cliffs in places where the backshore substratum is suitable. The low-shore vegetation type is a characteristic component of northern lowland lochs, and a limited range of emergents and floating-leaved species occur in sheltered bays and inlets, where fine sediments have accumulated. In contrast to the true lowland lochs, species characteristic of northern lowland lochs, such as reed, bulrush, bottle sedge, are tolerant of a wide range of water nutrient-status; they are, however, often associated with species characteristic of poorer water such as bogbean and bulbous rush. Water-lilies (mostly white) and floating bur-reed were virtually the only floating-leaved species observed. A final contrast with the true lowland lochs is that the backshore vegetation is typically low shrub, moor grass or coniferous woodland.

4.8.3   Moorland lochs

(a) Peaty lochs (Fig. 25)

Most of these are reservoirs above 200 m O.D. in Central Scotland. They are frequently characterised by the absence of a shore and by vertical peat-cliffs dropping in to water of over 1m depth. They have only a few emergents (except at the heads of shallow inlets) which are tolerant of oligotrophic conditions, and floating-leaved vegetation is confined to small patches of the floating bur-reed. Backshore vegetation is heath, low shrub or moor grass except in places where old improved pastures survive.

(b) Rock basin lochs (Fig. 25)

These moorland lochs (usually over 200m O.D. in the sample studied) are generally surrounded by steep, rough and often rocky slopes, also their shores are stonier than in the preceding category. They are, however, often broader with more irregular shorelines than valley lochs (see 4.8.4). Low-shore vegetation is a very common feature while emergents (such as bottle sedge) and floating-leaved vegetation (such as white water-lily) is restricted to small sheltered inlets. The backshore is usually a mixed, damp moor-grass or low shrub.

4.8.4   Valley lochs

Valley lochs (Fig. 26) are characteristically long and narrow, at relatively low altitudes and in deep valleys

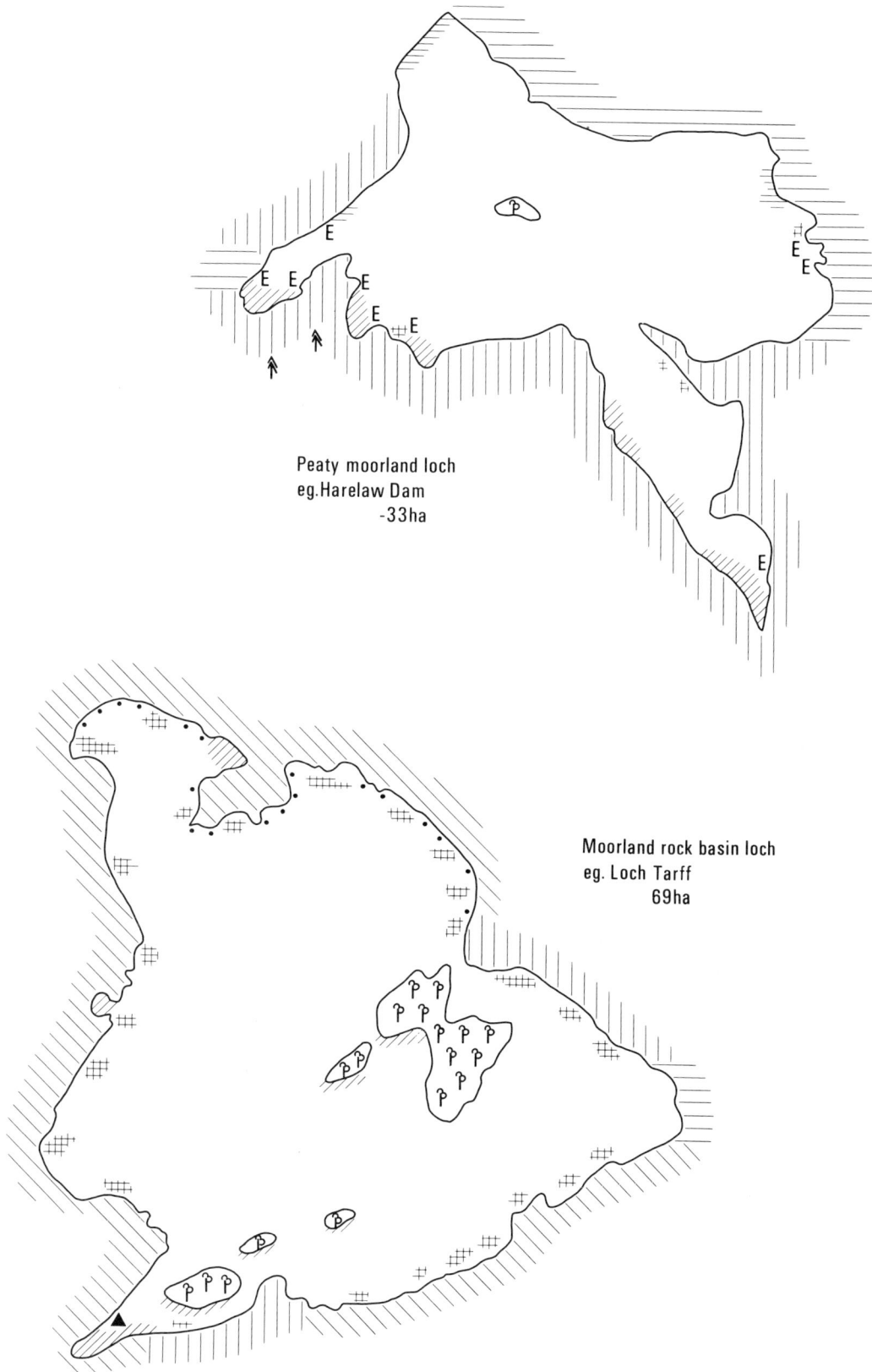

Fig. 25 Vegetation of loch types: peaty moorland and moorland rock basin

62

eg. Loch Lubnaig
- 216ha

E · E

eg. St. Mary's Loch
- 244ha

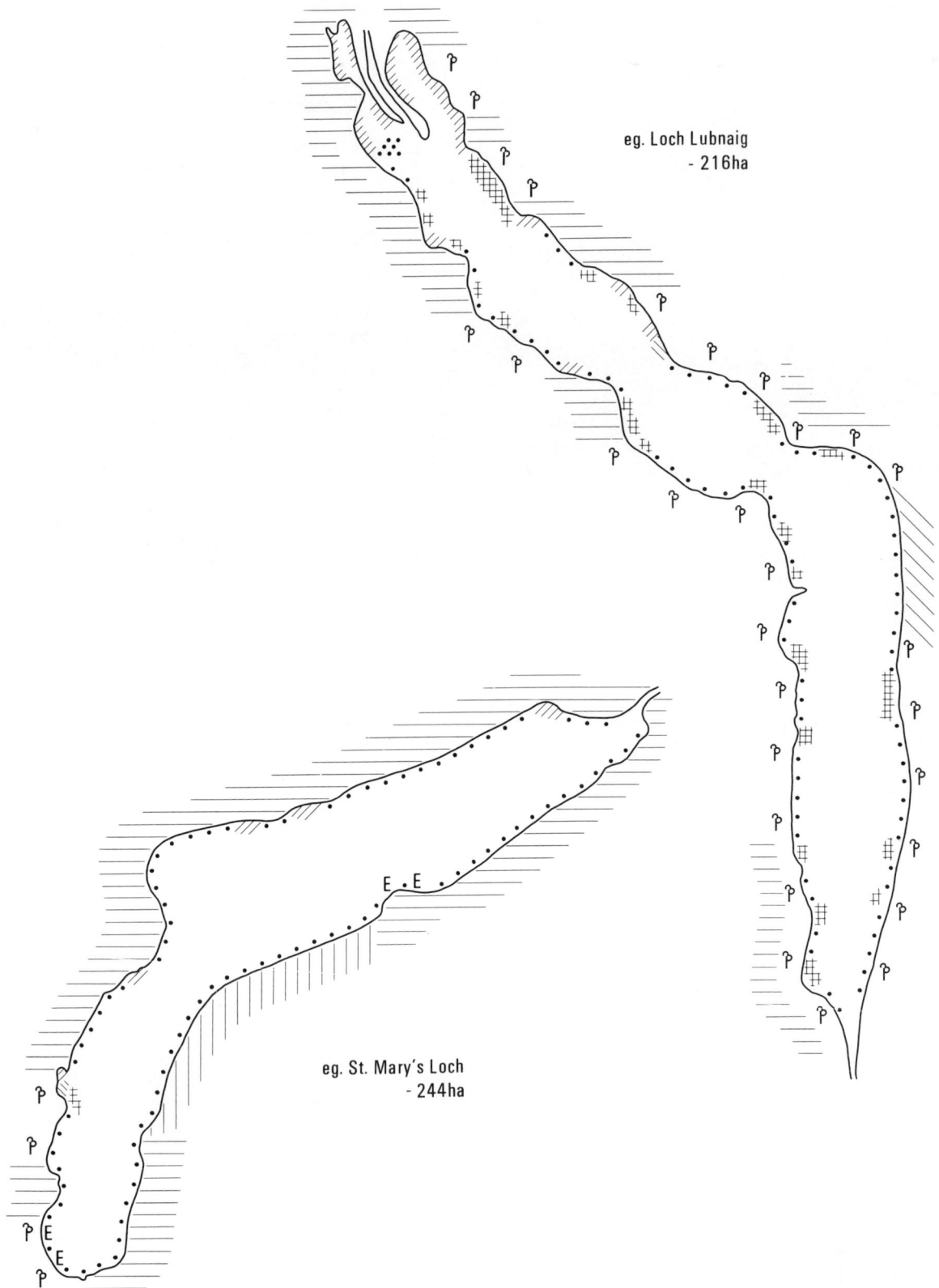

Fig. 26   Vegetation of loch types: valley lochs

surrounded by high steep slopes. These include many of the largest, as well as those most commonly regarded as typical Scottish lochs. They occur in both the Highlands and the Southern Uplands, and are among the most heavily used and highly developed for recreational uses. Low-shore or a sparse weedy vegetation is characteristic of their straight stony-bouldery shores; the latter are nearly always separated from the backshore by a cliff of varying height. Emergents (such as bottle sedge and common spike-rush) are restricted to larger sheltered bays, inlets and in-flow deltas. The backshore vegetation is usually short grassland, moor grass or low shrub but is often on ground too steep for recreational use except where delta-fans have formed at river mouths.

4.8.5 Transitional lochs

(a) Lowland-moorland lochs (e.g. Glanderston Dam NS 499562)

The majority of this type are in Central Scotland, between 150-200m O.D. and are surrounded, at least partly, by farmland, hence their nutrient status is probably intermediate between the poor moorland and the richer lowland lochs. Low-shore vegetation is uncommon, although, on mud-floored lochs adjacent to grazing land,

Canadian pondweed is frequently found and variable amounts of emergent vegetation, composed of species tolerant of a wide range of water quality, occur. The backshore is low grass or grass moor.

(b) Northern lowland – valley lochs (e.g. Loch Ruthven NH 620276)

These are lochs with relatively long, straight stretches of narrow stony shore alternating with large crescentic bays; surrounding slopes are generally steep and rough. Low-shore vegetation is sparse and/or patchy while emergents such as the bottle sedge and reed occur only in the sheltered bays.

4.8.6 Bare stony shored reservoirs

These reservoirs (e.g. Talla Reservoir NT 120214) are of varying size and shape where water draw-down can expose extensive bare shores and, sometimes, muddy floors as well. The shores are generally devoid of vegetation except where a higher percentage of fine material allows a sward of shoreweed to develop. Fluctuations of water level are too great to allow the establishment of emergents. Backshores are extremely variable.

Recreational impact on the lochside is dependent on the type or types of use to which it is put on the one hand, and the nature of the resource bases used on the other. This chapter analyses the general characteristics of recreational activities and facilities on the lochside which affect impacts. This is followed by a detailed discussion of the recreational use around the Trossach Lochs and Loch Ken in order to provide a basis for assessing the relative vulnerability of lochsides to recreational use.

The lochside is used for a wide range of recreational activities. Eighteen major activity groups have been identified in Scotland giving a total of twenty-seven individual activities (see Table 29). Curling and skating which are carried out so infrequently out-of-doors have been discounted. None of these activities is exclusive to the lochside but, viewed as a set of activities, they have certain attributes which distinguish them from similar individual or combined activities associated with other sites. Tables 31, 32 and 33 provide a systematised comparative analysis of the characteristics of lochside activities. These will be discussed in relation to the characteristics of all outdoor recreational pursuits (see Table 30).

5.1 Factors Influencing Lochside Recreational Activities

5.1.1 Resource base

Resource-based activities predominate around the lochside largely because the primary resource and attraction is the water. Table 31 shows that only in a few cases such as camping, caravanning and picnicking is there likely to be a significant development of user-oriented activities on the lochside. The wide range of activities associated with the lochside is closely related to the spatial coincidence of three types of resource base:

(a) land — the basic resource for camping, caravanning, picnicking, pony-trekking, walking, etc.

(b) water — the basic resource for boating, canoeing, sailing, water-skiing, swimming, etc.

(c) the biotic material — the basic resource for wild-fowling, nature study, and fishing; these activities being more complex in their requirements than those associated with the preceding resource bases.

Although there can be as many land-based as water-based activities associated with a lochside, these categories are not mutually exclusive and some, particularly those in (c) above, can be either land- or water-based. As has already been indicated in Chapter 2, the ratio of land- to water-based activities varies from one part of Scotland to another, and from one loch to another. One of the principal factors determining the balance, the characteristics of the resource bases, has already been discussed in Chapter 4.

5.1.2 Location

By the very nature of the resource base, lochside activities are outdoor pursuits. In contrast, however, to many similar activities, the division into rural versus urban location is not so clear-cut. In Scotland the loch has three possible outdoor recreational locations; rural lochs, suburban lochs, and urban-fringe lochs, the latter two being characteristic of Central Scotland. These may be completely, or almost completely, surrounded by land used for urban and/or industrial purposes or, characteristically, by derelict urban and/or industrial land. Some well-known examples are the Lochmaben lochs (near Dumfries), Glenboig Loch and Johnston Loch (near Coatbridge), and urban-park lochs such as Hogganfield, Duddingston and Linlithgow. Although lochside activities are all outdoor, they can in most cases be either urban or non-urban in location; even camping, caravanning and nature study occur, albeit to a limited extent, on the urban-park lochside.

5.1.3 Time

On the basis of observation of the lochs studied, it would seem that the duration and seasonality of lochside activities differ little from those pursued elsewhere. Camping and caravanning are the only two activities which exceed a day's duration and, of the remainder, about half the total are carried out on a seasonal basis because the biotic resource base is seasonally available, the activity is a competitive or a seasonal activity, or the necessary facilities are available only in the summer.

5.1.4 Effort

Passive (static) recreational activities predominate over active (mobile) activities on the lochside. The majority of these passive activities are closely though not entirely associated with water-based pursuits. However, the balance varies regionally in relation to variations in the resources throughout Scotland.

5.1.5 Organisation

Apart from orienteering and bank-fishing, all land-based activities on lochsides are non-competitive, and are fairly equally divided between formal and informal pursuits. In contrast, all water-based activities (except power-boating and sub-aqua), can be either informal or formal, competitive or non-competitive.

5.1.6 Distribution

Recreational activities have three main spatial characteristics, all of which are found on lochsides. These are linear such as those based on routes, trails or defined water-courses, as in the case of pony-trekking, nature trails, casual walking and competitive water-sports; concentrated, in that they are confined to certain parts of the lochside, usually because of their association with or dependence on particular facilities, such as moorings, jetties and piers, marinas; and dispersed activities which can be found anywhere (provided physical or other limitations do not inhibit their presence) since they are independent of facilities, as in the case of informal or 'wild' camping.

A comparison of the columns in Table 31 reveals little correlation between the characteristics of the various activities. Inspection of the rows, however, indicates that the following groups of activities have similar characteristics:

(a) formal-camping and caravanning;

(b) informal-camping and caravanning;

(c) competitive-canoeing, rowing, sailing and water-skiing;

(d) informal-canoeing, rowing, sailing, water-skiing, power-boating and possibly, sub-aqua.

The remaining activities show some degree of, but not complete, similarity. Only half of the twenty-nine activities fall into groups with the same characteristics. Hence, it must be concluded that the variety of possible activities and their attributes, rather than the activities themselves, tends to distinguish the lochside from other recreational sites. This is, as has already been noted, a function of the diversity of the resource base.

The characteristics which have been outlined above are intended to provide a flexible and adaptable basis for comparing and categorising groups of recreational activities in a number of different ways dependent on the particular purpose. For example, in zoning a loch to accommodate a number of recreational activities, spatial and temporal characteristics could be of prime importance.

5.1.7 Recreational facilities

The inter-relationship between recreational activities and facilities is a close but complex one; the characteristics of one can influence those of the other. Table 32 illustrates the nature of the relationships between recreational activities and facilities considered essential for a particular

Table 29

Recreational activities which commonly impinge on lochsides

	Activity	Type	Other features	
1.	Camping	Formal	Camp site	
		Informal	'Wild' camping	
2.	Canoeing	Competitive		
		Informal		
3.	Caravanning	Formal	Caravan site	
		Informal	'Wild' caravanning	
4.	Car parking	Formal	Car park, lay-by	Car parking is not a recreational activity, but since it is a prior stage to most
		Informal	Pull-off, roadside verges etc.	activities it has been included in the list
5.	Fishing	Bank		
		Boat		
6.	Nature trails			
7.	Nature study		Includes bird-watching and observation of plants and animals	
8.	Orienteering			
9.	Picnicking	Formal	Picnic site (with picnic table symbol sign)	
		Informal	No facilities	
10.	Pony trekking			
11.	Power-boating		Only informal power-boating has been found on lochs	
12.	Rowing	Competitive		
		Informal		
13.	Sailing	Competitive		
		Informal		
14.	Sightseeing		This may be carried out from a viewpoint, by walking, or from a car	
15.	Sub-aqua			
16.	Walking			
17.	Water-skiing	Competitive		
		Casual		
18.	Wildfowling		This comprises shooting of waterfowl. Shooting of grouse, which may occur around moorland lochs, is not related to the water body and is therefore not included.	

activity and those considered desirable but not essential. Specialised, essential facilities are required for only four activities. For twice as many, however, certain facilities are desirable; these are largely associated with water-based activities which can be pursued without facilities, but for which their presence is a definite advantage. On the whole, the water-based activities require the greater number of facilities because of the problems associated with launching, mooring and landing the wide variety of craft used on lochs.

Table 33 is a matrix designed to illustrate the nature of the relationship between groups of facilities (1-10) and facility characteristics (A-D). Definitions of recreation facility groups (1-10) are given in Appendix VI. Rows B to D outline some of the properties of the facilities in relation to their associated activities. Row B refers to the time when the facility is used, which may be throughout the activity (as in the case of a picnic table), or at the start and finish of the activity (as in the case of jetties and car parks). Row C refers to the effect of the facility on the spatial distribution of the activity; it may channel the users into a route such as footpaths, it may focus them at the start and finish of the activity, or it may confine them to a particular area such as a camp site. Some facilities formalise the activity in that they organise the participants (row D). Finally, in row E a rough indication has been given of the relative expenditure required to provide a facility, as low, medium or high.

5.2 Compatibility of Lochside Recreational Activities

While the diversity of recreational lochside activities is potentially as great as, if not greater than, in any other situation, the actual number and particular groups of activities which may occur depends on their compatibility. Compatibility in this context has been defined as "the extent to which two or more activities can be pursued in a given area; it is dependent on their ability to use either the same or adjoining sites at one and the same time, or to use a given area at different times". It is usual to consider the compatibility of land- and water-based activities separately. On the lochside they occur in such close proximity, particularly on the lochshore, that their spheres of activity may in many cases overlap to a greater or lesser extent. As a result, it is necessary, in this case, to evaluate the combined compatibility of activities occurring on adjoining stretches of land and water. The most important factors affecting the compatibility of lochside recreational activities are set out below:

(a) User requirements: activities with the same user requirements (such as camping, caravanning and picnicking) will tend to compete for the same space and will be mutually exclusive in that it is not possible to put a tent on the same piece of ground as a caravan at any one time.

(b) The nature of the activity in so far as it:
     – constitutes a danger or hazard to other ac-

Table 30

Characteristics of outdoor recreation pursuits

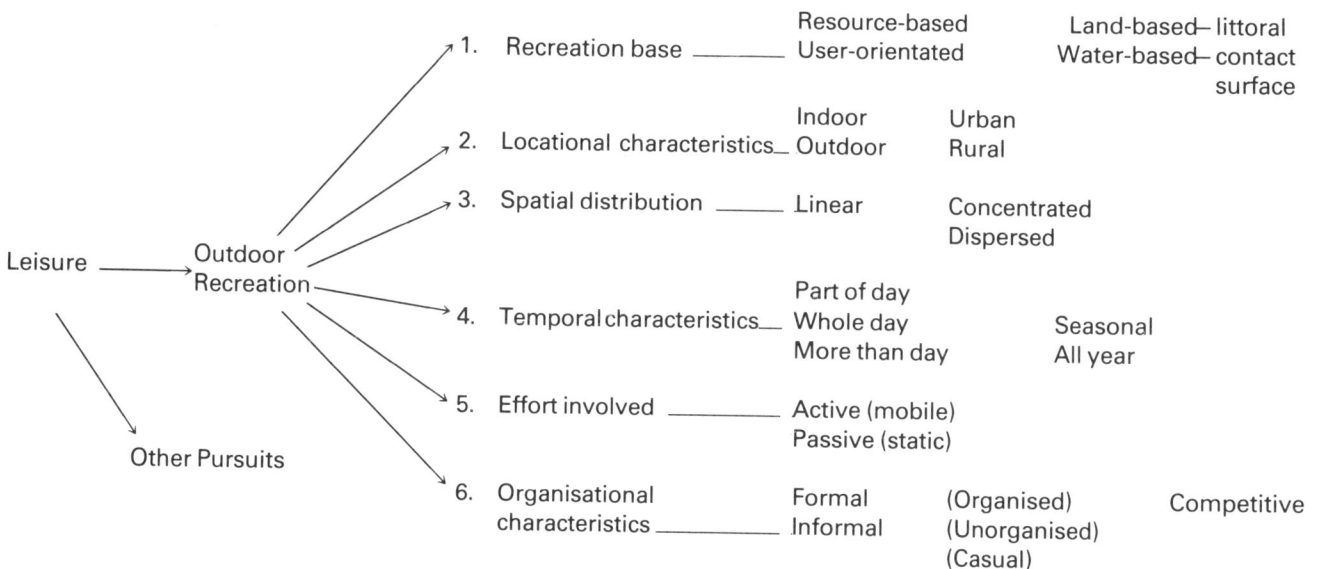

Table 31

Characteristics of lochside-based recreational activities

Activity	Resource based	User-orientated	Land based	Water based	Biological based	Indoor	Outdoor	Urban	Rural	Part of day	All day	1 day	Seasonal	All year	Active (mobile)	Passive (static)	Formal (organised)	Informal (cas. unorg.)	Competitive	Linear	Concentrated	Dispersed
Camping: formal	√	√	√				√		√			√	√			√	√				√	
informal	√		√				√		√			√		√		√		√				√
Canoeing: compet.	√			√			√	√	√	√	√	√			√		√		√	√		
informal	√			√			√	√	√	√	√				√	√		√				√
Caravanning: formal	√	√	√				√		√			√	√			√	√				√	
informal	√	√					√		√			√		√		√		√				√
Fishing: bank	√	√			√		√	√	√	√	√		√		√	√	√	√	√			√
boat	√			√	√		√	√	√	√	√		√		√	√	√	√	√			√
Nature trails	√		√		√		√		√					√	√		√			√	√	
Nature study	√		√		√		√		√	√	√		√	√	√	√	√	√				√
Orienteering	√		√				√		√	√	√				√	√	√		√	√		
Picnicking: formal	√	√	√				√	√	√	√				√	√	√					√	
informal	√	√					√	√	√	√				√		√		√				√
Pony trekking	√	√					√		√	√	√			√	√		√			√		
Power-boating	√			√			√	√	√	√	√			√	√		√					√
Rowing: compet.	√			√			√	√	√	√	√		√		√		√		√	√		
informal	√			√			√	√	√	√	√			√	√		√					√
Sailing: compet.	√			√			√	√	√	√	√		√		√		√		√	√		
informal	√			√			√	√	√	√	√			√	√			√				√
Sightseeing	√	√	√				√		√	√	√			√	√	√	√	√		√		√
Sub-aqua	√			√			√	√	√	√				√	√		√	√				√
Swimming	√			√			√	√	√	√			√		√		√					√
Walking	√	√	√				√	√	√	√	√			√	√		√	√			√	
Water-skiing: compet.	√			√			√	√	√	√	√		√		√		√		√	√		
informal	√			√			√	√	√	√	√			√	√		√					√
Wildfowling	√		√	√	√		√		√	√			√		√		√	√				√

*See **Glossary** for definition of terms*

Table 32

Matrix illustrating inter-relationships between recreational activities and facilities

Facility \ Activity	Camping: formal	Camping: informal	Canoeing: compet.	Canoeing: informal	Caravanning: formal	Caravanning: informal	Fishing: bank	Fishing: boat	Nature trail	Nature study	Orienteering	Picnicking: formal	Picnicking: informal	Pony trekking	Power-boating	Rowing: compet.	Rowing: informal	Sailing: compet.	Sailing: informal	Sightseeing	Sub-aqua	Swimming	Walking	Water-skiing: compet.	Water-skiing: informal	Wildfowling
Car park	√		√		√			(√)	√		√	√		√		√		√						√		
Lay-by		(√)		√		√	(√)	(√)		√			(√)		√		√		√	(√)	√	(√)	(√)		√	(√)
Picnic site												√														
Footpath									√														√			
Nature trail										√																
Viewpoint																				(√)			(√)			
Camp site	√	(√)																								
Caravan site					√																					
Accommodation	√	(√)	(√)	(√)	(√)			(√)	(√)	(√)	(√)	(√)		(√)	(√)	(√)	(√)	(√)	(√)	(√)	(√)		(√)	(√)	(√)	(√)
Toilets	√		(√)	(√)	√	(√)	(√)	(√)	√	(√)	(√)	√	(√)	(√)	(√)	(√)	(√)	(√)	(√)	(√)	(√)	(√)	(√)	(√)	(√)	(√)
Litter bins	√	(√)	(√)	(√)	√	(√)			√			√	(√)		(√)	(√)	(√)	(√)	(√)							
Fishing hut							√																			
Clubhouse			(√)				(√)	(√)					(√)		(√)	(√)	(√)	(√)	(√)			(√)			(√)	
Pier							(√)									(√)	(√)	(√)	(√)	(√)						
Jetty/slipway			(√)	(√)											√	(√)	(√)	(√)	(√)		(√)			√	√	
Moorings				(√)											(√)		(√)	(√)	(√)							
Boat storage			(√)	(√)				(√)								(√)	(√)	(√)	(√)					(√)	(√)	
Steamer								√												(√)						
Boats for hire				√												(√)	√		√		(√)				√	
Hide										(√)																

(√) = facility desirable for, but not essential to, the activity.   √ = facility essential to the activity.

tivities (such as wildfowling, power-boating, water-skiing);

- creates noise such as to make the pursuit of another activity such as bird-watching difficult or even impossible; and

- is a mobile, space-demanding, or linear activity which creates a direct or indirect hazard for itself and for other less mobile or static pursuits in the same area.

(c) The level of intensity of use (as expressed in terms of number of user-units combined with their speed of movement) which may cause a particular activity to be a danger or hazard to other activities;

(d) The size and shape of loch, which may in itself determine the degree of compatibility, because of the amount of water-space and lochside space available. Three aspects of the inter-compatibility of lochside-based recreation activities are illustrated in the matrices shown on Tables 34a, b and c.

Table 34a (Matrix 1) shows activities compatible in the same area at the same time. For water-based activities the same area means the whole area if the loch is small (less than 15 ha), or sections of it (if larger). In the upper triangle (land-based activities), the most important point to note is the incompatibility of wildfowling with any other

activity because of the danger involved. Formal camping, caravanning, and picnicking, together with nature trails, have a high degree of incompatibility because they have, to varying degrees, exclusive use of the land. Informal activities are, generally, more compatible. There is more potential conflict among water-based activities; this is because all competitive and power-boat activities require exclusive use of the water area. Casual rowing and boat-fishing are the only really compatible activities in this respect.

Table 34b (Matrix 2) shows activities compatible in adjoining areas at the same time. This is directly applicable to the lochside zone. Separation of land- and water-based activities lessens incompatibility; however, the very narrowness of the land/water interface may, in fact, increase the potential for conflict. If the latter is confined to the time during which the activity in question is being pursued, then the only activities likely to disturb others in adjoining areas are wildfowling, power-boating and water-skiing; the first conflicts with all other pursuits because of noise and danger; and the latter two disturb biologically-based activities and are in conflict with other water users because of the mutual dangers involved.

Table 34c (Matrix 3) shows activities taking place in the same area at different times. In this case the only conflicts

Table 33

Characteristics of lochside recreational facilities

Characteristic	Car park	Lay-by	Picnic site: fireplaces, seats, litter bins	Footpath	Guided trail	Viewpoint	Camping site	Caravan site	Accommodation	Toilets	Litter bins	Fishing hut	Club house	Pier	Jetty/Slipway	Mooring	Marina	Steamer services	Boats for hire	Hides
A. Facilities grouped according to function	1		2	3		4		5		6		7			8			9		10
B. (i) Facility used throughout activity			√	√	√	√	√	√					(√)					√	√	√
(ii) Facility used prior to and after activity takes place	√	√							√	√	√	√	√	√	√	√	√		√	
C. (i) Facility channels activity to a route			√	√														√		
(ii) Facility focuses start and finish of activity	√	√							√	√	√	√	√	√	√	√	√		√	
(iii) Facility concentrates activity in an area			√			√	√	√					(√)							√
D. Facility necessarily formalises the activity			√		√		√	√												√

(√) = facilities desirable for, but not essential to, the activity. √ = facilities essential to the activity.
Definitions of groups of activities are given in Appendix VI.

occur with formal activities (camping, caravanning, nature trails and picnic sites) where constraints of land ownership, capital investment in facilities and the ongoing nature of the activity make it unlikely that the sites could be used for alternative purposes in the short run. There is no conflict between the water-based activities.

A comparison of the three matrices indicates the considerable extent to which potentially conflicting activities can be lessened by space (Table 34b - Matrix 2) and time zoning (Table 34c - Matrix 3). However, the lochshore presents a rather particular condition which it shares to a certain degree with the sea-shore, and which is not entirely covered by the preceding matrices. Jaakson (1973) identified three types of lake activity: firstly, surface activities which include recreation that takes place on the water surface; secondly, contact activities in which the recreationist's body is in contact with the water; and thirdly, littoral activities which take place on land either adjacent to or within sight of the water. These three categories are not mutually exclusive. Their significance is in relation to compatibility; surface activities tending to be fastest and potentially least compatible, with the littoral activities being slower or sedentary and, potentially, more compatible with other activities. Littoral, in Jaakson's terms, would cover all land-based lochside activities and as such does not cope adequately with the problem of that most complex part of the littoral – the lochshore.

As the lochshore is a relatively narrow zone of variable width, subjected to alternating but irregular periods of inundation and exposure, it can be used at different

Table 34   Loch and lochside recreation compatibility matrices

Matrix compatibility table (Matrix 3). Activities listed in identical order on rows and columns; symbols: O = compatible, X = incompatible. The phrase "DOES NOT APPLY TO SAME AREA" is written across empty cells of the boating rows.

	CAMPING FORMAL	INFORMAL	CARAVANNING FORMAL	INFORMAL	BANK-FISHING	NATURE TRAIL	NATURE STUDY	ORIENTEERING	FORMAL PICNICKING	INFORMAL	PONY TREKKING	SIGHTSEEING	WALKING	WILDFOWLING	BOAT-LAUNCHING/MOORING	CANOEING COMPET.	INFORMAL	BOAT-FISHING	POWER-BOATING	ROWING COMPET.	INFORMAL	SAILING COMPET.	INFORMAL	SUB-AQUA	SWIMMING	WATER-SKIING COMPET.	INFORMAL
CAMPING FORMAL																											
INFORMAL	O																										
CARAVANNING FORMAL	O	O																									
INFORMAL	O	O	O																								
BANK-FISHING	O	O	O	O																							
NATURE TRAIL	X	X	X	X	O																						
NATURE STUDY	O	O	O	O	O	O																					
ORIENTEERING	X	O	X	O	O	X	O																				
FORMAL PICNICKING	O	X	X	X	O	X	O	X																			
INFORMAL	O	O	O	O	O	O	O	O	O	O																	
PONY TREKKING	X	O	X	O	O	X	O	O	X	O																	
SIGHTSEEING	O	O	O	O	O	O	O	O	O	O	O	O															
WALKING	O	O	O	O	O	O	O	O	O	O	O	O	O														
WILDFOWLING	O	O	O	O	O	O	O	O	O	O	O	O	O	O													
BOAT-LAUNCHING/MOORING	O	O	O	O	O	O	O	O	O	O	O	O	O	O	O												
CANOEING COMPET.																											
INFORMAL																O											
BOAT-FISHING																O	O										
BOATING																O	O	O									
ROWING COMPET.																O	O	O	O								
INFORMAL																O	O	O	O	O							
SAILING COMPET.																O	O	O	O	O	O						
INFORMAL																O	O	O	O	O	O	O					
SUB-AQUA																O	O	O	O	O	O	O	O				
SWIMMING																O	O	O	O	O	O	O	O	O			
WATER SKIING COMPET.																O	O	O	O	O	O	O	O	O	O		
INFORMAL																O	O	O	O	O	O	O	O	O	O	O	

MATRIX 1

O COMPATIBLE in same area at the same time

⊠ INCOMPATIBLE '' ie. no zoning required

MATRIX 2

O COMPATIBLE in adjacent areas at the same time

⊠ INCOMPATIBLE '' ie. space zoning required

MATRIX 3

O COMPATIBLE in same area at different times

⊠ INCOMPATIBLE '' ie. time zoning required

MATRIX 3

periods for water- and land-based activities respectively. The latter tend to predominate, since the exposed shore is at its greatest width during the summer season of peak recreational activity. It is at this period, however, that it becomes a zone of conflict between land- and water-based activities, because of the spatial overlap across a restricted area. Figure 27 illustrates the degree of conflict, and the extent to which recreational activities may or may not be incompatible on the lochshore. The point to be stressed is that not only is it a zone of maximum potential conflict, but that three conflicting directions of movement can cross each other and thereby greatly exacerbate mutual hazards and dangers as well as intensifying physical impacts.

5.3 Level or Intensity of Use

Recreational impact will also be dependent on what may be called the degree, level or intensity of use. This is, in turn, dependent on a complex interaction of factors among which the most important are those associated with:

(a) the number of people and vehicles directly or indirectly involved in certain types of recreation activity;

(b) the duration of the activity in terms of its seasonal and diurnal periodicity; the frequency with which it is pursued during these longer or shorter periods of time, and the length of time it is pursued in the course of any recreational period.

Table 35 summarises, in general terms, the duration (or temporal) characteristics of the main types of lochside recreational activities. Apart from coarse-fishing and bird-watching, which can be and are pursued all the year round (except in those waters where Sunday fishing is prohibited) all activities have a marked seasonal and diurnal periodicity. In most parts of Scotland, informal car-based activities can be, and are, carried on throughout the year, though winter activity, as elsewhere, is greatest during periods of good weather at weekends. The peak period is the short summer season with a diurnal peak of activity, in terms of numbers of people on site, commonly occurring between 14.00 and 16.00 hours. Water-based activities, with the exception of game-fishing and wildfowling, are almost entirely confined to the summer period, although their diurnal periodicity tends to coincide with that of the informal land-based activities; water-skiing and sub-aqua are popular evening pursuits. Also, competitive water-sports are weekend activities, which because of their visual appeal tend to attract, and thereby increase, the usual build-up of land-based visitors at this period.

Game-fishing and wildfowling differ in that these activities are controlled by close seasons designed to conserve breeding stocks. Further, spring and early summer months are more favoured for trout fishing because water temperatures and food supply result in more 'active' fish than later in the summer; whereas the most favoured months for salmon fishing vary depending on the timing of local runs of fish.

The frequency and length of time a lochside or part of a lochside is used for a particular purpose within a season is a vital factor in determining intensity of impact. However, the relationship between frequency and duration of use on the one hand, and numbers involved, on the other, is not easy to establish. Whether the impact-effect of a large number of people during one short time-period is the same as that arising from frequent short-duration use by a few people over a long time-period has not been established: it complicates an already complex issue. Also, where temporal compatibility allows a diverse activity-mix, the result may be either to lessen or to exacerbate the impact effects of single-activity use. At present, there is not even sufficient empirical data on which to base tentative hypotheses as to the effect of recreational compatibility (or of incompatibility as the case may be) on intensity of recreational impact on the lochside.

The problem of defining and describing level of recreational use, in standardised terms, is therefore extremely difficult, indeed, virtually impossible with any degree of accuracy. Not only are the number of variables involved considerable, but many do not lend themselves to quantitative measurement. In face of this problem, to which no satisfactory solution is available, various parameters can be used independently or in combination, dependent on the data available, and the use for which some index of level-of-use is required.

In order to provide a standardised basis for descriptive and, particularly, for comparative purposes, in relation to level or intensity of informal lochside or lochside-site use, it was decided to use two characteristics:

(a) Amount of Use
This is expressed as user-units per unit-area per unit-time. User-units have been taken as numbers of users plus numbers of vehicles. No attempt is made to differentiate between types of users (by age or activity for instance) because, except on very specialised sites, most users indulge in more than one impact-causing activity in the course of a recreation period. On informal sites, it has been demonstrated by user-surveys that over 90 per cent and often over 95 per cent of the parked vehicles are cars, therefore no attempt is made to equate cars, caravans or trailers. A unit-area is equivalent to one hectare (which allows comparison with other types of density), and unit-time equals a ½-hour for cumulative periods and ¼-hour for the peak period. This time-unit was selected because a very high proportion of short-duration visits at monitored sites were of this length, and between ¼ and ½ hour is taken as minimum for the pursuit of even the least time-demanding recreational activity. The calculation of user-units on this basis is possible. The main problem is the practical one of assessing or measuring the area of the recreational site at the time of the user census (records are few and difficult to acquire). It is necessary that both be measured at the same time, particularly when the use of the

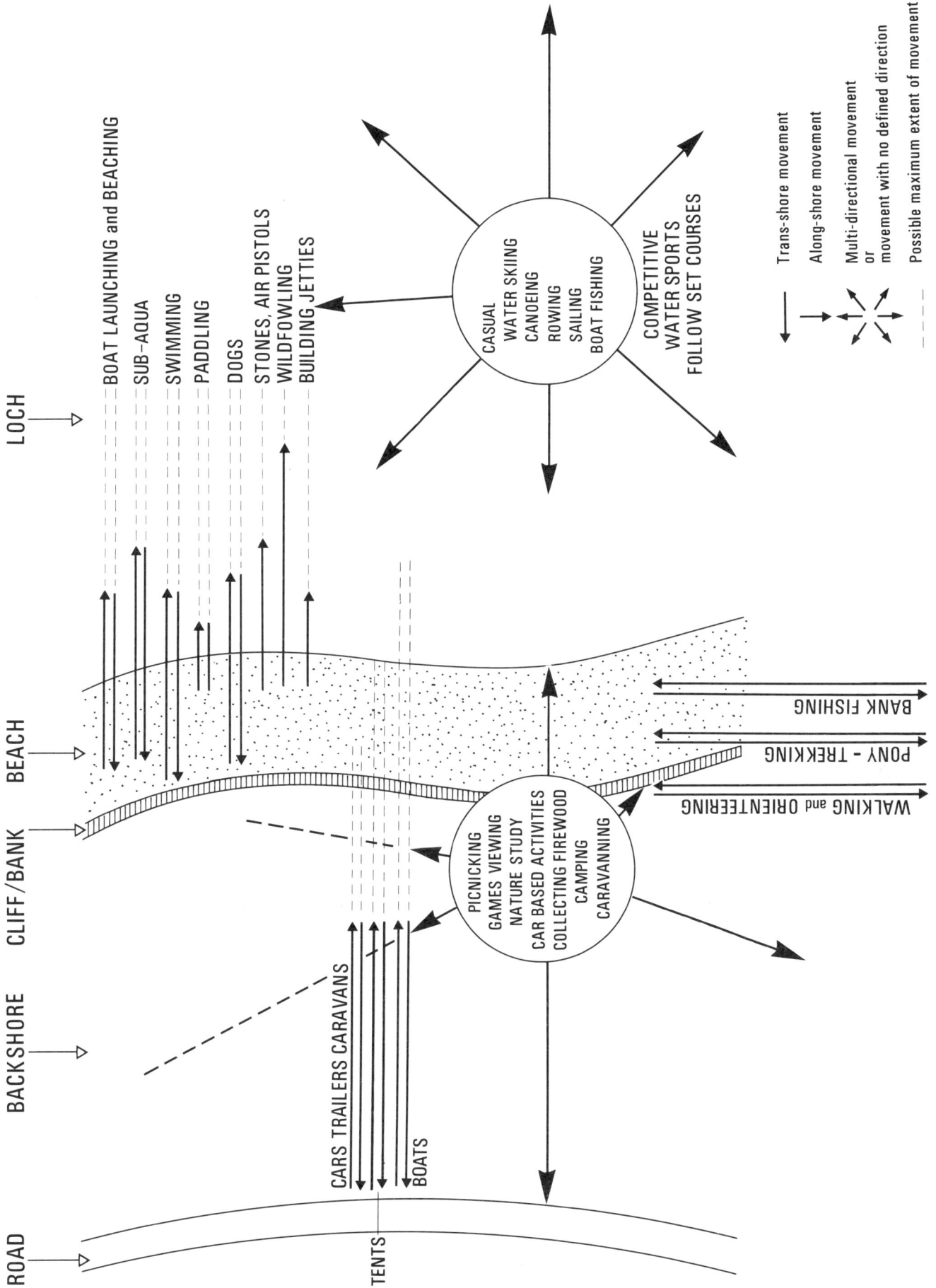

Fig. 27   Model of spatial distribution of recreational activities on the lochshore and backshore zone

lochshore is involved. Where no boundaries exist the delimitation of a recreation area is a further problem and again, on lochshores, the distribution of users may have to be used to define the boundaries of the recreation area. An index calculated on this basis will obviously have little absolute value. It does however provide, albeit a crude, estimate of the comparative density of use of particular sites at particular times.

(b) Duration of Use

This is also taken into consideration in assessing intensity of use. It is necessary to distinguish between short- and long-stay sites, since a rapid turn-over of visitors and vehicles can compensate for low density of numbers and vice versa.

This method was used to estimate the intensity of recreational use around the Trossachs Lochs and Loch Ken (see Section 5.4.4).

5.4 Characteristics of Lochside Recreational Use around the Trossachs Lochs and Loch Ken

The only readily available, detailed, and relatively up-to-date data about loch and lochside use is that in the *Loch Lomond Recreation Report* sponsored by the Countryside Commission for Scotland (Brown and Chapman, 1975). It presents information about the visitors and their patterns of activity in order to indicate the extent to which existing resources are used, the adequacy of various sites for visitor requirements, and existing conflicts between recreation and other land uses. Apart from this publication, no other data existed from which the type, distribution and intensity of lochside recreational activities could be assessed either at the national, regional or lochside scale.

The lochside data set referred to in Chapter 2 has gone some way to rectifying this deficiency at a national scale. However, more detailed information at the loch and lochside scale was required to verify subjective assessments, and two user-surveys were undertaken – one on

Table 35

Temporal characteristics of the main types of lochside-based recreational activities

Activities	Month (Jan Feb Mar Apr May Jun Jul Aug Sep Oct Nov Dec)	Diurnal periodicity (Peak time)	Frequency	Duration of activity
Informal car-based activities	— — — — — ——————— — — —	(14.00-16.00)	Ww; WHs	2 hours av.
Camping	— — — — — ——————— — — —	(18.00-10.00)	WHs	All night/part or all day
Fishing [1] salmon/ salmon trout	10th_____10th	Variable	WseHs Ww	Several hours whole day weekend trips
trout	14th_____7th			
coarse fish	———————————————			
Wildfowling	——————— ———————	Early am/or pm	W	½ – 1 day
Bird watching	———————————————————	Variable	W	½ – 1 day
Sailing [2]	— ——————— — —	(14.00-16.00)	WeHs	Several hours whole day
Canoeing [3]	— ——————— — —	(14.00-16.00)	s	1 – 2 hours
Water-skiing [4]	— ——————— — —	(14.00-16.00)		½ – 1 hour
Sub-aqua [3]	— ——————— — —	(14.00-16.00)		½ – 1 hour
Casual boating	— ——————— — —	(14.00-16.00)		½ hour – whole day

[1] Most days, particularly competitive fishing on Saturday; no Sunday fishing on game fish and on some coarse fish lochs; spring early summer favoured because fish are more active; [2] Competitive sailing; [3] Canoeing, sub-aqua at weekends; [4] water-skiing an evening activity in summer. W = Weekend; H = Holidays; w = winter; s = summer; e = evening.

the Trossachs Lochs in July/August 1975, and the other on Loch Ken in July/August 1976. Both are within the special study area (see Fig. 28), and detailed work on resource characteristics and recreational impact on the lochs had been carried out during the field season previous to the user-surveys in question. The main objective of these user-surveys was to make a simultaneous (between 11.00 and 17.00 hours) record of the numbers of people and vehicles, and types of activities, on all the main informal sites on four peak days in the period July-August, in order to provide a comparative assessment of levels of use and hence degree or amount of physical impact on the lochsides in question. (A more detailed explanation of methods of conducting this survey and the type of questionnaire used is given in Appendix IV).

5.4.1   Recreational development

The lochs surveyed (see Table 36) varied in size, physical

characteristics and setting, as well as in types and level of recreational use. Loch Ken provided a contrast with the Trossachs group, in being further away from Glasgow, as well as in the origin and development of its recreational activities. The Trossachs has long been a popular recreational area for people from Central Scotland, and more particularly from Clydeside. Although the scenic values of the Trossachs had been extolled before the publication of Walter Scott's romantic *Lady of the Lake* (1806), there can be little doubt that, after that poem's publication, the area acquired international renown. Recreation development also started early. The then Duke of Montrose improved what is still known as the Duke's Road north from Aberfoyle, and provided horse-drawn carriages for visitors who wished to follow the scenic route across the watershed between the Upper Forth Valley and the Loch Achray - Loch Venachar basin. The later extension of roads and branch railways within the area

Table 36

Trossachs Lochs and Loch Ken: selected physical and user characteristics

Loch	Size (ha)	Depth mean (m)	Depth max (m)	Mean range water-level (m)	Riparian rights	Fishing rights	Special recreational features	Other uses
L. Achray	79	11.3	29.7	1.2	F.C., 2 private	Callander District F.C.	Loch Achray Hotel, Trossachs Hotel, Water-ski club	Way-marked forest track along south-east shore
L. Ard	235	13.4	32.6	c0.91	F.C., multiple private	Ditto	Sailing club	—
L. Chon	106	9.1	22.9	c0.91	F.C., private	Ditto	—	—
*L. Drunkie (1)	48	11.8	29.6	5.9	F.C., private	L.C.W.B.	—	Compensation reservoir draw-off from L. Katrine
L. Lubnaig	216	6.4	44.5	1.2	F.C., 2 private	Sub-let Perth Fishing Ass.	—	—
Lake of Menteith	264	6.0	23.5	c0.61	F.C., D.O.E., multiple	Lake of Menteith Fisheries Ltd.	Inchmahome Island with Priory. Lake Hotel	—
*L. Rusky	20	7.0	14	?	Private multiple, private	Private + Angling club	Club house and private parking area; limited wildfowling and bird watching	Water supply
L. Venachar (1)	394	12.5	33.8	3.4	C.R.W.D., multiple private	Private	Venachar Water Users Ass.; loch zoned; fishing permits sold by Callander District and sailing club	Compensation reservoir
L. Ken (1)	464	6.4	15.2	1.8	Multiple private	Private		Hydro-electric power generation

(1) impounded; *lochs not included in user-survey;   L.C.W.B. = Lower Clyde Water Board.
F.C. = Forestry Commission; D.O.E. = Department of the Environment; C.R.W.D. = Central Region Water Department.

F

76

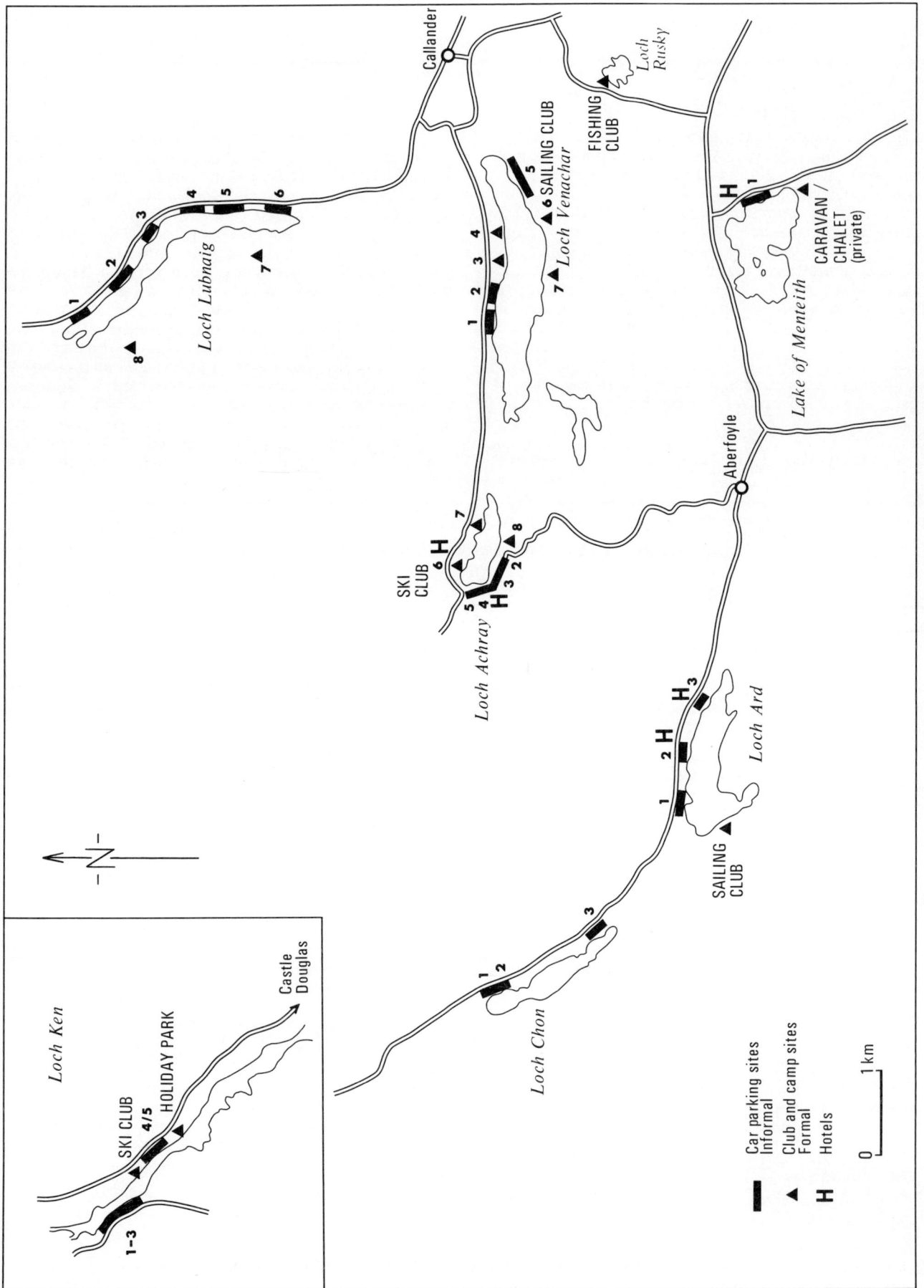

Fig. 28 Trossachs Lochs and Loch Ken: location of user-survey sites

assured the establishment of the Trossachs as a fashionable resort area by the latter half of the nineteenth century.

In contrast, Loch Ken's recreational development came nearly a century later. The present loch originated when this part of the River Dee was impounded in 1935 to create the largest of the reservoirs of the Galloway Hydro-Power Scheme. Situated in a quiet backwater, off the main road and rail routes, Loch Ken has experienced an upsurge of recreational use only within the last ten years or so. Two factors have contributed to this rapid development: firstly, improvements to the A75 (Dumfries-Stranraer) and to the A74 and M6 roads which facilitated travel from northern England to Galloway; and secondly, the growing demand from the north of England for new sites for coarse fishing and water-skiing, partly a result of excessive recreational pressure in the Lake District.

5.4.2 Relative importance of recreational activities

All the lochs fall into the rural category and, as Table 37 indicates, informal recreation is dominated by the more passive, land-based activities. The most frequently pursued passive recreation activity is always, and the second ranking activity is frequently, in Group 1 - 8 of this table. Walking, Group 9, is undoubtedly the most popular of the more active land-based pursuits. Around all these lochs, activities Groups 1 or 6 rank first for all except the most specialised sites, while the second ranking activity by site is either Groups 1 or 6 with 8 and 9 as the most popular alternatives.

In contrast, the proportion of groups involved in water-based pursuits is relatively small, usually less than 20 per cent; only on Loch Lubnaig and Loch Ken do they attain significant levels, which are 30 and 35 per cent respectively (see Table 38). In the case of the former loch, this was due to the ease of access from the road to the shore by cars, caravans and boat trailers at three large recreation sites. The development of two major facilities on Loch Ken, the Loch Ken Holiday Park (caravan site) and the Loch Ken Water-Ski Club, at the two points on the lochside where access from an A-class road is easiest, has accounted for the recent increase in recreational activity.

Table 37

Percentage total groups for which a given recreational activity was recorded on four peak days around Loch Ken, Trossachs Lochs and Loch Lomond

Group number	Recreational activities	L. Ken	L. Chon	L. Ard	Lake of Menteith	L. Venachar	L. Lubnaig	L. Lomond (1973)
1	Sitting in car	24.7	?	28.8	40.5	40.4	29.0	—
2	Picnicking in car	11.9	—	20.2	17.6	21.5	14.0	(15.0)
3	Picnicking near car	18.9	25.4	32.0	—	—	—	(40.0)
4	Picnicking over 50m from car	7.4	—	30.0	22.9	25.5	36.0	—
5	Playing games	7.6	13.2	14.1	16.8	23.9	19.2	(9.0)
6	Relaxing out of car	31.8	35.0	62.0	57.3	62.0	62.2	(64.0)(a)
7	Taking photos	18.7	14.1	32.0	19.8	32.4	43.8	—
8	Strolling near car	19.3	13.2	30.0	45.0	39.7	40.2	(2.3)
9	Walking some distance from car	19.5	28.3	43.6	38.9	46.4	—	(43.0)
10	Hill walking	2.7	14.1	—	—	—	—	—
11	Bank fishing (competitive)	1.3	—	—	—	—	—	—
12	Pony trekking	1.0	—	—	—	—	—	—
13	Bank fishing (casual)	9.7	—	—	—	—	10.3	—
14	Bathing/paddling	11.7	14.1	22.2	35.1	28.0	40.2	(26.0)
15	Boat fishing	2.2	—	—	—	—	—	(3.0) (b)
16	Casual boating	6.5	—	11.0	—	—	—	(3.0)
17	Competitive boating	1.5	—	—	—	—	—	—
18	Water-skiing	7.0	—	—	—	—	—	—
19	Sub-aqua	0.3	—	—	—	—	—	(2.0)
20	Bird watching	5.5	—	—	—	—	—	(10.0) (c)
21	Others	—	—	—	—	—	—	(25.0)

Loch Lomond figures from *Loch Lomond Recreation Report* 1975; (a) looking at scenery; (b) fishing unclassified; (c) nature study.

78

Bank-fishing and casual boating are the most frequently pursued water-based activites, except where such incompatible sports as power-boating, with or without water-skiing, take place. Also, water-based activities tend to be formal or semi-formal (for example, fishing organised by clubs and associations often in conjunction with a fishing-hut, or less commonly, a club-house). In this group, the use of Loch Rusky by a private fishing club virtually inhibits any other use of the lochside. Water-skiing and sailing focus on private club-houses and associated jetties located on Lochs Ard, Achray and Venachar; numbers involved, however, are relatively small and are regulated. On Loch Ken, the facilities provided by the two commercial concerns are foci for similar water-sports; no club membership is involved, pursuit of the activities is less formal, and greater numbers of people are indirectly associated with them. Probably the most important informal water-based activity is bank-fishing, without a permit, a common lochside activity throughout Scotland. In contrast, there are few formal, land-based activities.

5.4.3   Facilities

Recreational facilities available, especially around Loch Ken and the Trossachs Lochs, closely reflect the type of activities and, particularly, their low level of development (see Table 39). The most common are those associated with informal land-based recreation – Group 1 and Group 6 (though with a conspicuous lack of toilets) as defined in Table 33. The second most common Groups are 7 and 8, being facilities for berthing, launching and mooring, with associated club houses. All are facilities which are used prior to or after a particular activity, and which tend to focus and thereby concentrate informal as well as formal recreation activities. The result is that on the lochside dispersed informal recreation activities are less characteristic than on other sites. Also, linear activities are relatively limited: footpaths and tracks play a more im-

Table 38

Percentage groups of recreationists recorded who were involved in water-based activities

Activities	L. Ken	L. Chon	L. Ard	Lake of Menteith	L. Venachar	L. Achray	L. Lubnaig	L. Lomond
Water-based	34.7	15.0	19.6	8.7	14.3	10.1	30.6	(36.0)
Bank fishing	11.8	8.0	6.7	—	—	3.5	10.3	(3.0)
Casual boating	8.0	—	—	—	8.8	—	9.3	(3.0)
Sub-aqua	0.3	—	—	—	—	—	8.8	—
Water-skiing	8.5	—	—	—	—	—	—	(2.0)
Boat fishing	2.7	—	—	—	—	—	—	—
Competitive boating	1.8	—	—	—	—	—	—	—
Competitive bank fishing	1.6	—	—	—	—	—	—	—
Pleasure cruising	—	—	—	—	—	—	—	(2.0)

Source: Field Surveys 1975/1976 and Brown and Chapman (1975) *Loch Lomond Recreation Report*.

Table 39

Number of recreational facilities around Loch Ken, Trossachs Lochs and Loch Lomond

Recreational facilities	L. Ken	L. Chon	L. Ard	Lake of Menteith	L. Venachar	L. Achray	L. Lubnaig
Car parks	2	2	1	1	2	4	3
Jetty/pier	1	1	1	2	1	1	—
Club house	1	—	1	—	1	—	—
Picnic tables	—	—	—	—	—	—	—
Boat house	—	1	1	1	—	—	—
Log cabin	—	—	—	—	—	—	17
Hotels	—	—	2	1	—	2	—

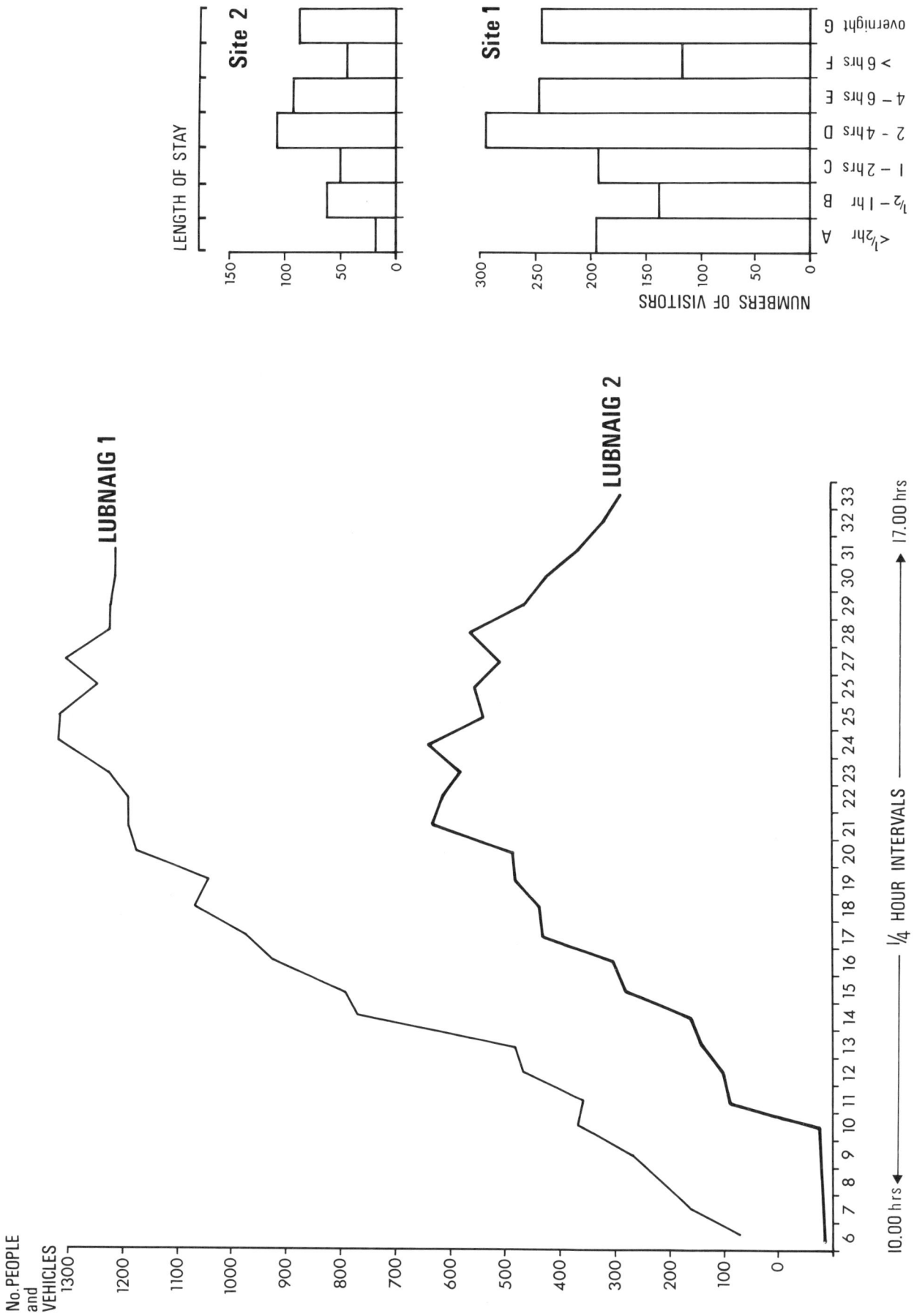

Fig. 29 Daily variation in numbers of people plus vehicles per ¼-hour interval and average length of stay on selected recreational sites on Loch Lubnaig

80

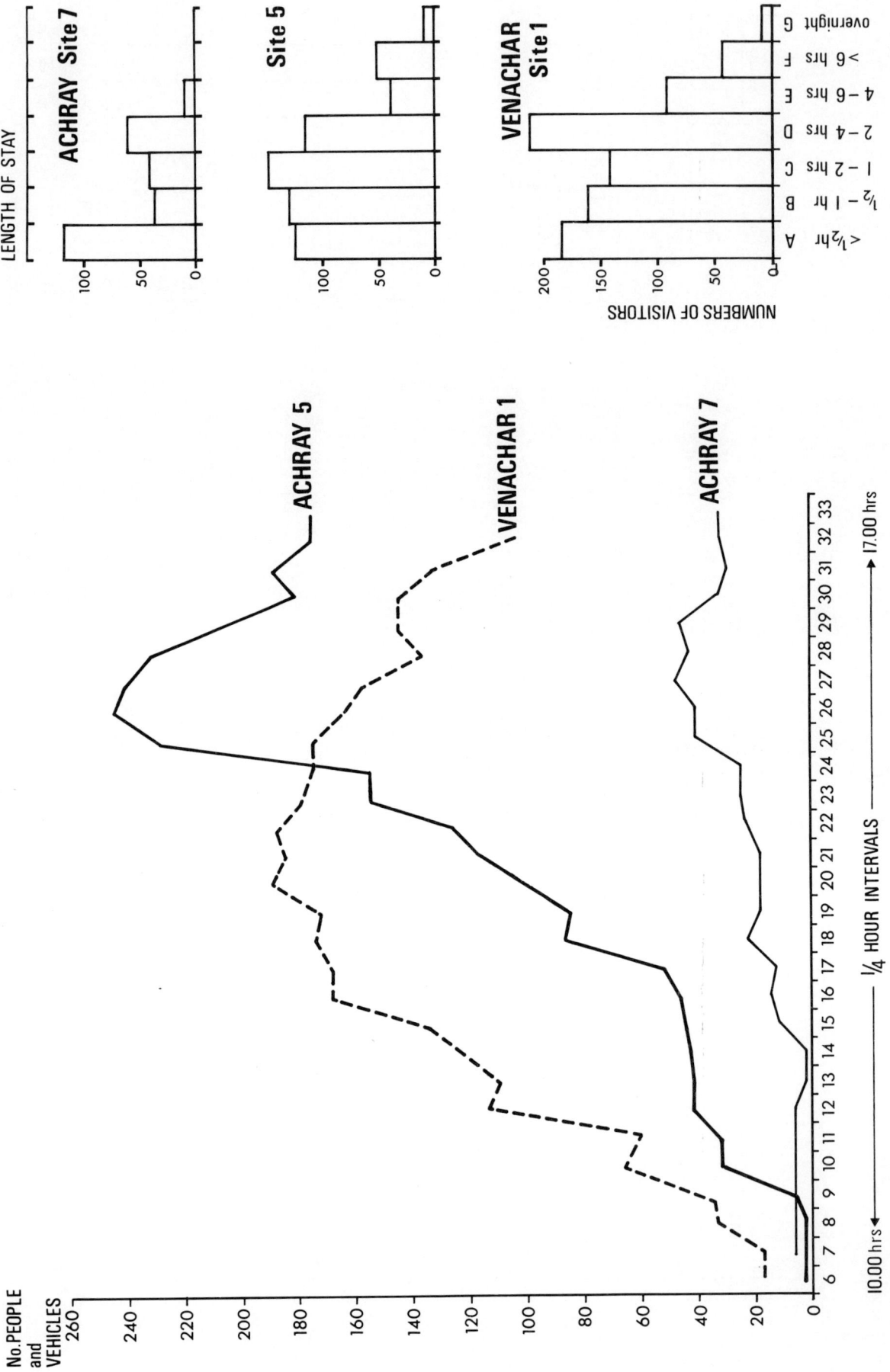

Fig. 30 Daily variation in numbers of people plus vehicles per ¼-hour interval and average length of stay on selected recreational sites on Lochs Venachar and Achray

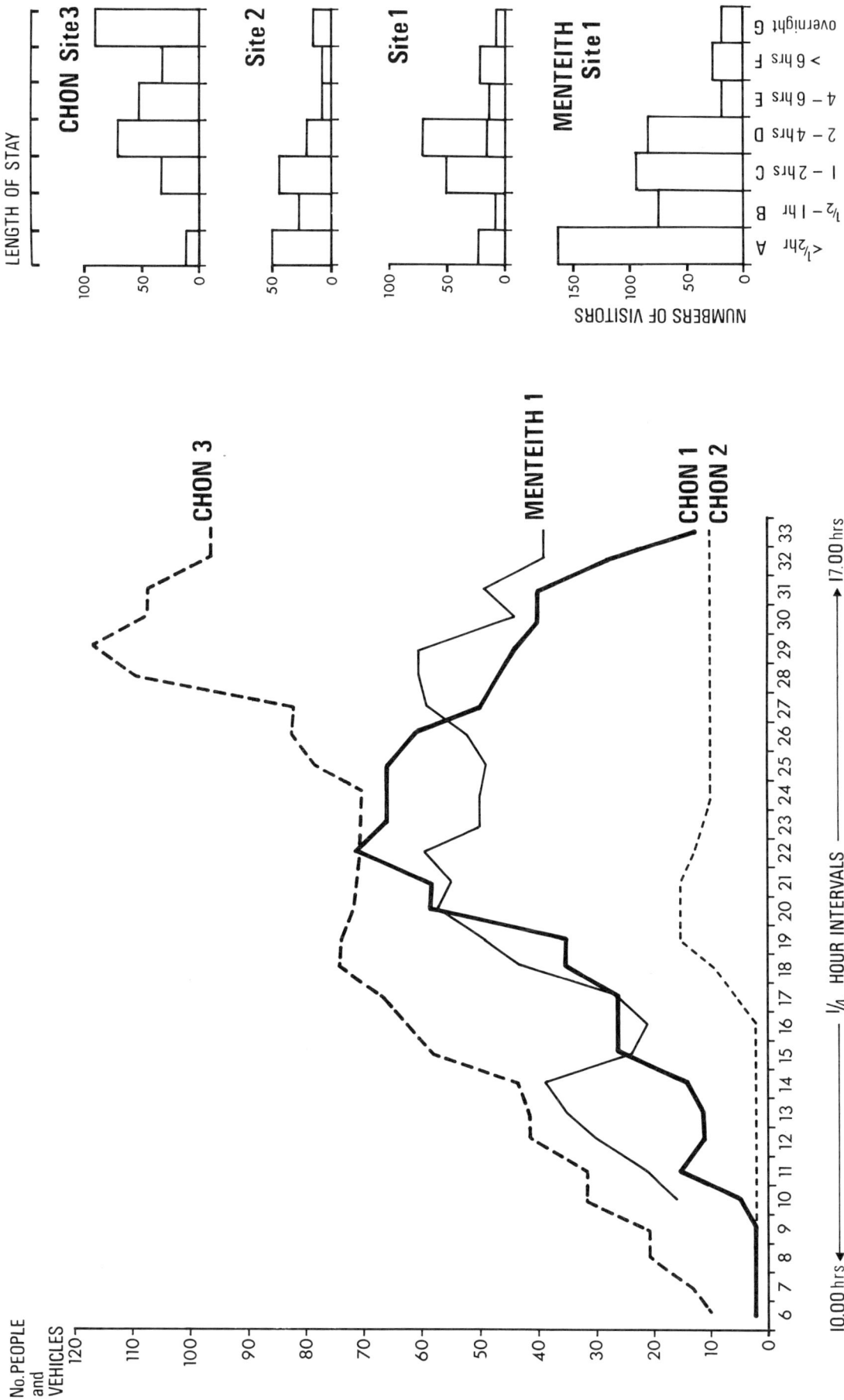

Fig. 31   Daily variation in numbers of people plus vehicles per ¼-hour interval and average length of stay on selected recreational sites on Lochs Chon and Lake of Menteith

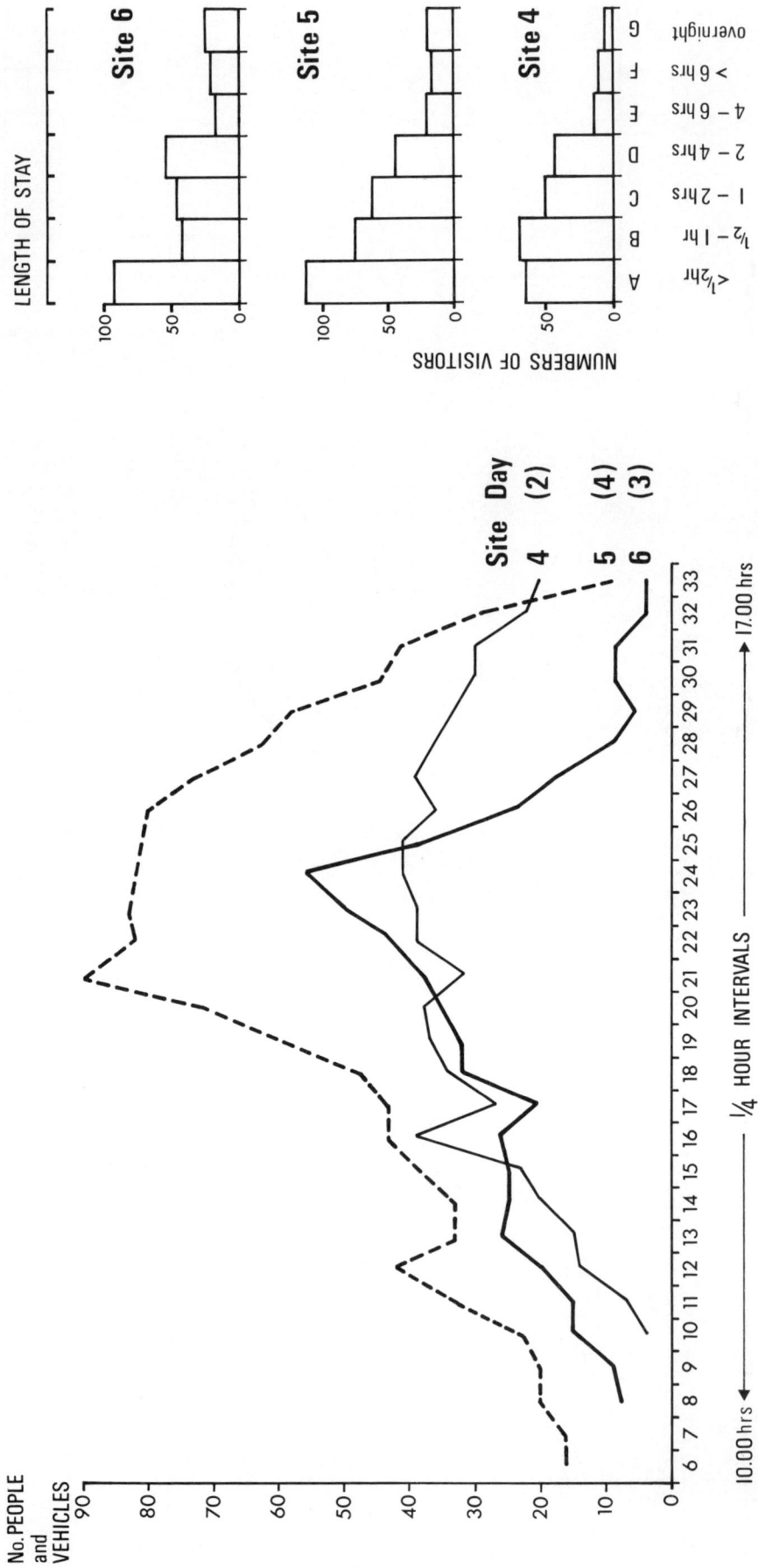

Fig. 32 Daily variation in numbers of people plus vehicles per ¼-hour interval and average length of stay on selected recreational sites on Loch Ken

Table 40

Intensity of use in terms of density of people and vehicles, and turn-over (length of stay) at selected lochside recreational sites

Loch and site	Peak numbers people + vehicles	Approx. area (ha)	Peak density (nos/ha/ ¼ hr)	Length of stay	Formal or informal sites	Peak day and weather
Loch Ken 4	41	0.4	110	Short	Open large formal car park with easy access to shore	Sunday 18/7/76 Dry, overcast, cool fresh SW wind
5	90	1.2	75	Short	Verge and off road parking with easy access to shore	Sunday 22/8/76 Dry, sunny, warm ESE wind
6	55	0.9	60	Short and moderate	Small car park	Monday 2/8/76 Cloudy, occasional showers, cool
Trossachs Venachar 1	231	0.8	288	Short and medium	Formal car park with easy access to shore	Sunday 3/8/75 Very hot, clear skies, slight heat haze
Chon 1	74	0.6	79	Medium	Forestry Commission car parks and picnic areas	Sunday 6/7/75 Very warm, clear skies, good visibility
2	21	0.6				Monday 21/7/75 Cool, continuous rain, low persistent cloud
3	116	4.5	27	Medium and long	Off-road parking ('wild' camp site)	Sunday 6/7/75 Very warm, clear skies, good visibility
Achray 5	224	0.8	305	Short and medium	Forestry Commission car park landward side of road	Sunday 3/8/75 Very hot, clear skies, slight heat haze
7	47	0.2	235	Short	Walled lay-by, good viewpoint	Sunday 3/8/75 Very hot, clear skies, slight heat haze
Lubnaig 2	639	1.8	360	Mixed	Informal off-road parking: access for people and vehicles to shore	Sunday 6/7/75 Very warm, no wind, clear skies, slight heat haze
1	1135	1.2	1096	Medium and long	Informal off-road parking: access for people and vehicles to shore	Sunday 3/8/75 Very hot, clear skies, slight heat haze
Menteith 1	60	0.7	86	Short and medium	Informal off-road parking: access to shore	Sunday 24/8/75 Cool, sunny, occasional showers

Short = less than 1 hour; Medium = 1 – 4 hours; Long = over 4 hours

84

portant role as short links between facility and activity areas than for direct recreational use.

Table 41

Rank order of intensity of use of selected recreational sites around Trossachs Lochs and Loch Ken

Loch and site		Density people + vehicles/ha	Turn-over	Intensity of use
L. Lubnaig	1	1096	Long stay	Very high
L. Lubnaig	2	360	Mixed	
L. Achray	5	305	Short/medium	High
L. Venachar	1	288	Short/medium	
L. Achray	7	235	Short	
L. Ken	4	110	Short	
Lake of Menteith	1	86	Short/medium	Moderate
L. Chon	1 & 2	79	Medium	
L. Ken	5	75	Short	
L. Ken	6	60	Short/medium	
L. Chon	3	27	Medium/long	Low

5.4.4 Intensity of recreational use

On a number of sites around the Trossachs Lochs and Loch Ken, intensity of use was calculated, using the method described in Section 4.3, for the peak ¼-hour on the peak day of the four days on which user-surveys were undertaken (see Figs. 29 to 32 inclusive). In the Trossachs, the peak day was Sunday, 3rd August, 1975, for four of the six lochs; on the remaining two lochs the differences in numbers on the 3rd August and their peak days (Chon, 6th July and Menteith, 24th August) were so small that it was decided to calculate peak densities of people plus vehicles on the 3rd August for all the Trossachs sites. This day was the warmest Sunday in the months of July and August that year. Many of the sites surveyed, especially those on Lochs Venachar and Lubnaig with easy access to the shore, were used at this peak period to maximum physical capacity. The variations between sites on Loch Ken, however, were such that peak densities had to be calculated for different days (see Table 40).

Even on the basis of a relatively limited number of sites (limited because of the difficulties of measuring areas with any degree of accuracy), a wide variation in intensity of use (see Tables 40 and 41) is revealed. If sites are ranked in order of density, as in Table 41, four orders of intensities of use emerge, providing a comparative scale against which other lochs can be judged. Lubnaig 1 was the most intensively used site observed in the course of two full summers' fieldwork. Only numbers of people recorded on Loch Morlich (peak: 1,600 in 1974) are higher, but the width and length of beach there is very much greater, and it is not accessible to vehicles. The ranking also provides a comparative scale by which relationships between intensity of use, vulnerability of site and amount and type of impact can be compared.

Table 42

<h2 style="text-align:center">Lochside impact – type matrix</h2>

ACTIVITIES / DISTRIBUTION / NATURE OF IMPACT

DISPERSED	POINT/CONCENTRATED	LINEAR	BACKSHORE	BANK or CLIFF	LANDWARD	LOCHWARD	ACTIVITIES	COMPACTION (FEET)	ROLLING (WHEELS)	CHURNING (HOOVES)	ABRASION (BOATS)	PICNIC FIRES	DIGGING	BUILDING
	√	√	√				CAMPING	√				√		
	√	√	√				CARAVANNING	√	√			√		
	√	√	√	√		√	BANK-FISHING	√				√	√	√
	√		√	√			PICNICKING	√				√	√	√
			√				CAR-PARKING		√					
		√	√	√	√	√	NATURE TRAIL/STUDY	√						
		√	√	√	√	√	CASUAL WALKING	√						
		√		√	√	√	ORIENTEERING	√						
		√			√	√	WILDFOWLING	√			√			
	√						ALL NON-COMPETITIVE BOATING ACTIVITIES	√	√					
	√	√		√	√	√	BOAT LAUNCHING+BEACHING	√	√		√		√	√
	√				√	√	PADDLING/SWIMMING	√						
	√	√	√	√	√	√	SIGHT-SEEING	√	√					
		√	√	√	√	√	PONY-TREKKING			√				
		√	√	√	√	√	CASUAL RIDING			√				
	√		√	√	√	√	CASUAL GAMES	√					√	√

DISTRIBUTION   **NATURE OF IMPACT**

RESOURCE CHARACTERISTICS

Affected by recreational impacts •

Affecting recreational impacts
    encouraging   +
    discouraging   –

COMPACTION	ROLLING	CHURNING	ABRASION	PICNIC FIRES	DIGGING	BUILDING	
+	+	+	+	+	+	+	SHORELENGTH ACCESSIBLE
+	+	+	+	+	+	+	" WIDTH
– •	– •	– •	– •	–	+ •	–	SHORE MATERIAL : MUD
+ •	+ •	+ •	+ •	+	+ •	+ •	SAND
+ •	+ •	+ •	+ •	+	+ •	+ •	GRAVEL
–	–	–	–	+	– •	+ •	COBBLES
–	–	–	–	+	–	–	ROCK in SITU
+ •	+ •	+ •	+ •	+ •	+ •	+ •	BEACH
+ •	+ •	+ •	+ •	+ •	+ •	+ •	SHORE VEGETATION – LOW SHORE
– •	–	– •	– •	–	–	–	EMERGENT
+ •	–	•	– •	+ •	+ •	+ •	BACKSHORE / SHORE CLIFF
+ •	+ •	+ •	+ •	+	+ •	+ •	BACKSHORE SLOPE : FLAT/GENTLE
+ •	–	+ •	–	–	•	•	MODERATE
– •	– •	– •	–	–	•	•	STEEP
+ •	+ •	+ •	+ •	+ •	– •	– •	BACKSHORE VEG. : SHORT GRASS
– •	–	– •	–	– •	– •	– •	LONG "
+ •	– •	+ •	–	•	– •	– •	MOOR "
– •	–	–	–	– •	– •	– •	SHRUBS
– •	– •	– •	– •	+ •	– •	– •	TREES
+	+	+		+			BACKSHORE-LENGTH-ACCESSIBLE
+	+			+			-WIDTH
+	+		+	+	+	+	FAVOURABLE ASPECT
+	+		+	+	+	+	SHELTER
+	+	+	+	+	+	+	VISIBILITY FROM ROAD

6.1 Factors Affecting Recreational Impact

Effects of recreational use will be dependent on the type, intensity, and spatial distribution of the activities involved. The susceptibility of any site to recreational impact is dependent on a complex of inter-related factors, some of which may be intrinsic to the resource, others extrinsic. Figure 33 illustrates the most important factors, and their inter-relationships, which may affect recreational impact on the lochside. While none of these is peculiar to the lochside *per se*, the understanding of the way in which they operate is basic to the aims of this report. Impact is dependent on factors associated with resource, user and location characteristics.

6.1.1 Resource and user characteristics

Figure 33, which shows intra- and inter-relationships between these factors, is designed to be self-explanatory. Some points, however, should be noted: first, the importance of the distribution of users and vehicles, activities, and facilities across (transverse) and/or along (linear) the lochside; second, the size (area), bio-physical characteristics and the relative vulnerability or durability of the lochside resource which must be considered in relation to the three levels or scales indicated which are the lochside component, lochside type and loch type; and third, extrinsic characteristics of the resource and the users. Among the latter, the factors which exert the greatest effect on the resource-user system are set out below:

(a) The type and number of tenure types relating to land, water or game rights affect access to both land and water and, hence, the extent to which the potential for lochside recreation use has been or can be fully exploited;

(b) The development and management of either user or resource characteristics which may affect the site's capacity for one or more types of water and/or land-based uses and its relative susceptibility to impact;

(c) Other non-recreational uses of the lochside which are important in that they may, as in the case of urban-type developments such as buildings or roads, protect the lochside or inhibit access to it. In the case of cattle grazing, and sand or gravel working, such activities can create conditions which may be confused with recreational impact. They may also initiate damage which will make the site more vulnerable to impact, or combine with recreational activities to intensify the impact.

6.1.2 Locational characteristics

The locational characteristics which affect recreational use and impact are factors extrinsic to the lochside, but are not indicated on the diagram.

They include:
(a) proximity of users to the lochside in terms of distance from major and minor vehicular roads and from centres of population of varying size;

(b) ease and type of access from a motorable road: in this sense 'ease' is determined by absence of either physical or legal constraints on movement from the road to the lochside; type of access refers to whether it is single-point, multiple-point, partial or completely linear access;

(c) visibility of the lochside from the road, particularly as seen from a moving car.

All these factors, singly or together, affect the volume of users and the intensity of impact on a particular lochside.

6.2 Types of Recreational Impact

Several recreational activities may effect the same type of impact, for example walking or playing games results in trampling. Therefore, it was decided that a discussion of the various types of impacts occurring on lochsides, rather than of the effects of individual activities, was a more rational basis for analysis of the processes involved. The impact-type matrix in Table 42 has been designed to aid in the analysis and identification of the:

(a) principal types of physical/biological impact associated with the lochside recreational uses considered in Chapter 5;

(b) locational and distributional characteristics of these impacts;

(c) reciprocal relationships between the resource-base and the recreational impacts.

On the basis of the processes involved, five main types of direct (or first order) impact can be identified, one or more of which may be associated with one or more lochside recreational activities. These are set out below.

6.2.1 Trampling by human feet and rolling by vehicle wheels

The impact of human feet and of vehicles are closely related. They are associated, either directly or indirectly and to a greater or lesser degree, with virtually every type of recreational activity, hence together they constitute the commonest and most widespread type of impact on the lochshore and other outdoor recreational sites. Their effects have previously been studied in a variety of habitats (Liddle, 1973 and Leney, 1974) many of which constitute components of the backshore. It has been established that these impacts lead to the elimination of less resistant herbaceous plants, and a proportionate increase in those more tolerant of trampling; to the reduction and eventual destruction of the ground-vegetation cover and the underlying organic layer; to the compaction of the soil and to the exposure of shrub and tree roots.

Trampling and rolling by vehicle wheels may occur over a wide area or may be concentrated along linear routes followed by walkers and/or vehicles. Such tracks can, over time, develop into deep surface ruts with compacted soil.

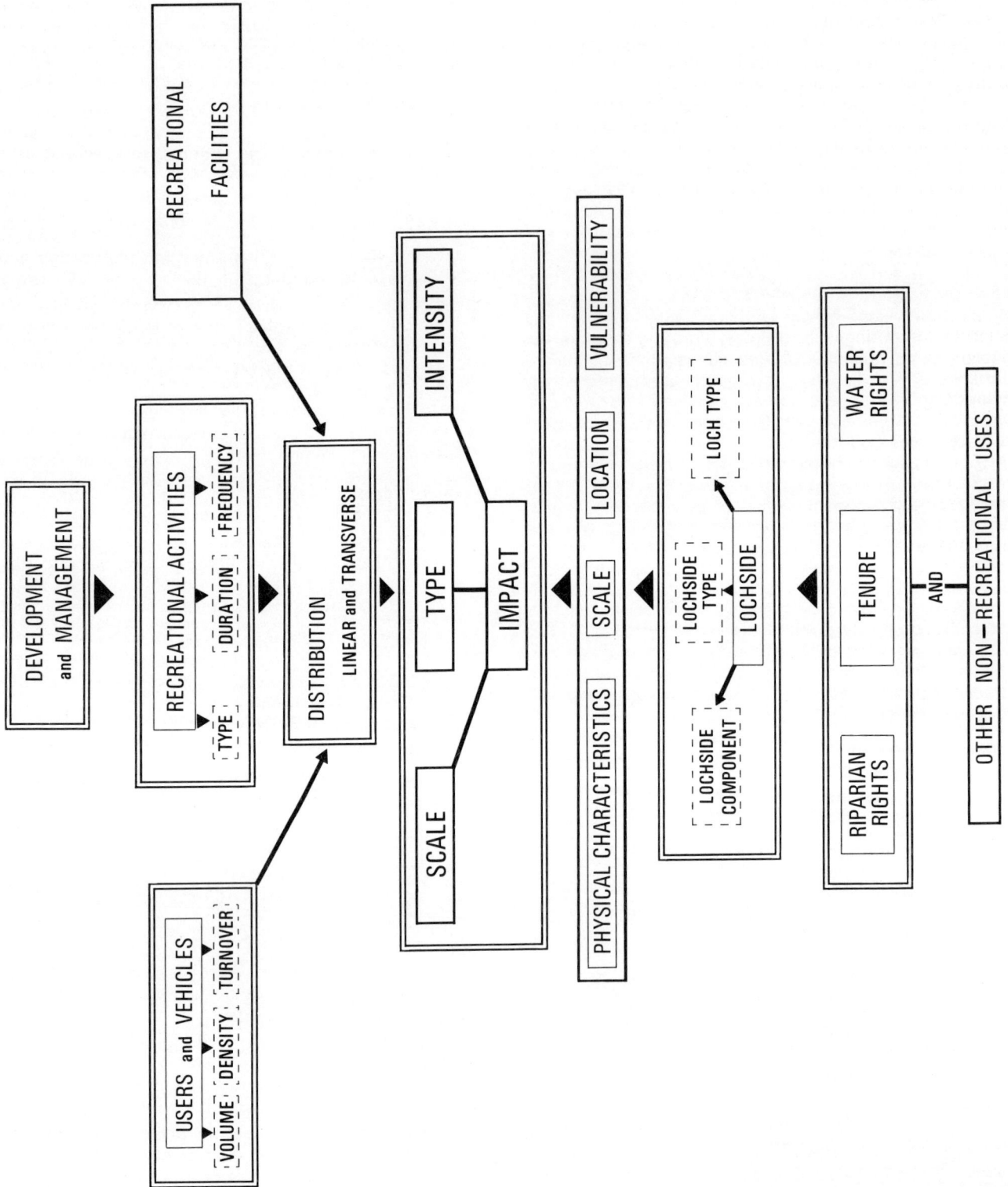

Fig. 33   Factors affecting scale, type and intensity of lochside recreational impact

Where there is a sufficient slope, surface run-off water may be channelled into the ruts to cause accelerated gully erosion. On very steep slopes, impact may result in down-slope slumping of surface vegetation and soil, with or without associated gullying.

Backshores tend to be subjected to a greater intensity of impact from trampling and rolling than the other lochside components, because of their attractiveness and/or suitability (see Appendix V) for particular recreational activities and because access is easier for vehicles. Where sufficient space is available for parking, backshores with a short grass or open tree-cover on a flat to gentle slope (Profiles 2a and 3a: Fig. 21) are the most attractive for nearly all car- and land-based activities.

Such sites experience the greatest intensity of impact, but except on formal sites the impacts associated with car-parking, caravanning and camping tend to be fairly widely dispersed over the site in use. On steeper slopes and on lochsides where the vegetation cover reduces penetrability to and/or visibility of the lochshore (Profiles 2b, 3b, 4b, 5b, 6b and 7b: Fig. 21), impact is concentrated along tracks many of which provide relatively short links between formal or informal parking areas, or caravanning or camping sites on the backshore and the main recreational area on the shore.

Several recent publications (Liddle 1975 a, b and Liddle and Greig-Smith, 1975) have summarised the studies which have been carried out on the effect of trampling on various types of vegetation; these have been concerned mainly with short grassland, sand dunes and heaths. The impact of trampling and vehicles on lochshores, however, has not previously been studied in detail, and consequently an understanding of the processes involved is still conjectural. Trampling occurs on the shores used for recreation, while vehicle rolling is less widespread, being confined to shores accessible to vehicles and across which it is possible for them to be driven. This requires the absence of a cliff of more than 15 cms between the backshore and the shore, and of such obstacles as might damage and/or obstruct vehicles on the shore.

Shores most attractive as recreational sites include those with sparsely vegetated or bare, dry beaches, particularly when composed of sand, gravel or mixed sand and gravel, and those with a continuous, or more commonly discontinuous, cover of low-shore vegetation on mixed sand/gravel/mud overlying a firm substratum, or less frequently cobbles and boulders.

The effect of impact is to reduce and eventually eliminate the low-shore vegetation, to disturb the beach material by destroying beach ridges, and generally to spread and flatten the surface material in such a way that the effects of wave-sorting are much less obvious. Compaction of loose surface material results in either closer packing of fine and coarse material, or the embedding of coarse material into the finer, muddy matrix below to form a harder and more even surface, often with a gentler gradient than before. The

particular effects of impact, however, vary with the nature of the shore material; these and the consequent vulnerability of the shore to impact will be discussed more fully in the following chapter.

In contrast, swampy wetland shores with emergent reed and sedge beds have a limited recreational value. Land-based activities tend to be confined to specialist users such as bank-fishers, wildfowlers and bird-watchers whose impact is not only less, but becomes concentrated along seasonal or permanent Indian-file tracks, which are between 25 and 50 cms wide. These footpaths frequently follow the inner margins of emergent reed/sedge beds parallel to the shore. Only in extensive beds such as may form at loch heads (for example Castle Semple Loch and Loch Lubnaig) are tracks formed at right angles to the shore; on such sites, if the level of use is high, multiple paths develop as those which become bare and muddy are abandoned in favour of new routes.

6.2.2 Digging
Digging is an impact associated with a wide range of activities undertaken on the lochside. Digging is particularly associated with:

(a) all general car-based activities and has two main motives: firstly, children's play-activity; and secondly, the need to find or create shelter for people relaxing in the open and for their picnic fires;

(b) bank- and boat-fishing where small surface excavations are made to form a fireplace or a temporary fish store.

Digging is a widespread activity particularly in the lochshore and in cliffs or banks at the landward edge of the shore. The impact of play-digging is frequently reinforced by excavations, which can assume cave-like proportions, at the base of a cliff to provide shelter for people and their picnic fires. The result of digging is to cause overhanging banks and cliff under-cutting; processes accelerated by the loosening, and accidental dislodgement, of large boulders which are common constituents of backshores composed of unsorted glacial detritus. Combined with trampling on steep slopes, this is the dominant process in bank- or cliff-retreat and the consequent reduction of the backshore area; where the latter is wooded, the undermining and the eventual fall of those trees along the cliff or bank edge is inevitable.

6.2.3 Building (or construction)
Using either deliberately excavated or loose surface material, building (or construction) on the shore is as common as digging. It tends, however, to be confined to the lochshore and more particularly to beaches with material of a size that can be easily excavated or moved by recreational users. Large stones, cobbles and small boulders are used to construct rough, informal jetties extending out from the water's edge a few metres into the loch. These are used as an aid to bank-fishing or for

informal boat-launching and berthing. Other shore constructions include fireplaces and a variety of small dams or enclosures, arising from play activity, the general outcome of such activities being a superficial redistribution of beach material. Along the waterline, particularly of highland and upland lochs with stony shores and a fairly large range of water-level, building may result in a concentration of large cobbles and boulders in a series of rough jetties at right-angles to the water's edge.

6.2.4 Abrasion (or scouring)

Abrasion is the main type of impact effected by the launching and/or beaching of such craft as fishing and casual rowing boats, with or without outboard motors, some light sailing dinghies, and less frequently canoes and power-boats. Abrasion, resulting from the pulling up or pushing out of a boat, is usually highly concentrated in a narrow zone on either side of the current waterline and to a distance off-shore which is dependent on the gradient of the loch floor. However, although most abrasion takes place during the summer low-water period, fluctuations in water-level in the recreation season seem to occur frequently enough to cause the abrasion-zone to move across practically the maximum possible width of a beach. On lochs where fishing is organised, the initial launching of boats is frequently concentrated at one or more points. These may be permanent jetties or piers or, often, rough berths composed of two parallel dry-stone walls which project into the water, and into which the boat can be pulled when not in use during the season. From these, and other informal sites, boats must be pushed or dragged across the beach and off-shore into deeper water. Informal beaching and re-launching occur frequently during the course of casual boating and boat-fishing. Both launching and beaching take place on a wide range of shore types, with only the large boulder beaches and the wide emergent reed/sedge beds on deep organic muds being avoided to any degree.

The physical impact of abrasion from boats results in the breakage of emergent vegetation, the scouring of the shore and loch-floor, with consequent removal and elimination of low-shore vegetation, and even of the sub-surface perennial rhizomes of the more firmly anchored emergent plants. A common feature of lochs used for boating are completely bare areas 2 to 3 metres wide alongside formal or informal launching-cum-beaching points on shores which elsewhere have well-developed low-shore and/or emergent vegetation. A characteristic feature of the latter is what has been called 'boat nests', which are distinct parallel-sided gaps usually about two metres wide in otherwise continuous reed or sedge beds. These correspond to points where continued abrasion has created and, presumably, maintained a completely bare, unvegetated area in an otherwise dense stand of reeds or sedges. In some places, the emergent beds have been deliberately cut to facilitate the launching and berthing of fishing boats.

6.2.5 Burning

The remains of picnic and camp fires are a widespread and characteristic feature of many proximal Scottish lochsides. Lochs which do not reveal evidence of fire sites can normally be assumed to have been subjected to a very low level of recreational use. Picnic fires are set by shore-based visitors, and are numerous around heavily used lochs near to large urban areas, such as Loch Lomond, where the use of the lochside for summer-evening barbecues is a long-established tradition. The remains of picnic fires set by boat and bank-fishers are more widespread, reflecting the longer diurnal (and/or nocturnal) duration of this activity. The origin of such fires can be identified by the nature of the litter associated with them.

Evidence of picnic fires is widely distributed and can be found throughout the lochside zone with the form of the fireplace varying with the particular lochside component. Circular, charred patches, sometimes in slight surface depressions, are common on open low grass or moorland backshore sites with flat to moderate slopes; irregular hollows occur at the base of trees, cliffs, or banks; while circular stone-rimmed depressions are characteristic of the lochshore. Once initiated, a fireplace, if it persists from one season to the next, will probably continue to be used by successive groups of people.

The impact of fire-sites on bare beaches appears to be minimal. On low grass and moorland, surface vegetation is gradually killed, and continued use, combined with the removal of the organic layer will lead to the eventual exposure of the underlying mineral soil. The impact of fires is more severe on trees. Trees growing along the inner edge of the shore, on the bank or cliff behind or on the backshore, are used for shelter and fires are set on the leeward side at the base of their boles. The result is charring of the bark, then fire-scars which lengthen and deepen with use to form the characteristic triangular fireplaces cut deeply into the bark and wood, and finally leading to complete penetration of the trunk, at which stage all that may remain of the tree is a charred, dis-embowelled stump. An indirect (second order) impact associated with picnic fires is vegetation damage, caused particularly by the breakage of tree branches and, where exposed, roots to provide the necessary firewood. Fires tend to be concentrated at the inner edge of the shore, and particularly at the base of the backshore bank or cliff, because of the requirements for wood and shelter.

6.2.6 Other types of recreational impact

Other types of recreational impact which it is difficult to systematise are listed below:

(a) Deliberate mechanical breakage of vegetation.
This has its greatest and most obvious effect on trees where branches and roots are broken off to provide firewood. Such damage also occurs as the direct or indirect effect of children's play.

(b) Plant picking.
It is difficult to make any meaningful statement

about the intensity or effects of this impact. On backshores, the effects of plant picking are comparable to those on other similar habitats found in other locations which are used for general recreation purposes. On the lochshore, reeds and particularly reedmace are sought for decorative purposes, but their habitat tends to protect them from severe impact. Impact on water lobelia, whose attractive blue flower projects a few centimetres above water near the shore, is probably greater, and where it occurs at the southern limit of its range (for example, around the Trossachs lochs) this may, together with other impacts, have contributed to its relative paucity in this area.

(c) Water pollution.
As reflected in the visual quality of the water, pollution is, in comparison to many inland water-bodies in England and other overseas countries, very low and highly localised. In fact, an outstanding characteristic of all but the most heavily-used, small, suburban and urban park lochs is the relative cleanliness and clarity of the water. Eutrophication would appear to be minimal. Only one small rural loch with a well-developed fringing algal scum was found, and this was surrounded by farmland on to which a large quantity of pig-slurry was being regularly sprayed. A few other lochs had patches of green algal scum. Yellowing of reed canary-grass around some of the more heavily-used urban park and urban-fringe lochs was also observed, and it has been suggested (Hamilton, *per. comm.*) that this may be due to incipient eutrophication particularly during very warm summer months when evaporation is high.

(d) Litter.
The presence of litter provides one of the first, easily-observed signs of recreational use. The amount of non-biodegradable litter on lochshores would appear to be out of proportion to the amount of recreational use on many lochs. This is particularly the case on lochshores used for picnic fires; the highest proportion of which, as judged by the litter-composition, could be attributed to either bank or boat fishers. In addition to the unsightliness of litter, broken glass is one of the main hazards on heavily-used beaches, and on other shores more lightly used by fishers.

6.3 Types of Non-recreational Impacts

One of the problems associated with any attempt to assess the effect of recreational activities on the lochside is that of identifying other non-recreational impacts which may take place at the same time, produce effects difficult to distinguish from those produced by recreation, or ameliorate or exacerbate recreational impacts.

Of these, the most important are the natural wind-generated waves and breakers which operate to a greater or lesser extent on and around all lochs, and about which

no data was available in Britain when this survey commenced. Studies in North America, Sweden and Australia have been undertaken on inland water-bodies, but these are of such a scale (for example, The Great Lakes) as to be inapplicable to the very small Scottish loch or reservoir. Fundamental empirical work on wave and breaker characteristics was undertaken during the course of this project on a number of Scottish lochs. The data collected (albeit limited in view of the scale and complexity of the problem), allows an attempt to be made to answer with more confidence than previously (Tivy, 1974) some of the questions raised about the nature of the inter-actions between the impact of the natural wave and recreational activities.

6.3.1 Loch waves and breakers

Impact on the lochshore is effected by breaking waves, and the size and speed of wave movements is directly related to the fetch. On lochs where wave recordings were made, the smallest recorded waves were produced by a light breeze (6 mph/Beaufort Force 2) and an estimated minimum duration of ten to fifteen minutes. The highest wind-speed recorded with an all-day duration was 37 mph (Beaufort Force 7 – moderate gale bordering on fresh gale, 39-40 mph). However, once waves are initiated, duration of wind speed is not so significant as the length of fetch: the longer the fetch, the greater the wave development and thus the greater (or more efficient) will be the potential wave-energy. Detailed studies were undertaken on twenty-two beaches to analyse the possible relationships between fetch, on the one hand, wave and breaker characteristics and beach form and function on the other (see Table 43). On these, the maximum fetch ranged from 1,053 m on Loch Tarff to 33 km on Loch Ness. Unfortunately, maximum fetch cannot be directly correlated with loch length as measured for the data set.

The longest uninterrupted straight-line distance across water is usually very much less than the overall loch length because of promontories, islands, and loch-shape; loch-length and fetch approximate most closely on lochs with particularly straight shore-lines, such as Loch Ness. Loch Lomond, the second longest loch (66.1 km) in Scotland has a maximum fetch of only 10.8 km. Also, on lochs with long fetches (particularly in the case of the long, narrow, valley lochs so characteristic of Highland Scotland), the chance of waves running along this line for long periods of time is less than across broader highland or lowland lochs with much smaller fetches. This is because any deviation of wind-direction from the maximum fetch on a long narrow loch can cause an extremely rapid decrease in length of fetch. On the other hand, lochs of any size, in either lowland or upland conditions, which lie in open-ended troughs or valleys, which funnel wind from a particular direction, may experience high wind-speeds along their line of maximum fetch for a longer period than on the large lochs in other situations.

The efficiency of waves as agents of material transport and of erosion increases as they break and move across a

G

92

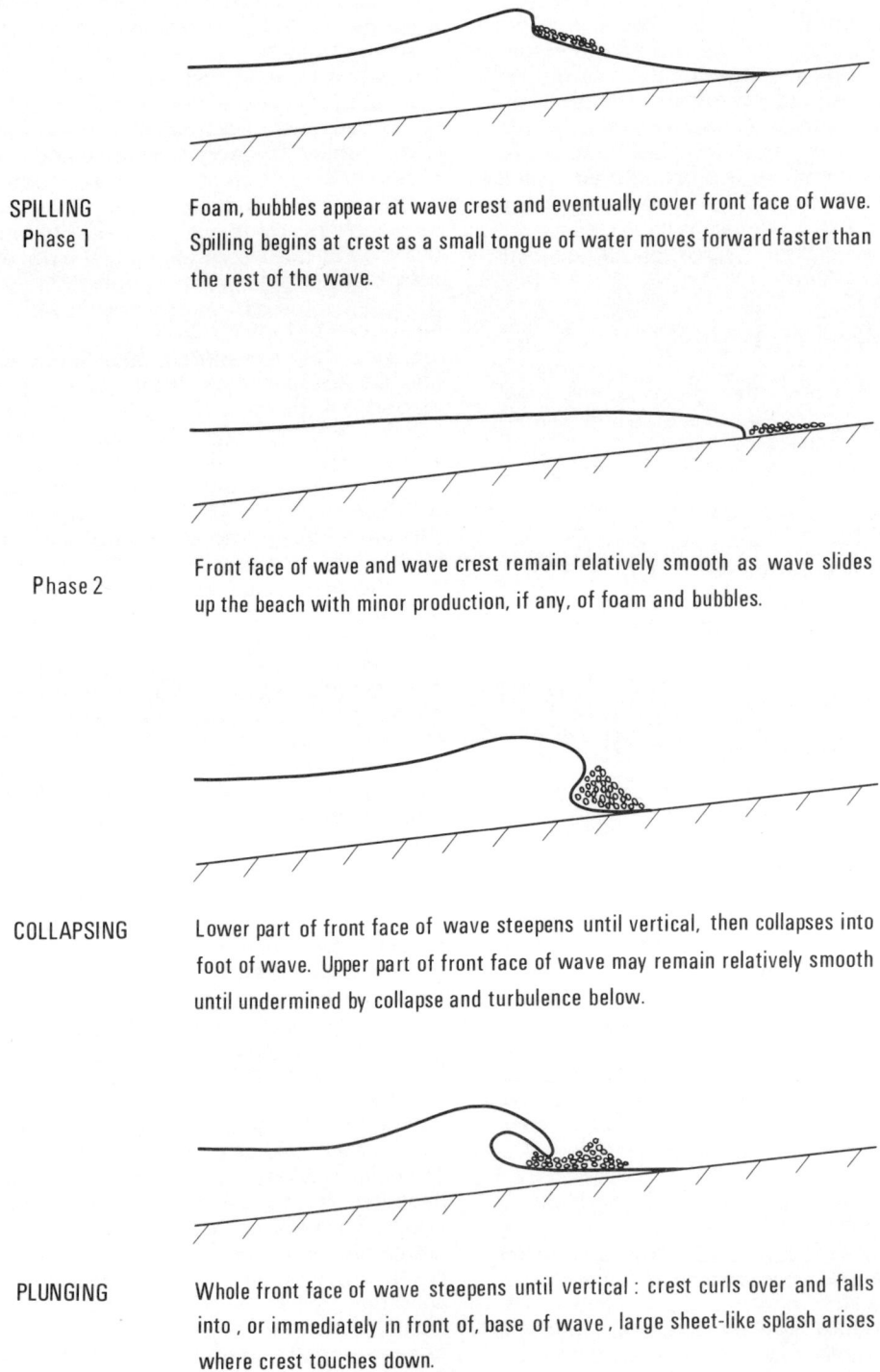

SPILLING
Phase 1

Foam, bubbles appear at wave crest and eventually cover front face of wave. Spilling begins at crest as a small tongue of water moves forward faster than the rest of the wave.

Phase 2

Front face of wave and wave crest remain relatively smooth as wave slides up the beach with minor production, if any, of foam and bubbles.

COLLAPSING

Lower part of front face of wave steepens until vertical, then collapses into foot of wave. Upper part of front face of wave may remain relatively smooth until undermined by collapse and turbulence below.

PLUNGING

Whole front face of wave steepens until vertical : crest curls over and falls into , or immediately in front of, base of wave , large sheet-like splash arises where crest touches down.

Fig. 34   Loch wave breaker-types

beach as swash and backswash. Three main breaker types have been identified (see Fig. 34): spilling, collapsing and plunging.

Spilling breakers are the least effective agents of transport. The wave-form is retained, it breaks gently and such transport of fine material as occurs does so in the final swash stages of the breaker. Collapsing breakers are much steeper, break more suddenly, and the front face of the wave collapses as it advances up the slope of the beach: the effect is similar to throwing a bucket of water up the beach. These are the most efficient agents of transport of larger material – the size being dependent on the initial wave amplitude; they may also play some part in bringing beach material into suspension. Plunging breakers form more rapidly still and most of the water falls vertically onto the beach where the wave is checked. These are not particularly efficient agents of transport since most of their potential wave energy is dissipated in a mass of foaming swash. The resulting sudden increase and decrease of water-pressure has the effect of sucking mineral particles up into the water just behind the breaker. As a result, plunging breakers are much more efficient at loosening material which, if very fine, may then be put into suspension or, if coarser, may become available for transport by larger waves.

The collapsing breaker is the commonest type to have been observed and recorded on the lochs investigated. Data gathered to date suggests that plunging breakers may only occur for 20-30 per cent of the time and collapsing breakers for 50-60 per cent. Spilling breakers are not very common. The main reasons are that the average lochshore slope is not steep enough to generate plunging breakers; the fetch is too short on the majority of lochs observed (indeed on the majority of all Scottish freshwater lochs); the beach material may be so coarse that when waves break much of the water percolates into the beach causing what is known as breaker-wastage, and these shoaling waves never become steep enough to plunge.

If this hypothesis is universally applicable then it follows that loch waves are only moderately effective agents of erosion. The most marked wave-erosion takes place under conditions where large, high-energy, plunging breakers occur near the backshore and can erode its landward edge, either by the force with which air may be compressed in cracks or crevices, or by the force with which the material lifted by the breaker hits the cliff. The collapsing breaker, most characteristic of the Scottish loch, is primarily a transporting agent. By its action, material is moved backwards and forwards across the beach, and under certain circumstances, and dependent on the size of the beach-material, it can have a horizontal abrasive action. However, because of the initially small wave-size, this action is on a very limited scale compared with that on a marine beach. In contrast to the sea coast, the breakers only affect a small part of the loch beach at a time, because of the seasonal rather than tidal variations in water-level. Furthermore, high water-levels are not necessarily coincident with periods of high wind speeds.

On natural lochs, storms accompanied by wind and heavy rain may occur before the consequent rise in water-level. Also, the time during which the relatively inefficient loch-wave can effect erosion of the landward margin is limited.

It must be borne in mind that a relatively high proportion of proximal Scottish lochs have had their levels raised and regulated (see Section 2.2). In many of these impounded lochs, where shore slopes may have been artificially steepened (by protective embanking for instance), the range of water-level is substantially modified. Raising of the water-level and/or steepening the shore-slope tends to increase the frequency of plunging breakers and to bring both collapsing and plunging breakers nearer to the landward edge for more protracted periods of time. Under these circumstances, the erosive power of the loch wave could be increased without any change in length of fetch or of wind speed.

A more immediate need was to try to identify the loch-size threshold below which the impact of the natural wave on the lochside is limited or negligible. Evidence of wave-action is, as has already been noted, most easily detected on mobile beaches. It was, therefore, decided to see if it were possible to distinguish 'true' beaches (the result of wave erosion, transport and deposition alone) from those where other processes were or had been dominant. The threshold for the formation of true beaches could be expressed in terms of fetch, beach slope, or loch-size. The lochs studied in detail suggested that, on those with the shortest fetch (for example, Loch Tarff), much of the exposed shore, though composed of unconsolidated material, is the loch-floor exposed by summer draw-down of water. The most gentle slope found was on Loch Garten where most of the shore was probably also exposed loch-floor. The true beach with the most gentle slope was found at Rowardennan (Loch Lomond) where there is a fetch of 2,010 m – but this is on the largest loch. Of the twenty-two beaches surveyed, the smallest loch with a true beach is Loch Morlich (121.7 ha) where the surveyed fetch is 1,672 m. This data suggests that true beaches are unlikely to be found on lochs, with a surface area less than 120 ha, a maximum fetch of c. 1,500 m and shore slopes of less than 0.05%. On the basis of area alone this accounts for about 85% of all proximal lochs over 5 ha. Where beaches occur below this threshold, it is likely that processes other than wave-action, such as the impact of recreational activities and/or one of the non-recreational activities discussed in the following part of this chapter, have been dominant in their production. It is also possible that a shore slope of 0.05% is the minimum threshold slope for the production of collapsing breakers which are the most effective (though not the most efficient) natural agents of transport and erosion.

In assessing the effect of wind-generated loch-waves on the stability of the lochside, two other inter-relationships must be considered. The first is the possible effect of water-based recreational activities on loch-waves and the extent to which they may reinforce or ameliorate their impact. Boats tend to generate short-period waves that decay rapidly away from the line of movement; such waves can be produced by either power-boats or non-power-boats. High-speed power-boats used for racing, water-skiing or pleasure boating are most efficient when planing (riding on their 'hull-step') which is usually a third to a half of the length of the boat. Waves produced by power-boats increase rapidly in size during acceleration until planing occurs when they decrease, often by as much as half the preceding maximum size. Non-power-boats do not plane and wave-size increases to a maximum which is related to operating speed.

From observation on the Scottish lochs investigated and more detailed studies of ships' waves in the Oakland Estuary, U.S.A. (Sorensen, 1966), the following generalisations can be made:

(a) Larger vessels such as steamers and large cabin-cruisers tend to move relatively slowly through lochs; their waves decay rapidly and, when they reach the shore, do not appear to differ significantly from what could be expected to occur naturally. The largest vessel operating on inland waters in Scotland is the *Maid of the Loch* on Loch Lomond during the season end of May to mid-September. When sailing, she generates two sets of bow and stern waves on each side, which produce, in the narrowest part of the ship's course (between Inversnaid on the east and Tarbet on the west shore) breakers comparable to those produced by a winter storm. This occurs, however, for approximately three minutes twice a day when, normally, the loch-level is at its lowest. The only loch other than Lomond with an appreciable number of cabin-cruisers is Loch Ness, but in relation to the area or perimeter, the number of boats involved is very small. In any event, wave-impact is negligible on the cobble shores which fringe much of Loch Ness.

(b) Smaller powered vessels move faster and, particularly when fishing, come closer in to the shore. Their wakes usually break on-shore and appear to differ quite markedly from prevailing or expected waves. They produce waves similar to those common in a short winter storm.

(c) Power-boats (other than hovercraft which do not generate a wake) produce a wake comparable to a cruiser or launch of between half to three-quarters their size. The breaking waves do not appear to differ greatly from those produced naturally, though the greatest disturbance occurs in acceleration and deceleration areas when the vessel is below planing speeds.

(d) Sailing boats produce wakes whose breakers are almost impossible to identify.

Thus, the greatest potential modification of the natural wave and breaker is by small power-vessels incapable of planing. These boats tend to operate in conditions when the lochs are more or less calm and when natural wind-generated loch-waves are not so efficient at damping-out boat-wakes. Wakes from large, slower-moving vessels over-ride prevailing wave patterns, but have a smaller impact on the shore. Boating, at the level which occurs on most Scottish lochs, does not appear to reinforce the erosive effect of the natural wave. The effect of the increased turbulence on bottom-rooted plants and fine organic and inorganic sediments is more conjectural, as the studies on the Broads have shown (Moss, 1977). This and the extent to which impact on the submerged and emergent vegetation may cause second-order wave impacts on-shore is a problem that requires more detailed study.

The second inter-relationship is the effect that land-based recreational and non-recreational activities on loch shores may have on wave impact. Recreational impact from trampling and vehicle movements tends to smooth out beach profiles, to compact mixed material into a fairly durable pavement, to spread material across the beach, and in some cases to make the beach-slope steeper at or near the water-line. The possible implications of these changes (dependent on the scale and duration) are as follows:

(a) A given rise in water-level will lead to waves breaking further up the beach.

(b) Water-percolation into the beach may be reduced following compaction, and since more water is retained in the breaker it becomes more efficient as an agent of transport and/or erosion.

(c) On beaches with material of a relatively uniform size-composition, the material is not so closely packed; it remains mobile but becomes churned and pitted by trampling. In this case, the uneven surface checks the rate of swash, may increase the rate of water-percolation into the beach material, and hence reduce breaker efficiency.

(d) A beach on which material is disturbed by recreation becomes prone to remodelling by wave action. The material becomes more susceptible to wave action and hence more effective as an agent of abrasion or erosion.

(e) Steepening of the beach slope at or near the water line by recreational impact may result in waves breaking a little further off-shore and hence being less effective agents of erosion.

The effect of lochside recreational facilities on wave impact is more difficult to assess since it varies with type of facility and the nature of the site. Structures on the water, such as pontoons, buoys, and nets, will tend to dampen wave action. Generally, however, structures projecting into the water such as piers, slipways, and even informal fishing jetties, will have the greatest direct effect. They will refract or reflect wave action, protecting the beach and encouraging deposition of beach material. If the off- to on-shore transport of material is oblique to the shore-line, an artificial promontory could cause the beach to be gradually depleted of its material. However, structures on or near the beach will have little direct effect, but channelling or focusing activity on a beach may alter or control the impact on the beach and hence on wave action.

6.3.2 Sand and gravel winning

Sand and gravel winning from lochsides on a large-scale commercial basis is uncommon, with the extraction on the backshore at Mid Ross (NS 361862) on the south-west of Loch Lomond a notable exception. Sporadic sand and gravel extraction from loch beaches, however, is a fairly common practice. Its main use is for the construction or repair of dirt roads or for building protective banks around the lochside itself. Continual removal of material from the same beach (as at Sallochie (NS 390950) and Cashel (NS 394940) on Loch Lomond) can result in a flattening of the beach profile with a concomitant steepening of the landward slope at the junction between the beach and backshore; eventually, if excavation reaches the level of the water-table, standing water will occur on the beach. Vehicles used to remove material may themselves cause severe localised compaction. In less extreme cases, reduced beach-slope may result in waves breaking further up a beach and thereby increasing the risk of landward cliff or bank erosion. Where removal continues, causing depletion of the beach material, the appearance of the shore suffers and its recreational value declines. Natural

recovery may not take place, or may only do so very slowly, for the process of beach building with sediment supplied from off-shore is a much slower process on the lochside than on the coast.

6.3.3 Domestic animals

Other types of non-recreational impact can be caused by animals, particularly cattle (and, to a much lesser extent, sheep) grazing and watering around lochsides. This occurs mainly where rough grazing or permanent pasture fringes the loch or where an access point is provided to allow cattle to reach the shore from backshore fields.

Cattle by reason of their numbers, weight and tendency to concentrate at a particular part of the shore where it is sandy or muddy, but firm and not too rocky, have the greatest physical impact. Unless the slope is very steep and a path has already developed, cattle do not follow well-defined linear routes in the same way as sheep: they tend either to spread out or to bunch together. The effect of their movement down the bank or cliff separating the

Table 43

Surveyed beaches ranked in order of length of maximum fetch

Loch	Grid ref. beach site	Loch length (km) (shore)	Fetch (m)	Exposure	Relief development
1. L. Tarff	NH 430099	4.2	1053	W	0.078
2. L. Garten	NJ 975184	2.5	1112	SE	0.070
3. L. Garten	NJ 967185	2.5	1153	SW	0.059
4. L. of the Lowes	NT 238204	5.0	1204	SW	1.453
5. L. of the Lowes	NT 238204	5.0	1257	SW	2.063
6. L. Insh	NN 838045	5.0	1580	SW	0.151
7. St Mary's L.	NT 246223	18.0	1605	NE	0.170
8. St Mary's L.	NT 258228	18.0	1638	SW	0.235
9. L. Morlich	NJ 972097	5.5	1672	WSW	0.066
10. L. Ruthven	NH 636284	8.4	1875	NW	0.035
11. L. Ruthven	NH 633280	8.4	1995	SW	0.195
12. L. Lomond	NN 359985	66.1	2010	NW	0.169
13. St Mary's L.	NT 247223	18.0	2441	NW	0.359
14. St Mary's L.	NT 266238	18.0	3425	SW	0.231
15. St Mary's L.	NT 267238	18.0	3485	S	0.231
16. L. Lomond	NN 326077	66.1	5560	SE	2.633
17. L. Lomond	NN 326078	66.1	5970	NE	0.392
18. L. Lomond	NN 353954	66.1	5970	NE	0.535
19. L. Lomond	NN 409925	66.1	6210	SW	0.567
20. L. Lomond	NN 396930	66.1	7690	S	0.484
21. L. Ness	NH 596350	90.2	33185	SW	0.347
22. L. Ness	NH 600378	90.2	33365	SW	0.277

backshore and the shore is to cause a characteristic stepped-slumping with large blocks of turf becoming displaced downwards and eventually detached from the backshore. Movement on the shores results in the churning or puddling of fine organic and/or inorganic sediments. These combined effects on the backshore and shore are easy to distinguish from recreational impacts. There are lochside sites now used for recreation which must in the past have been subjected to grazing, as well as those, less numerous, which have reverted from recreational back to agricultural use. Unfortunately, past changes in land use are difficult to establish with certainty, and it is impossible to assess their impact in retrospect.

Domestic animals tend to be selective grazers, and it has been suggested that heavy grazing of certain lochshores has resulted, on the one hand, in reduction and elimination of reed (Spence, 1964) and, on the other, in an increase in the proportion of the more resistant common sedge in the wet grasslands which so frequently border lochsides.

6.3.4 Wildfowl

Lochs, particularly those with undisturbed stands of emergent vegetation, are the feeding and/or nesting areas of a variety of birds. These add considerably to the loch's scientific and recreational value, and the presence of such habitats accounts for the number of lochs where the whole or parts of their shores are nature reserves or have been notified as Sites of Special Scientific Interest by the Nature Conservancy Council. Many lochs carry large, semi-resident populations of the mute swan (*Cygnus olor*), a bird long-protected by royal decree. On several of the lochs surveyed, the number of swans appeared to be high in relation to the size of the water body, and the emergent vegetation showed evidence of impact as a result of nesting, trampling, churning, and grazing by these large heavy birds. Nowhere, however, was the impact of swans as severe as that noted in Germany by Sukopp (1973); he recorded that one pair of swans can destroy or damage 90 square metres of sedges or reeds by their nesting, trampling and feeding. Deliberate reduction of swan populations has been necessary in order to preserve the emergent fringe around the Havel Lakes in Berlin.

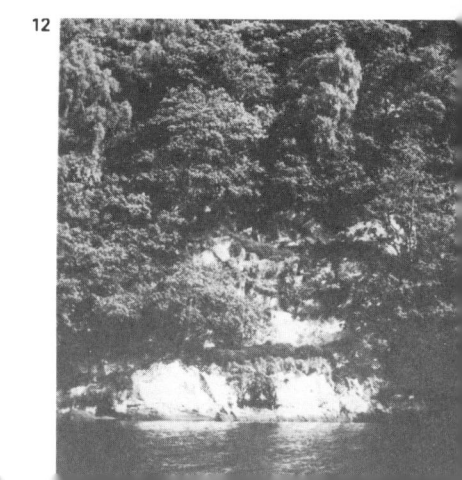

PLATE 1 SITE TYPES

Fig. 1   Loch of the Lowes. Site-type 1: wetland
           (foreground: low; middle-right: high). Also
           site-type 6a in background.
Fig. 2   Loch Achray. Foreground, site-type 2a: gentle
           slope and forbs. Middle left, site-type 2c:
           gentle slope and woodland.
Fig. 3   Loch Ruthven. Foreground, site-type 3a:
           moderate slope and forbs. Background, site-type
           3b: moderate slope and shrubs. Also, sandy
           beach: B.
Fig. 4   Loch Lomond. Site-type 4c: steep slope and
           woodland.
Fig. 5   Loch Ruthven. Site-type 5a: very steep slope and
           forbs, and site-type 5b: very steep slope and
           shrubs. Also, boulder beach B.

Fig. 6   Loch Achray. Site-type 5c: very steep slope and
           woodland.
Fig. 7   Loch Awe. Site-type 6a: extra-steep slope and
           woodland.
Fig. 8   Loch Chon. Site-type 7b: undulating slope
           with shrubs. Also, wetland fringe: W at
           water's edge.
Fig. 9   Loch Ken. Site-type 8: low embanked shore.
Fig. 10   Loch Ness. Site-type 9: high embanked shore.
Fig. 11   Loch Lomond. Site-type 10: rock outcrop < 15°
Fig. 12   Loch Ard. Site-type 11: rock outcrop > 15°

1

2

3

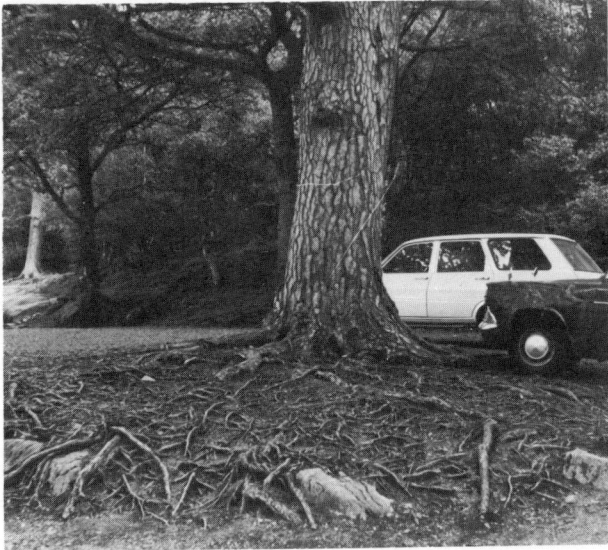

4

PLATE 2
DAMAGE TO
LOCHSHORE TREES

Fig. 1   Loch Lomond. Fire damage to oak trees.

Fig. 2   Loch Lomond. Root exposure: giraffe effect.

Fig. 3   Loch Lomond. Root exposure by trampling and car parking.

Fig. 4   Loch Morlich. Root exposure: octopus effect.

Fig. 5   Loch Lomond. Cliff retreat at Luss.

Fig. 6   Loch Lomond. Alder and oak with shoreline retreat.

5

6

1

2

3

4

5

6

PLATE 3   IMPACTS ON THE LOCHSHORE

Fig. 1   Loch Earn. Numerous jetties at camp-site.
Fig. 2   Loch Ruthven. Cattle watering.
Fig. 3   Duddingston Loch. Geese grazing.

Fig. 4   Loch Morlich. Informal fishing stances.
Fig. 5   Loch Lomond. Boat beaching and backshore abrasion.
Fig. 6   Loch Lubnaig. Slumping at steep, eroded bank.

1

2

3   4

PLATE 4   RECREATIONAL ACTIVITIES

Fig. 1   Loch Lubnaig. Parking on the backshore.

Fig. 2   Loch Ken. Bank fishing.

Fig. 3   Loch Morlich. Intensive beach use

Fig. 4   Loch Ness. Wild camping.

Fig. 5   Loch Ness. Lochside picnicking.

Fig. 6   Loch Lubnaig. Canoeing and spectating at the Trossachs Water Festival.

5   6

Up to this point, the Report has been mainly concerned with the identification of the nature, intensity and extent of recreational impacts currently taking place around proximal Scottish lochsides. In relation to the scale of the resource base, the impacts would appear to be relatively slight and highly localised. However, if present trends in the recreational use of the countryside are maintained, and the numbers of visitors from south of the Border and overseas countries continue to increase, so too will the use of the Scottish lochside. Pressure on already heavily used or even over-used sites will intensify and impacts will be initiated around other, as yet undeveloped, proximal lochs. Even within the past five years, the increased use of more remote and relatively untouched lochs, just a little off the main trunk roads, has been noticeable. The most striking examples among those studied are Loch Ken, and other Galloway lochs, and the Whitebridge group of lochs (Tarff, Duntelchaig, Mhor, Ruthven) to the south-east of Loch Ness and including the south-east side of Loch Ness itself.

It is therefore necessary, in the interests of both short and long-term developments, to assess the vulnerability of the lochside to impacts, that is, the extent to which part or the whole of a lochside is or may be susceptible to environmental changes resulting from the direct or indirect impact of recreation. This assessment of vulnerability is necessary in order to:

(a) establish base-line environmental standards against which the effect of current impacts can be compared, and from which indices of environmental deterioration can be formulated; from the latter, critical thresholds in the process of deterioration can be identified;

(b) aid the longer-term planning of lochs generally, as well as the possible development and management of unexploited lochs and lochsides, and for the drawing up of guidelines for the assessment of their liability to deterioration in the foreseeable future.

7.1 Concept of Vulnerability

Although the concept of vulnerability is basic to the understanding of the use of both renewable and non-renewable resources, it is a complex, multi-dimensional and relative attribute which is impossible to measure in any known absolute or precise terms. Degrees of vulnerability of lochside components, sites, or entire lochsides can be ranked on the basis of semi-quantitative and/or qualitative descriptions, but they cannot be summed-up to arrive at a measure of total vulnerability. Rank numbers when used are not real values, and ranking applied at one scale can rarely be applied at another, as will be explained later. Vulnerability of proximal lochs is dependent, at any one point in time, on a complex set of interacting variables some of which are more amenable to standardised description than others, and some of which have a wider range of variation than others.

7.1.1 Factors influencing vulnerability

Both intrinsic and extrinsic factors influence vulnerability:

(a) Intrinsic factors are related to the physical-biological characteristics of the resource, the most important of which are:

(i) the initial attractiveness of the site irrespective of its suitability for one or more types of recreational activities;

(ii) the relative durability (or resistance) of the physical and biological components of the site to impact.

(b) Extrinsic factors are not related to the nature of the resource, they are independent of it; but, in so far as they affect recreational use, they may intensify or ameliorate the resulting impacts and hence the vulnerability of the site. They include:

(i) location of the lochside relative to markets;

(ii) accessibility, both *de facto* and *de jure*, of the lochside;

(iii) visibility of the loch and/or lochside from the road;

(iv) management, including restriction as well as promotion of recreational use.

The size of the loch is often regarded as an extrinsic factor. For the purpose of vulnerability assessment, it is perhaps more meaningful to regard it as a dimension within which the other factors operate. This is significant because the vulnerability of a lochside often has to be assessed by a different intrinsic or extrinsic factor dependent on the scale of investigation or level of planning.

The objects of this Chapter are firstly, to analyse in various ways the vulnerability of lochside components, sites and entire lochsides, and secondly, to examine the extent to which intensity of use, type of impact, vulnerability, and existing intensity of damage, are inter-related on those lochsides surveyed and, more particularly, on those for which the necessary data on recreational use is available, for example, Loch Ken, and the Trossachs Lochs. Finally, this Chapter aims to summarise these inter-relationships in the most appropriate and convenient manner.

7.2 Vulnerability of Lochside Components

No standardised method has been adopted to assess vulnerability of the lochside components at varying scales. It was not possible to analyse all the components involved in the same manner or degree of detail, but beaches and shore vegetation, for the reasons already noted, were measured more precisely and on a finer scale than site-types. In the case of site-types, the vulnerability of the components was assessed on the basis of both qualitative and quantitative comparisons of a wide range of used and unused sites. The form and composition of lochshore vegetation and beaches, for which little or no previous data existed, were surveyed and measured more accurately. Lochshore, and some closely related back-shore vegetation communities, were subjected to experimental impact (running, walking and a limited amount of boat-launching and beaching), and the experimentally-induced damage was compared with existing conditions on recreationally used sites (see Appendix III). Finally, attempts were made to measure the rate of erosion and retreat of lochshore cliffs of varying height and composition, on a number of heavily used sites, to test qualitative estimates of their comparative vulnerability.

The components and shore-features, two or more of which comprise a lochside-site are:

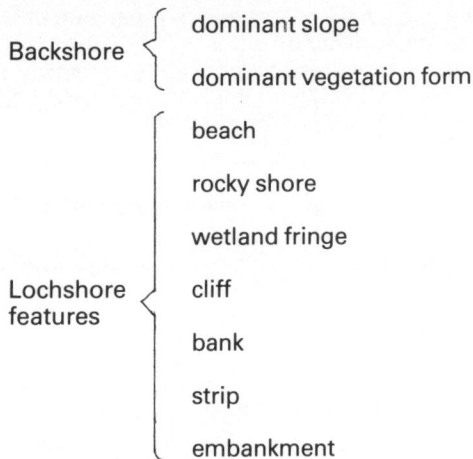

Backshore {
  dominant slope
  dominant vegetation form
}

Lochshore features {
  beach
  rocky shore
  wetland fringe
  cliff
  bank
  strip
  embankment
}

(see Glossary for definitions of terms).

On the basis of the data recorded, the relative vulnerability of each feature has been ranked in Table 44. However, for the purpose of this Report, only the vulnerability of the three principal, and most frequently recurring, lochside components — the backshore, the backshore/shore cliff and the shore — will be considered in detail.

7.3 The Backshore

The backshore is the most complex feature, since its description is dependent on the combination of two constituent components — slope and vegetation. Insufficient data is available to rank the relative durability of the vegetation (one of the most important indices of vulnerability) and it has had to be assumed, from observation of lochsides, that the steeper the slope under a given type of vegetation and soil, the more vulnerable it will be to recreational impact.

Table 45 is an attempt to assess the comparative vulnerability of the components of the proximal backshore in terms of their attractiveness for the visitor as seen from the roadside, and their susceptibility to one or more of three levels of damage.

Of the main types of backshore-site, the most vulnerable are those characterised either by short grass vegetation with a gentle to moderate slope, or open woodland (through which the loch can be seen from the road), on similar slopes with a comparable substratum. These two sites combine high attractiveness for nearly all or most land-based activities (and more particularly for car-parking, picnicking and general relaxation) and low to moderate durability of the resource base.

The short-grass site can withstand a fairly high degree of impact before noticeable deterioration occurs, and light impact by trampling can bring about an increase in flowering herbs. Under moderate pressure from trampling and vehicles, a shift in the relative balance of constituent species occurs. Trample-resistant plants such as common bent (*Agrostis tenuis*), crested dog's-tail (*Cynosurus cristatus*), plantains (*Plantago major* and *Plantago lanceolata*), white clover (*Trifolium repens*) and the creeping buttercup (*Ranunculus repens*) tend to replace the less durable species; in addition, there is a noticeable absence of flowering plants (see Tables 4 and 5 in Appendix III for full lists). It has been suggested that the presence of non-flowering white clover and creeping buttercup may indicate a critical threshold between moderate and heavy use and of over-use in terms of sward deterioration. If impact is not too heavy and seasonal recovery of the grass is possible, a durable self-perpetuating sward of trample-resistant species, capable of withstanding a moderate amount of pressure, is gradually established. This is characteristic of picnic sites which do not receive undue vehicle impact. Heavy impact by trampling and vehicle movement results in the appearance of a high percentage of annual meadow-grass (*Poa annua*) and of bare ground. Modifications to short grassland attributable to recreation as observed during this survey are similar to findings recorded by Leney (1974).

Goldsmith (1974) has suggested that the absence from a sward of one to three of the following species can be used to identify a critical threshold of use. These species are bent, crested dog's-tail, meadow fescue (*Festuca pratense*), perennial rye-grass (*Lolium perenne*), meadow-grasses (*Poa annua* and *Poa pratense*) and white clover. However, not all of these plants are constants in Scottish backshore grassland; on Scottish sites, the dominance of annual meadow-grass and a high proportion of bare ground are the most widespread and most distinctive indicators of damage. Table 46 shows species characteristic of grassland under various levels of use on sites around the Trossachs lochs. On heavily used informal car parks, annual meadow-grass is the most durable grass; it occurs only rarely in grazed areas, except at points where cattle concentrate and puddle the ground, and it is also common on paths, picnic and camp-sites. In addition, it has a considerable ability to recover even under heavy impact. This has been strikingly demonstrated by the seeding of a bare area near the marina on Castle Semple Loch (NS 358591). A mixture of hard-wearing grasses, mostly meadow-grass and perennial rye-grass, were sown in May 1975, and despite very heavy wear during the summer, an almost complete grass-cover had been established by the end of August in the same year.

Table 44

Ranking on basis of high (H), moderate (M) or low (L) relative vulnerability of lochside site-type components

1. Dominant slope	0 - 5° L	6 - 10° L	11 - 15° M	16 - 20° H	20° H		
2. Dominant vegetation form	Wetland L	Short grass L	Long grass M	Moor grass L	Low shrub M	Shrub (dense) L	Trees (open) H
3. Beach	Boulders L	Cobbles L	Sand L	Mixed gravel M		Mud H	
4. Rocky Shore	L						
5. Cliff	High organic L	Low organic M	High mineral M	Low mineral L			
6. Bank – (see relevant combination of slope and vegetation form)							
7. Strip – (see relevant combination of slope and vegetation form)							
8. Embankment	Stone L	Un-consolidated M					

Table 45

Backshore components: relative vulnerability in terms of attractiveness and ease of damage

H = high   M = moderate   L = low

Slope category	Vegetation-physical component	Attractiveness	Plant species change	Ease of damage/ development of bare ground	Gullying or slumping
Less 10°	Wetland	L	M	L	L
	Shore grass	H	H	L	L
	Long grass	M	M	M-L	L
	Moor grass	M-H	L	L	L
	Low shrub	M	L	M-L	L
	Shrub (dense)	L	L	M-L	L
	Tree (open)	H	M-H	M	L
11 - 20°	Short grass	H	H	M	M-H
	Long grass	M	M	M	M-H
	Moor grass	M	L	M	M-H
	Low shrub	M-L	L	M	M-L
	Shrub	L	L	M	L
	Tree	M-H	M-H	H	M-H
Over 20°	Short grass	M-H	H	H	H
	Long grass	L	M	H	H
	Moor grass	M	L	H	H
	Low shrub	L	L	H	H
	Shrub	L	L	H	H
	Tree	L	M-H	H	H

In short grass areas, one of the most vulnerable parts is the regularly used path near to and paralleling the shore. Continual wear can result in such a path becoming bare, and depressed below the level of the surrounding surface. During periods of high water-levels, this bare ground can be subjected to abrasion and undercutting by wave action, and if the narrow strip separating the path from the shore is broken, accelerated erosion can occur.

Second only in attractiveness to short-grass sites is open woodland. The vulnerability of the ground flora is very similar to that of the short-grass vegetation discussed above. However, the relative resistance of the trees makes this type of backshore (compared with other sites with similar slope and substratum) very resistant. Durability, combined with size and longevity makes the tree a particularly good indicator of the level of impact to which a site has been subjected. In so far as it is also a common element on lochside banks and on many lochshore sites, the relationship between tree-damage and intensity of impact will be discussed at the end of this section.

The remaining types of backshore are all less attractive because of high gradients and/or density of the vegetation cover. Where slopes are steep impact tends to be concentrated on pathways which become potential erosion-lines. In cases where rock outcrops occur *in situ*, these backshores are very durable.

Two backshore vegetation types were investigated in more detail because they are intermediate in character between the backshore and wetland shores: these were dominated either by the common sedge or by meadow-sweet. Common sedge is characteristic of the wetter short-grass sites on gently sloping ground. During the summer, when loch levels are low and the site dries out, it is both attractive and remarkably durable. It showed little change even under high intensities of experimental trampling. Meadow-sweet sites are moderately attractive during the flowering season, because of their dense cover of sweet smelling flowers, but are very vulnerable to impact on account of a habitat somewhat damper in the summer than that of the common sedge. Paths with bare ground can develop fairly quickly during one season under moderate levels of use. However, by the beginning of the following season bare ground has virtually disappeared, except where subjected to heaviest experimental running intensities, but the new vegetation is shorter than that not subjected to experimental impact (Rees and Tivy, 1977).

7.4 The Lochshore

The assessment of the vulnerability of the lochshore to the effects of recreational impact posed greater problems than in the case of either the backshore or main backshore/shore discontinuity. This was because of the lack of any empirical data about the relationship between natural impact from wave-action and that from recreational activities around lochshores. In this respect, a fundamental distinction must be made between shores with sparse or no vegetation and those with wetland fringes.

7.4.1. Shores without vegetation

Shores without vegetation are both attractive and durable. While it has already been established that the beach is the most attractive type of shore for recreational purposes, it is less certain that the concept of durability and vulnerability can be applied to it in precisely the same way as before. Beaches composed of well-sorted sand or gravel appear to suffer little from recreational impact, as they remain mobile, and although their surfaces may be disturbed by use, they recover their natural form rapidly during periods of rising and falling water-levels in autumn through to spring. Finer material drawn down-shore during the spring and summer as the loch level falls is replenished when the water rises again in the winter.

Beaches composed of mixed mud, sand and coarse gravel are only slightly less attractive than the more homogenous beaches, and despite the fact that they are more susceptible to recreational modification, they are almost as durable. The impact of vehicle wheels, for example, compacts the beach material and thereby reduces its mobility. It also causes the coarser particles to become firmly embedded in the sandy/muddy matrix beneath, forming a shingle-pavement which actually increases the durability of the beach. This pavement can resist a high intensity of continued use. It may remain intact until, with increasing use, it is broken into, after which it can be stripped fairly rapidly by the abrasive effect of breaker-swash and back-wash.

The consequent flattening of the profile on these beaches means that waves may break further up the beach (see Fig. 35). In summer, this might accelerate the lochward removal of beach material; in winter it could conceivably bring the breakers nearer the backshore cliff, without necessarily returning mobile material to the beach. Hence, the risk of increased wave erosion of the cliff tends to be counteracted by a reduction in the amount of material available for transport and thus for abrasion and erosion. Furthermore, cliff erosion by recreational impact may produce more debris than can be dealt with by the breakers, hence the base of the cliff may be protected from wave erosion by a bank of loose material.

7.4.2 Shores with fringing wetland vegetation

Of the shores investigated, those with some form of wetland vegetation account for as large an area as those with bare beaches. The recreational value of shores with this type of vegetation is low, but because of the wetland characteristics of the habitat, it is attractive to ornithologists and wildfowlers. However, because of increasing pressures on, and the decline in areas occupied by, wetlands generally, the ecological and scientific value of lochside swamps is relatively high.

The vulnerability of the constituent wetland species to recreational impact is dependent on the following factors:

(a) The attractiveness in terms of height, colour and general appearance of the vegetation cover as viewed from the road, or of individual species, such as the bulrush;

(b) The ability of the plants to maintain growth, or even to be stimulated by recreational impact, during the growing season;

(c) The relative resistance of leaf and root systems to direct impact: short, flat and rosette leaves are less vulnerable than tall upright ones; fibrous less so than flaccid and fleshy leaves; and deeper less so than shallower rooting systems;

(d) The ability of plants to recover after an impact. Indications of relative recoverability can only be given for one season's use at varying intensities. There can be no guarantee that ranking on the basis of long-term (20 years) use would be similar. It was not possible to establish a critical threshold of use when plants failed to recover;

(e) The nature of the substratum: vegetation growing in a dry soil has, in general, a greater resistance to impact than that in saturated conditions; Burton (1974) has estimated that damage to terrestrial vegetation on wet soil is twice that when it is dry. In this respect shore species are, therefore, particularly vulnerable when the substratum is not only saturated but is composed of soft organic mud. On firmer sites, a layer of plant litter frequently builds up, which protects the ground surface from continued impact by trampling. This has particular relevance for reed and sedge beds where impact by a few bank-fishermen in winter, when water levels are high and the site very wet, may do more damage than a greater number of fishers using the same area in the summer.

The three principal types of lochshore vegetation described in Chapter 4 vary in vulnerability in terms of attractiveness of the habitat for recreation and durability of the constituent species:

(a) Emergent communities – unattractive and susceptible to damage: vulnerable;

(b) Floating-leaved communities – avoided by recreationists: durable;

(c) Submerged/low-shore communities – attractive and of variable durability: vulnerable on sites which are not stony.

However, the nature and distribution of impact varies between these groups, as does the relative vulnerability of the main species of which they are composed. Table 47 summarises the relative vulnerability of the main wetland species subjected to recreational impact. A species with low vulnerability, but showing little change, has low recovery rates; the most dramatic changes are obviously those involving high vulnerability and high recoverability. The empirical and experimental methods used and the

data on which this ranking is based is published elsewhere (Rees and Tivy, 1977 and in press). In contrast to other rankings used in this Report, quantitative values based on the ratio of the difference in plant growth before and after impact on an experimentally trampled site, to the difference in growth during the same period on a control site, can be assigned for both vulnerability and recoverability. However, these are relative figures whose principal value lies in the standardisation of the qualitative descriptions of high, moderate and low.

Table 47 illustrates the much greater vulnerability of low-shore vegetation compared to emergent vegetation. Although shoreweed spreads very rapidly and has a moderate potential for recovery, it is always very vulnerable to impact. When dry, it becomes desiccated and its growth-rate decreases, and when wet it is more vulnerable to damage both by wheeled vehicles and trampling. Shoreweed can be eliminated by continuous trampling and abrasion from the launching and beaching of boats. The bulbous rush is somewhat more resistant to recreational impacts. A high proportion of this species, particularly along the more obvious lines of movement, is probably one of the best indicators, known to date, of a high level of use of this type of site. On all sites except those with a fine sandy substratum, the low-shore vegetation cover tends to be naturally discontinuous to a greater or lesser degree. Impact is therefore unlikely to reduce the recreational value. Indeed, it could be argued that it is enhanced either directly or indirectly on these shore sites. The exception is where a continuous, dense grass-like sward of shoreweed occurs on sandy sites. In this case, high vulnerability makes for rapid wear and elimination of the vegetation cover, when it is subjected to a relatively small amount of use. This effect has been noted on a previously unused area recently opened up to picnickers. Also, such shore swards are sufficiently uncommon to have a scientific value, and to warrant either protection from use, or conservation management.

Emergent vegetation is, for reasons already discussed, much less vulnerable to the more spatially restricted and specialised impacts to which they tend to be subjected. Observation and experimental testing however indicate a range of variation in degree of vulnerability among the commonest species (see Appendix III). Reed and reed sweet-grass beds are the most vulnerable to impact and bare-ground tracks can be formed rapidly through them under relatively modest intensities of use (see Table 48) particularly by bank-fisher men and wildfowlers. However, ability to recover is such that by the middle of the next season all trace of the previous year's track had disappeared.

Adjacent to footpaths through stands of reed canary-grass, bank-fishing stances are commonly found. Here, as near informal picnic sites and boat-beaching points, the community is more open, and the number of weedy species increases with the level of use.

The commonest component of Scottish lochshore sedge-beds, bottle sedge, is much less vulnerable to impact than the reeds; bare tracks were produced only at the highest intensities of experimental impact, particularly in sedge beds on soft organic sites. Recoverability, though good, was more erratic and much less dramatic than in the reed beds. The common spike-rush and water horsetail which are also found in considerable amounts around Scottish lochshores are, likewise, less susceptible to damage, and recoverability is reasonable. However, because of their low vulnerability, their recoverability ratio is also low. The tall reedmace and bulrush are the least vulnerable because of their tough tissues and deep water habitat, which protects them from most forms of recreational impact, except stem breakage by boats running into or very occasionally through them.

The degradation of reed and sedge beds probably has more immediate and serious consequences than in the case of the low-shore vegetation. There is a more dramatic visible and scenic change, and the loss of nesting areas and cover for wildfowl and other birds reduces the recreational value of these sites and, in some cases, of the whole loch. More important, however, is the possible effect of the destruction of reed and sedge beds on impact by loch-waves and breakers. Emergent and floating-leaved plants dampen waves and stabilise loch-floor sediments. Their effect on the former, in this respect, will be proportional to their density and width. Observation indicates that even small breaks, either in the form of boat-nests (for details see Appendix III) or tracks, in these wetland fringes, allows waves and breakers to be more

effective agents of transport. The fine substratum is progressively disturbed and the stability of the adjacent stands of reeds and sedges weakened by the removal of material at a rate greater than it can accumulate. Gradual elimination of reed/sedge beds will be accompanied by removal and redistribution of bottom sediments and a concurrent deepening of water and complete inundation of a greater area than before. Both changes make successful re-establishment of the littoral wetlands increasingly difficult. There has been no adequate monitoring anywhere in Britain of the processes involved in the decrease of lochshore emergent fringes, or the effect of their removal on the wetland habitat. Figure 36, however, illustrates the influence of different forms of land use, including very heavy recreational use, on the littoral vegetation of the Havel Lakes, which in 1959 became the main recreation area for West Berlin (Sukopp, 1971).

7.5 The Backshore/Lochshore Discontinuity
Characteristic of many natural lochsides is a sharp break of slope at the back of the shore, referred to as a bank or cliff. It can vary from a few centimetres to several metres in height and can be formed of unconsolidated mineral matter, peat, or solid rock. It is usually, though not invariably, coincident with the landward limits of mobile beach material. For the recreationist, it forms a discontinuity between the two major lochside habitats; the aquatic or semi-aquatic of the shore and the terrestrial of the backshore. Furthermore, it is an obstacle to the

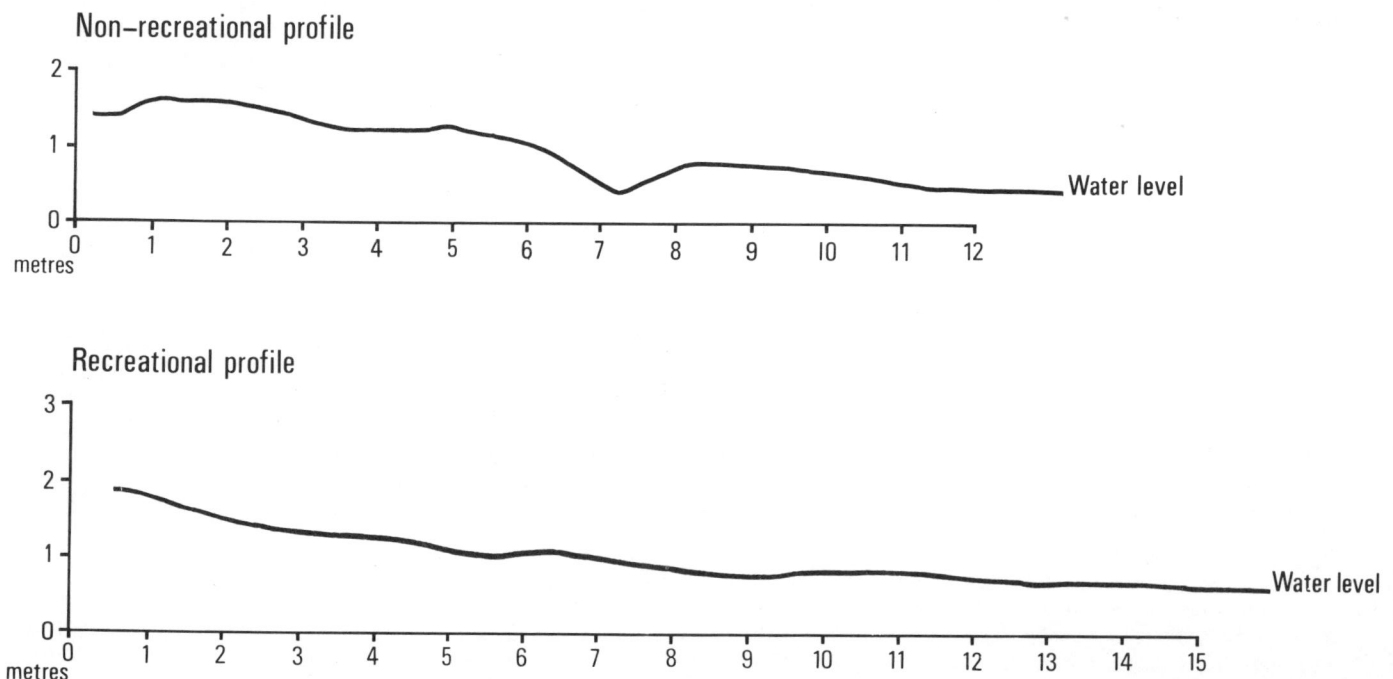

Fig. 35   Recreational and non-recreational lochshore beach profiles

Table 46

Percentage frequency of species characteristic of grassland under various levels of use on recreational sites around the Trossachs Lochs

Recreational sites	Bare ground	Poa annua	Lolium perenne	Trifolium repens	Agrostis tenuis	Holcus lanatus	Ranunculus repens	Cynosurus cristatus	Ribes species
Trampled car park (Heavy use)									
L. Lubnaig upper	40	72	11						
	40	41	13	—	20	8	—	—	—
L. Lubnaig lower	44	42	32	6	—	—	—	—	—
	60	38	15	2	—	—	—	—	—
	31	45	20	37	—	2	—	—	—
	37	40	12	—	—	—	—	—	—
L. Chon (F.C.)	96	24	—	—	—	—	—	—	—
	50	43	15	—	6	—	—	—	—
	83	48	—	—	—	—	—	—	—
	12	31	21	11	34	—	—	—	—
	55	45	36	—	—	—	—	—	—
L. Venachar east	85	25	—	—	—	—	—	—	—
Trampled (Moderate use)									
L. Lubnaig upper	5	81	15	32	—	—	—	—	—
L. Venachar west	14	26	15	1	30	—	13	10	—
	21	46	27	7	15	—	3	2	—
	8	9	30	4	60	—	11	2	—
L. Venachar east	10	6	27	18	36	24	9	—	—
	30	32	45	10	7	—	—	7	—
L. Chon (F.C.)	5	1	—	14	46	50	34	—	—
	2	4	—	14	60	35	13	—	—
	5	—	12	—	67	2	—	—	—
L. Chon (boathouse)	14	1	14	4	13	—	—	23	—
Little disturbed (Light use)									
L. Lubnaig upper	—	—	—	—	12	95	—	—	—
L. Lubnaig lower	—	—	—	—	50	—	—	—	40
	—	—	—	—	—	62	—	27	—
	—	—	—	—	—	81	—	—	10
	—	—	—	—	26	40	—	—	41
	—	—	—	—	31	72	24	—	12
L. Chon (F.C.)	2	—	—	—	30	43	28	—	—
	—	—	—	—	29	69	—	—	—
	—	—	—	15	28	47	—	—	—
L. Chon (boathouse)	—	—	—	—	29	24	30	—	—

F.C. = Forestry Commission

Table 47

Relative vulnerability and recoverability of lochshore wetland vegetation in relation to recreational impacts

Type of vegetation and main plants	Attractiveness	Durability	Vulnerability	Recoverability	Main type impact
Emergent					
Reed	L	L	H	H	Footpaths and trampling
Reed sweet-grass	L	L	H	H	
Reed canary-grass	L	(variable)	(variable)	M	
Bottle sedge (firm site)	L	L-M	L-M	M-H	Wildfowl
Bottle sedge (soft site)	L	L	L	M-H	
Common spike-rush	M	L	L	L	Boating
Horsetail	L	L	L	L-M	
Reedmace	L	H	V.L.	N.D.	
Bulrush	L	H	V.L.	N.D.	
Floating-leaved					
(all types of plants)	L	H	L	N.D.	Boating
Submerged (all types)	L	H	L	N.D.	Boat-beaching and paddling
Low-shore					
Water lobelia	H	L	H	L-M	Diffuse trampling, boat-beaching and flower-picking
Dense sandy: shoreweed	H	L	H	L-M	
Stony: shoreweed	H	M	M	L-M	Sometimes vehicles and boat-trailers
Bulbous rush	H	H	L	L-H	
Wetland backshore					
Common sedge	H	M-H	L	L	Diffuse trampling and boat-beaching
Meadow-sweet	H	M-H	H	M	Paths and boat-beaching

Vulnerability ratios

Low (L)　　　= 1: +1 to −1
Moderate (M) = 1: −1 to −2
High (H)　　 = 1: below −3

Recoverability ratios

Low (L)　　　= 1: less than +3
Moderate (M) = 1: +3 to +7
High (H)　　 = 1: over 7
N.D.　　　　 = No Data

movement of people and/or vehicles from one part of the lochside to the other. Cliff-height classes based on observation are given in Table 49.

Cliff durability is dependent mainly on height and the material of which it is composed, particularly the susceptibility of the material to mechanical erosion. The attractiveness of this cliff for recreational purposes is related to the amount of shelter or bottom-edge it provides at its base, and/or the amount of space for overviewing the shore and loch at the top-edge. In other words, the attractiveness of the cliff is related to its top or base, rather than the cliff-face itself. The latter is mainly a focus for such play-activities as digging, scrambling and sliding.

Vulnerability of the cliff then is related more to its lack of durability, and to its location between the backshore and the lochshore, than to its attractiveness for recreation.

The vulnerability of the principal types of cliff based on selected height and composition classes is analysed in Table 50. The slightly lower vulnerability of the peat cliffs is a function not only of their lesser attractiveness, but also

Table 49

Cliff height classes

Cliff	Height	Degree of limitation to movement
Low	0 - 30 cm	Above 30 cm height deters cars but not boats or boat trailers
Moderate	31 cm - 1 m	No obstacle to majority of people; difficult for boats and boat trailers
High	Over 1 m	Obstacle for boats and boat trailers; difficult to step off

Table 50

Vulnerability in terms of attractiveness (A) and durability (D) of main lochside cliff-types

Height	High (over 1 m)		Moderately high (31 cm - 1 m)		Low (Less than 30 cm)	
Material	(A)	(D)	(A)	(D)	(A)	(D)
Organic (peat)	L	L	M	L	M-H	M
Inorganic (mineral)						
Sorted	H	L	H	L	H	L
Unsorted	M	L-M	H	L-M	H	L

H = high   M = moderate   L = low

Table 48

Footpath development in three types of reed beds; type of path is that produced by known intensity (monitored) of actual recreational use

Vegetation community	Intensity of use experimental impact	Type of path produced
Reed (*Phragmites communis*)	(Total number passages across per season) 84	No defined path-level of use probably inadequate to maintain path
	128	Light use by bank fishermen
	256	Moderate use by bank fishermen
	512	Heavy use by visitors in general
Reed sweet-grass (*Glyceria maxima*)	84	Light use by bank fishermen
	128	
	256	Moderate use by bank fishermen
	512	Heavy use – typical of park loch pathways
Reed canary-grass (*Phalaris arundinicea*)	84	Light use by bank fishermen
	128	
	256	Moderate use by bank fishermen
	512	Heavy general use

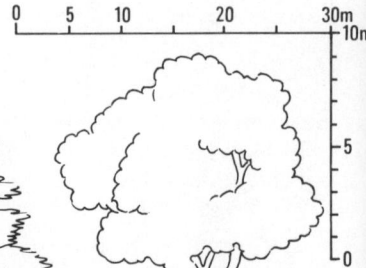

1. Natural condition of shore

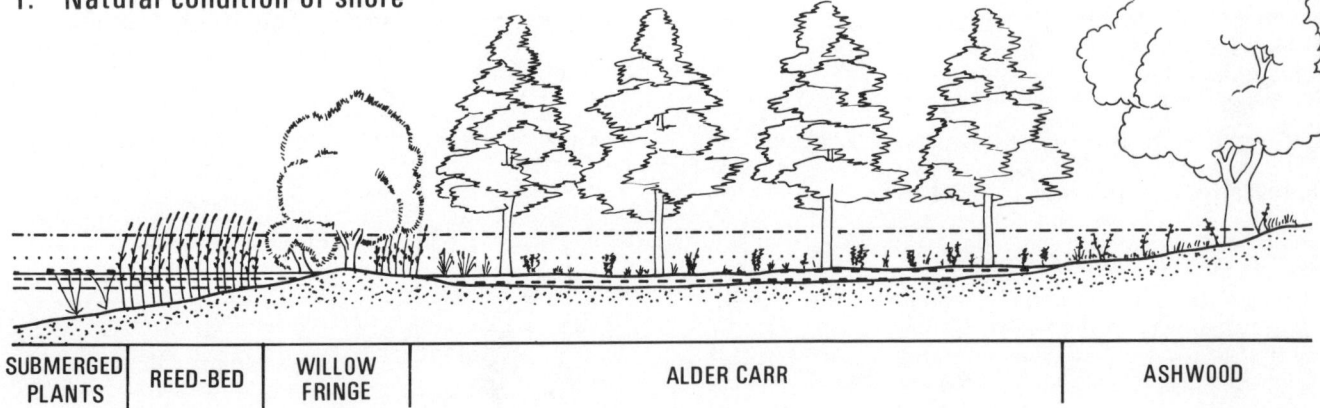

| SUBMERGED PLANTS | REED-BED | WILLOW FRINGE | ALDER CARR | ASHWOOD |

2. Meadow

| SUBMERGED PLANTS | REED-BED | WILLOW FRINGE | REED CANARY-GRASS | REED SWEET-GRASS | REED CANARY-GRASS | DAMP MEADOW | REED CANARY-GRASS | DAMP MEADOW |

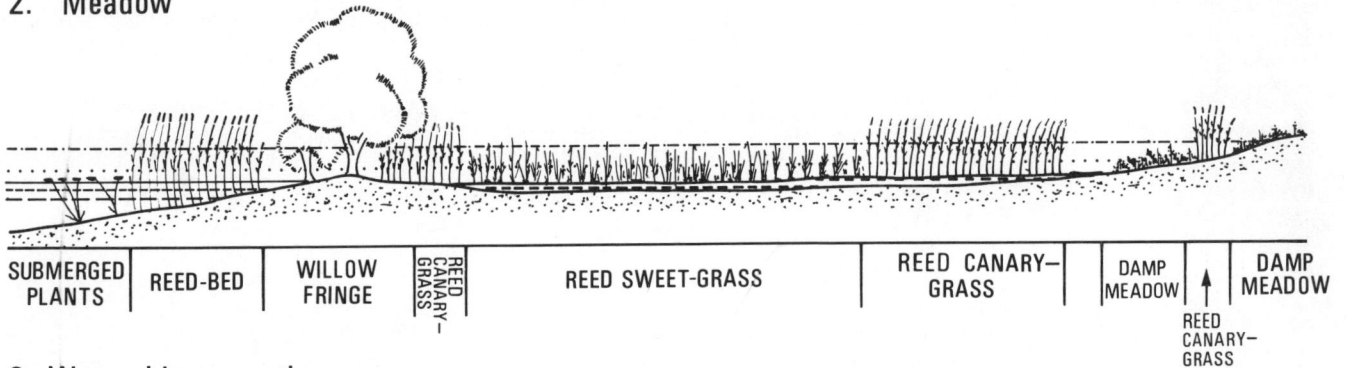

3. Waterside recreation

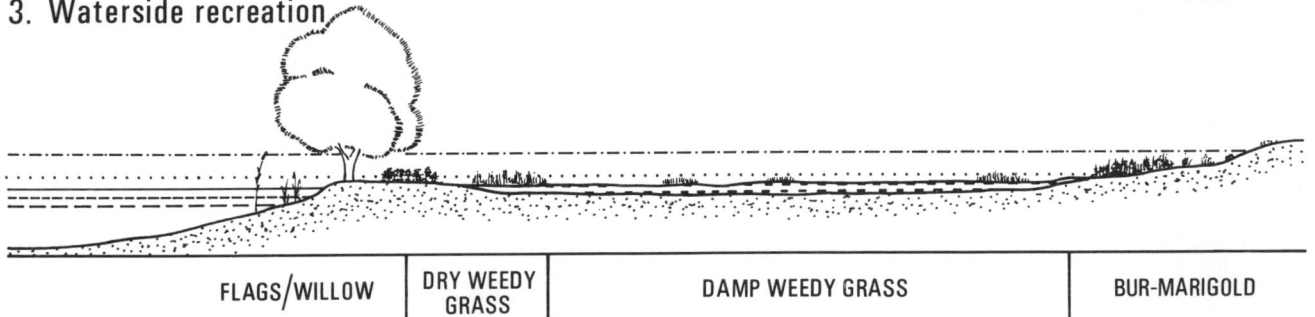

| FLAGS/WILLOW | DRY WEEDY GRASS | DAMP WEEDY GRASS | BUR-MARIGOLD |

4. Long-duration waterside recreation

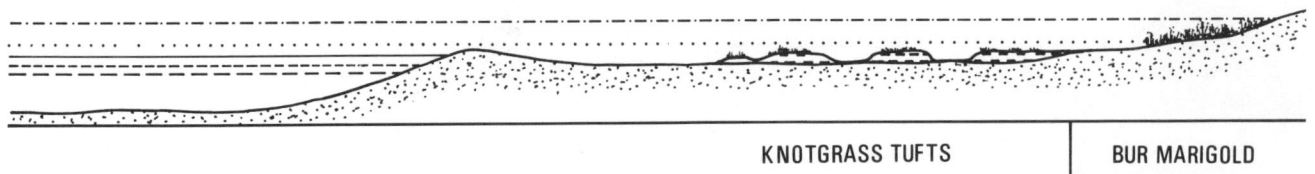

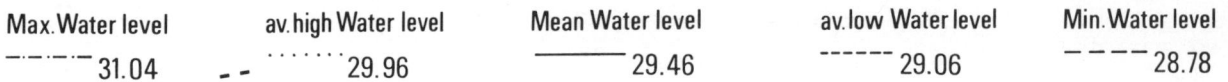

| KNOTGRASS TUFTS | BUR MARIGOLD |

Max. Water level	av. high Water level	Mean Water level	av. low Water level	Min. Water level
31.04	29.96	29.46	29.06	28.78

Fig. 36   Impact of recreational land use on Havel Lakes, Berlin (Sukopp, 1971)

1. HIGH PEAT CLIFFS

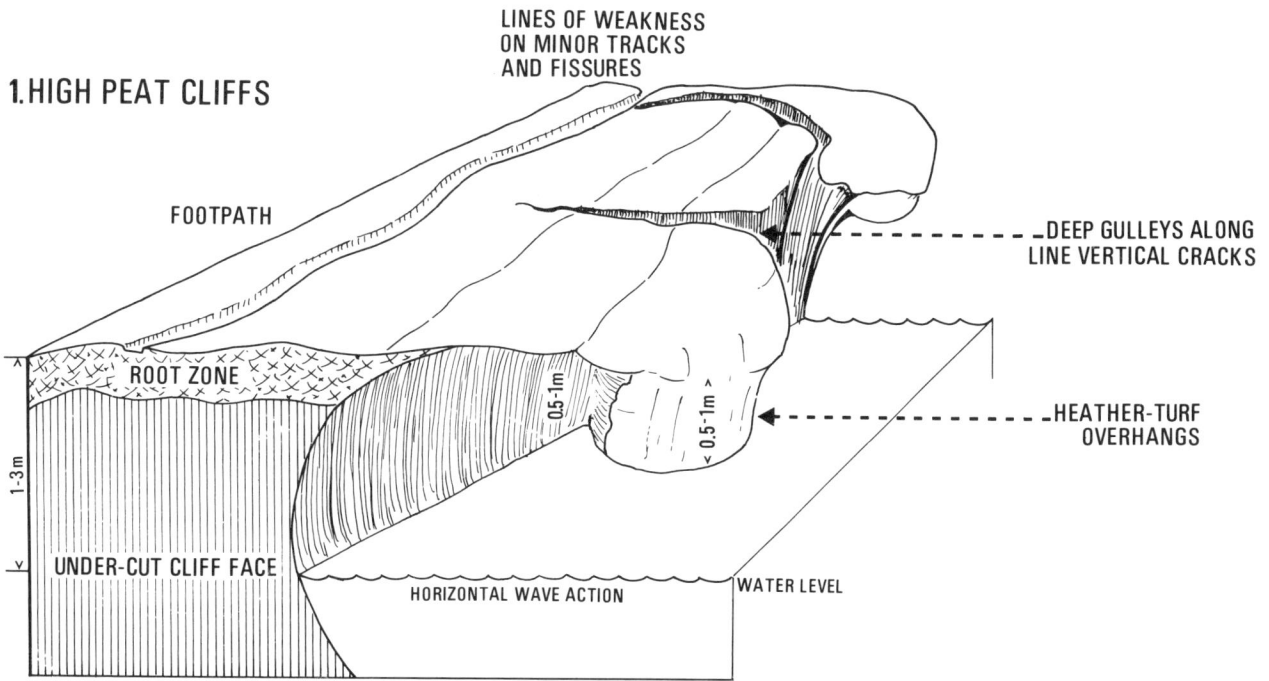

LINES OF WEAKNESS
ON MINOR TRACKS
AND FISSURES

FOOTPATH

DEEP GULLEYS ALONG
LINE VERTICAL CRACKS

ROOT ZONE

0.5-1m

< 0.5-1m >

HEATHER-TURF
OVERHANGS

1-3m

UNDER-CUT CLIFF FACE

HORIZONTAL WAVE ACTION   WATER LEVEL

2. LOW PEAT CLIFFS

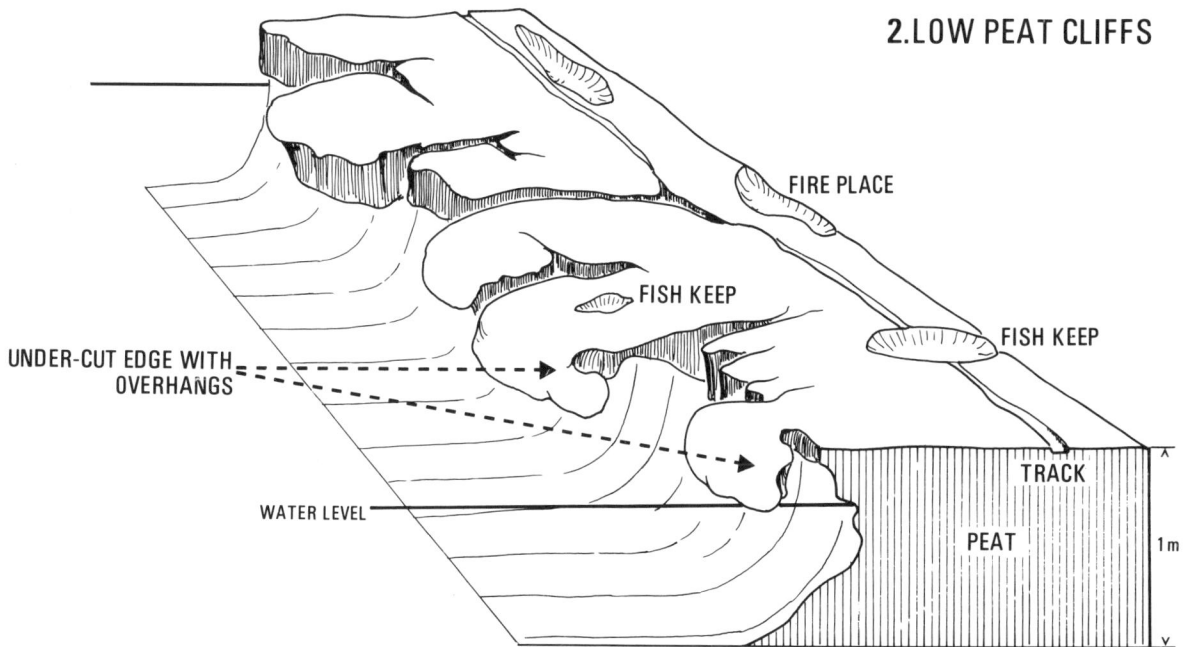

FIRE PLACE

FISH KEEP

FISH KEEP

UNDER-CUT EDGE WITH
OVERHANGS

TRACK

WATER LEVEL

PEAT   1m

3. LOW MINERAL CLIFFS

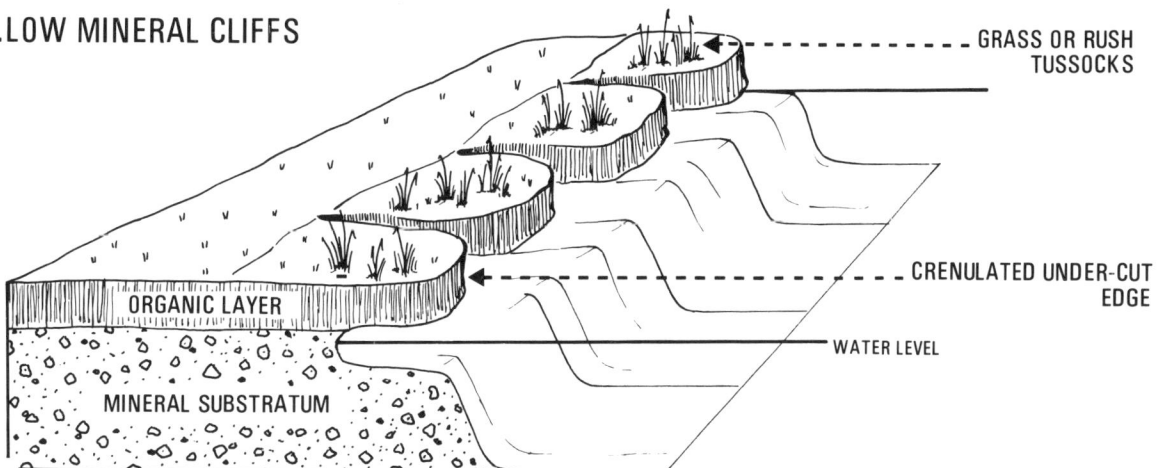

GRASS OR RUSH
TUSSOCKS

CRENULATED UNDER-CUT
EDGE

ORGANIC LAYER

WATER LEVEL

MINERAL SUBSTRATUM

H

Fig. 37   Effects of recreational impact on three types of water-edge cliff

of their location. Where there is no shore, the water-level fluctuates vertically at the cliff-face, and impact from recreational activity is confined almost entirely along the top edge; any impact along the base of a cliff-face is from wave action. Where there is a shore, the recreational impact may be from both above and below, and wave impact on the cliff-face is usually of secondary importance or even insignificant.

Three types of cliff at the water's edge have been distinguished and effects of different types of impact are analysed diagrammatically in Figure 37. Despite a liability to heavy impact from bank-fishermen and general users, the water-edge cliffs are much less vulnerable than those at the back of the shore. High cliffs are less attractive, particularly if composed of well-sorted sandy material, and they are often of low durability because of their

Table 51

Maximum cliff retreat measured on selected types of lochside sites

Location	Black Loch (nr. Slamannan)	Lochend Loch (Coatbridge)	Loch Lubnaig (Trossachs)	Loch Venachar (Trossachs)	
Period	Mar 77 - Oct 77 (7 mths)	Oct 75 - Oct 76 (12 mths)	Oct 75 - Oct 76 (12 mths)	Apr 75 - Mar 76 (12 mths)	Mar 76 - Oct 77 (6 mths)
Cliff-type	High peat (Water edge)	Low peat (Water edge)	High mineral (Backshore)	Low mineral (Backshore)	
Max. change cliff height	−0.5 m	−0.5 m	−1.45 m	−2.7 m	−1.7 m
Relative intensity of impact	Moderate	Very high	High	Very high	

Table 52

Criteria used to define and rank intensity of lochside erosion damage

Scale	Intensity of damage	Main shore/backshore cliff or bank	Minimum threshold conditions
4	Severe	High cliff	− Anthropogenic caves and cliff overhang − Frequent, deep, gully dissection of cliff-face or − Severe slumping of cliff-face
3	Bad	High cliff	− Continuous cliff-face − Pronounced cliff undercutting and slumping − Frequent, pronounced path erosion on cliff and on backshore
2	Moderate	Moderately high-low cliffs and high banks	− Discontinuous cliff-face − Incipient undercutting and/or slumping − Pronounced path erosion, cliffs, banks and backshore
1	Low	Low to high water-edge cliffs and moderately high-low banks	− Crenulated cliff edge, marked cliff overhang or slight bank undercutting − Incipient path and gully erosion on banks

vertical unvegetated faces. Low cliffs, although attractive and frequently subjected to heavy use above, are relatively little affected by wave-action. Small waves breaking at the face or at the base have little erosive potential, either in terms of their energy, or of the material they can transport. Their main action is to loosen unconsolidated mineral material or soft peat which results in a slightly undercut, crenulated edge, usually protected by a well-developed turf and root mat above. Bank retreat is a gradual process, even under recreational impact, and is effected mainly by unstable, undercut grass or heather sods breaking away from the low cliff edge. The most serious breaks, with accelerated erosion, occur where boat-beaching or launching has breached the cliff and created a small embayment (enlarged boat-nest) with an incipient beach at the base of a cliff pushed back from the water's edge.

Mineral cliffs at the back of the shore are the most vulnerable because of their high degree of attractiveness and relatively low durability. Such cliffs vary not only in height, but also with the nature of the constituent material. If the latter is compacted or indurated, it can be much more resistant; if it overlies an impermeable rock or clay-layer, at or just under the water-table, impeded downward drainage can lead to saturation and instability of the cliff face. Low cliffs are the most vulnerable as they can be rapidly destroyed by the passage of people, and particularly vehicles, across them. Destruction of surface vegetation inevitably accompanies this process with the result that the bare shore-zone is extended inland at the expense of the back-shore.

In the case of the moderately high and high mineral cliffs, retreat of the cliff face takes place by a process of undercutting and slumping along the face, and of gully erosion along established lines of movement down the cliff. Retreat tends to be more rapid the higher the cliff, but the rate can also vary depending on whether the backshore is tree-lined or wooded. Also, it takes place at the expense of the often limited backshore area and its associated tree-cover. In the latter instances, cliff-retreat leads more gradually to the undermining of trees which either become progressively unstable, recumbent and eventually fall (hence increasing cliff instability), or which become detached from the backshore and persist as a decimated remnant fringe in front of the retreating cliff. Retreat of low mineral cliffs can lead to an increase in the width and amount of beach material supplied by eroded material. In contrast, however, the retreat of high mineral cliffs, particularly when composed of a high proportion of large cobbles and boulders, can lead to an accumulation of debris, either at the back of the shore or, in extreme circumstances, over its entire surface, with a consequent decrease in its recreational value.

Rate of cliff-edge retreat can be used as a measure of relative vulnerability of this component. In order to rank the vulnerability of the main types of cliff identified, measurements were made of the difference between the position of the top cliff-edge at the beginning and end of

one or two seasons. The time and resources available only permitted a limited number of measurements. The results given in Table 51 are generally consistent with the observations above.

The backshore/lochshore discontinuity, particularly if it is a cliff, is the most vulnerable of the three components. It is the weak-point of the lochside on which the vulnerability of the whole lochside site may depend. Since its vulnerability to erosion is high, degrees and types of cliff-damage can be used as one of the most important diagnostic characteristics of intensity of lochside erosion damage (see Table 52).

7.6 Trees as Indicators of Recreational Impact
A relatively high proportion of Scottish lochsides, particularly in the Highlands, have some type of tree cover. This comprises either small stands of open mixed deciduous woodland, narrow linear tree fringes or coniferous plantations. The recreational significance of these trees is related to their amenity value, and to their role as stabilisers of what is frequently an unstable or potentially unstable site. As previously indicated, trees are peculiarly resistant to recreational impact; they can tolerate and exhibit a wide variety of types and degrees of damage before being completely destroyed. The main types of easily identifiable tree-damage are discussed below:

(a) Crown damage.

   (i) Branch breakage and deformation: if breakage continues, stunted, sharply-angled branches, with bulbous joints and elbows are common;

   (ii) Dead, leafless branches particularly around the periphery of the otherwise living, summer-green crown of deciduous trees: at a more advanced stage of damage the proportion of dead branches may exceed that of living ones;

(b) Trunk or bole damage.

   (i) Fire-scars: these vary in intensity and are one of the most obvious and striking types of tree-damage. Other types of damage include the abrasion and removal of bark from an area of varying width and extent by rubbing, notching, cutting and carving the bole, and bark stripping;

(c) Root damage.

   (i) Root-mat exposure: with increasing impact the entire area occupied by the lateral root systems of individual trees becomes exposed at the surface of the ground;

   (ii) Undermined roots: these occur where there has been partial or nearly complete removal of the soil and subsoil around and under the root-mat;

   (iii) Vertical root-stilting: these occur when at least 10 cms of soil and subsoil have been removed and the upper part of the vertical roots becomes exposed; in extreme cases, the trees are left standing on stilts;

(iv) Root breakage and deformation: these produce malformed or deformed trees. Continual breakage of some young trees (such as alder, willow, hazel (*Corylus avellana*), birch (*Betula* species) and oak (*Quercus* species)) can stimulate suckering and produce a coppice-form; more severe damage results in stunted growth and in extreme circumstances will destroy the tree.

Neither coppicing nor stunting alone, however, are reliable indicators of recreational impact, because they may be the result of varying environmental conditions or of other non-recreational processes. Fire scars, in association with other evidence of recreational impact on the tree and/or its immediate habitat, provide more reliable indicators; indeed, repeated charring of the bark and bole alone may result in the eventual death of the tree. More commonly, burning of the base of the bole is combined with root-breaking and undercutting of the bole and root-mat. The latter causes increasing instability, particularly on steep, unstable slopes, and finally recumbent, up-rooted, and fallen trees.

Tree species, however, vary in their vulnerability to impact. In this respect, the most significant difference is between the alder and other types of broad-leaved deciduous trees. The former can tolerate and survive a greater degree of impact for a comparatively longer period of time than other deciduous trees. It is, together with various species of willow, one of the most characteristic trees or shrubby trees of the Scottish lochside (McVean, 1956).

Alders grow best in damp habitats and can, in winter months at least, tolerate relatively long periods of partial inundation. Optimum conditions for germination and growth, however, occur where there is a marked seasonal fluctuation of the water-level, but where the capillary-fringe is high enough to ensure that the surface layers of the soil remain sufficiently moist for some 20-30 days in the period April-June. The tree has extensive, relatively shallow lateral and very deep vertical roots. Hence it can tolerate and survive not only alternating periods of desiccation and submergence but also inherently unstable soils and sub-soils such as are characteristic of many lochshores. It is also relatively fire-resistant, and can withstand moderate amounts of browsing by cattle and deer, but normally escapes depredation by sheep because of the wetness of its habitat. In addition, the alder coppices very readily, and cutting or breaking of its branches can stimulate the growth of suckers.

For the recreationist, two main types of alder wood can be identified on the Scottish lochside. The first are dense thickets of alder-willow on level, very damp to wet sites on the landward side of the more extensive reed/sedge beds. These woods are impenetrable and often unsafe under foot, and are therefore usually avoided, even by fishermen. The second type is the alder-fringe or remnant alder-fringe (often composed of no more than a single line of trees) which parallels the lochside around the inner edge of the shore. Whether this extends inland or not depends on the presence and height of the bank or cliff-discontinuity between the shore and the backshore.

The alder-fringe is subject to a greater variety and intensity of both recreational and wave-impact than other trees, because of its location and habitat. Where the fringe has suffered a moderate amount of recreational impact, the trees grow isolated one from the other on remnants of what must have been a more complete and extensive turf cover; which previously may have been up to a metre above the present lochshore level. On more heavily used sites, all that may remain of the alder-fringe is a line of stunted fire-scarred trees, from which all lateral roots have been stripped, supported by vertical roots exposed to some 2 metres above the bare lochshore surface. There are many lochshores from which the alder-fringe has disappeared completely.

Other lochside deciduous trees are less vulnerable to damage because of their location on the backshore. Ash (*Fraxinus excelsior*) and beech (*Fagus sylvatica*) with shallow root systems are more susceptible than oak and birch, which are among the commonest constituents of the backshore woodlands. Coniferous trees, particularly the Scots pine (*Pinus sylvestris*), have even more tenacious roots. However, because of the concentration of impact at the junction between the shore and the backshore, alder will tend to reveal earlier, more intensive, damage than more vulnerable backshore species. In addition, badly damaged alders, for the reasons outlined above, will tend to persist longer than some of the less tolerant and less naturally stable deciduous trees, particularly ash.

Individual trees could not therefore be used as indices of varying intensities of damage or of site vulnerability, because of this variability in tolerance of and persistence in the face of damage. Instead, an attempt has been made to define and rank intensity of tree damage on the basis of a combination of one or more types of tree species with or without alder present. Table 53 outlines the minimum threshold conditions selected to define four categories of tree-damage.

7.7 Vulnerability of Lochside Site-types

As already indicated, the location of recreational activities tends to vary from one lochside site-type to another. However, the amount of use of continuous backshores and shores does not necessarily coincide because of variations in their relative attractiveness. This is well illustrated in Table 54. The use of both components will also depend on ease of access. On the backshore, this is often restricted by the nature of the vegetation to one or more heavily used paths, while ease of movement along the lochshore may be unimpeded. There are also sites where vegetation does not hinder access, but where water-rights are such as to restrict the use of the shore. In general, however, bare shores or those with sparse and/or low-growing vegetation constitute the principal lochside recreational focus. Hence, in the majority of sites, the backshore is subject to less recreational use than the lochshore. Those lochsides where bare shores are combined with backshores of short grass or open woodland are subject, in general, to the heaviest recreational use (see also Table 27).

While the vulnerability of a given lochside is dependent upon the nature of its components, the former cannot be estimated quantitatively from the sum of its parts; the durability and/or attractiveness of one component may increase or reduce that of another with which it is associated on a given site-type. Hence, it was decided to rank site vulnerability on the basis of the comparative attractiveness and/or durability of the main backshore components with those of the lochshore and the main cliff-discontinuity between the shore and the backshore. The presence and nature of a shore, albeit limited in extent and duration, can, as has already been demonstrated, attract or discourage people from the water's edge. The presence of a particular type of shore, or the absence of a shore, may increase or decrease the vulnerability to recreational impact of an adjacent backshore. The cliff is the main weak-point on which the vulnerability of the whole site may depend. Tables 56a and 56b indicate the relative attractiveness and durability of each type of cliff, shore, and backshore, according to the scale given in Table 55.

Combinations of A-C indicate the most vulnerable sites, of G-I the most durable, regardless of attractiveness; while D, E, G, H, indicate those which are both attractive, or moderately so, and reasonably durable. Table 56 systematises basic data in such a way as to facilitate comparisons. The ranking of lochshore and backshore types with the presence of one of four types of cliff, aids in the identification of those sites which are more attractive to recreationists, and are reasonably resistant to their activities.

Durable sites whether attractive or not, include all those with a low organic cliff, except for those with steep backshore slopes. These are usually sites with no shore, or a very limited, most often bare, shore at the cliff foot.

Moderately vulnerable sites may be divided into two categories: firstly, those in which at least one component is reasonably durable, and one more vulnerable; and secondly, those in which all the components are inherently durable, but where vulnerability is increased because the sites are very attractive. In the first category are sites with a low organic cliff and a steep backshore; those with low-shore and emergent vegetation with a low inorganic cliff, particularly when the backshore is also steep; and those where a high mineral cliff occurs between a bare shore and a gently sloping backshore. Sites in the second category are those where bare sandy or stony shores are separated by a low inorganic cliff from a gently sloping backshore.

Table 53

Criteria used to define and rank intensity of tree-damage around Scottish lochsides

Scale	Degree or intensity of damage	Key species	Minimum threshold conditions
4	Severe	Pine (with or without other conifers), Beech, Ash, Oak (with or without alder)	(a) Advanced root damage (root exposure and undermining with at least incipient stilting in majority of trees other than alder, or instability such that more than 50 per cent of the trees are recumbent or thrown). (b) Fire scars on more than 50 per cent of trees. (c) High proportion dead and/or broken branches.
3	Bad	Alder, Oak and/or Birch	(a) Degree of damage to alder as in 4 above. (b) Oak: bad root damage (root extensive exposure and undermining + instability), and either bad trunk damage or branch damage or instability.
2	Moderate	Alder-willow and/or Oak-birch	(a) Alder: bad root damage (25-50 per cent trees with fire scars) and branch damage. (b) Others: moderate root damage (exposure and incipient undermining) and branch damage.
1	Slight	Alder	(a) Moderate root damage. (b) Slight branch damage (coppice and stunted forms). (c) 25 per cent trees with fire scars.

112

Very vulnerable sites are those where high inorganic cliffs occur, especially with low-shore vegetation and a moderate to steeply sloping backshore, or those with similar shore and/or backshore characteristics, but with a high organic cliff. The latter tend to be less attractive to visitors.

7.8 Lochside Vulnerability

So far, vulnerability has been discussed only at the scale of the site-type and its components, and in terms of their attractiveness for and durability to recreational impact. However, the actual or potential recreational use of any site-type and the consequent direct or indirect impacts cannot be realistically assessed in isolation; it must be considered in the context of the whole lochside of which it is a part. The vulnerability of the lochside, as distinct from any part of it, will depend on the scale, variety, and pattern of site-types and the proportion and location of that part of the lochside proximal to a motorable road.

Table 55

Scale of attractiveness and durability

Scale	Description of attributes
A	Attractive but easily damaged by recreational use
B	Moderately attractive but easily damaged
C	Not attractive but easily damaged
D	Attractive and moderately resistant
E	Moderately attractive and moderately resistant
F	Not attractive but moderately resistant
G	Attractive and resistant to recreational use
H	Moderately attractive and resistant to recreational use
I	Not attractive but resistant to recreational use

Table 54

Proportion of lochshore to backshore sites used for recreation on lochside types surveyed

Lochside type	Shore type	Percentage of sites where shore is used for recreation	Percentage of sites where backshore is used for recreation
Bare/weedy	Sandy, bare	94.8	94.8
Backshore	Rocky, bare	100	100
	Stony, bare	80.7	65.1
	Stony, weedy	29.6	22.2
Low-shore	Shoreweed, dense	—	—
Backshore	Shoreweed, discontinuous	48.0	25.0
	Shoreweed/water lobelia	—	30.8
	Shoreweed, open	83.7	65.1
Emergent	Common spike-rush/shoreweed	75.0	62.5
Backshore	Common spike-rush	30.0	36.6
	Reedmace	33.3	83.3
	Reed	53.3	62.5
	Reed sweet-grass	50.0	50.0
	Reed canary-grass (dominant)	36.8	36.8
	(open/mixed)	100	69.2
	Branched bur reed	33.3	100
	Bulrush	35.0	50.0
	Water horsetail	—	75.0
	Bottle sedge	65.9	36.3

The method devised to categorise loch-types on the basis of the variety and pattern of dominant site-types (see Chapter 4) reveals the very considerable range of variation within the sample of lochs studied in the field. It can also

Table 56a

Relative attractiveness and durability of main cliff types

Cliff type	Scale
Low (less than 1 m) organic	I
High (over 1 m) organic	C
Low (less than 1 m) inorganic	D
High (over 1 m) inorganic	B

Table 56b

Relative attractiveness and durability of shore and backshore types

Lochshore and backshore types	Scale
Lochshore types:	
Low shore vegetation	A
Emergent (soft site)	B
Emergent (firm site)	E
Bare shore	G/H
Backshore types:	
Less than 10° slope	
Wetland	I
Short grass	G
Long grass	E
Moor grass	H
Low shrub	E/F
Shrub	F/I
Trees (open)	D
10°-20° slope	
Short grass	D
Long grass	E
Moor grass	E
Low shrub	E/F
Shrub	F
Trees (open)	B
Over 20°	
Short grass	A/B
Long grass	C
Moor grass	B
Low shrub	C
Shrub	C
Trees (open)	

For key see Table 55

be used as a guide to the assessment of the relative vulnerability of lochsides. For instance, the greater the proportion of the loch perimeter occupied by one or two sites attractive for the more popular car-based, land- or water-oriented, recreational activities the greater will be the vulnerability of the lochside. Conversely, the less attractive site-types, particularly those with very steep slopes and/or a dense shrub or tree cover, which reduce visibility and/or accessibility, will give the lochside a measure of self-protection. The relationship between variety of lochside site-types and vulnerability, is, however, complicated by the continuity or discontinuity of the dominant site-types. The effect on lochside vulnerability then becomes even more complex; it will depend on the scale and durability of the intervening site-types and may in some cases increase, and in others reduce the lochside vulnerability. Further, as in the case of the site-type, vulnerability may be affected one way or the other by the proportion of the lochside occupied by one of the shore features noted in Figure 28 (Chapter 4). A beach, a strip or, to a lesser extent, a rocky shore may provide access (particularly for walking, pony-trekking and bank-fishing) around more of the loch than would be provided by the backshore alone. This again may increase the vulnerability of less-attractive, accessible, but otherwise less-durable site-types, and hence of the lochside. Other features such as a wetland fringe, high cliffs, banks or steep rocky slopes at the water's edge can very effectively cut one part of the lochside off from another. This may have the effect of isolating otherwise attractive and/or easily damaged site-types, thereby making the lochside as a whole less vulnerable to impact. Generalisation beyond this level is, however, not possible, and the variability of the physical character of the lochside is such that any attempt to rank lochside vulnerability on the basis of the small sample of lochs studied would be of doubtful practical value. Within the five basic loch-types identified, the range of lochside variation is dependent on the number of divisions (1-23) of the dominant site-type, together with the seven shore features, one or all of which can be found around a given loch. Table 57 illustrates the range and type of intrinsic lochside vulnerability within the sample studied.

Lochside vulnerability, however, is also dependent on a number of extrinsic factors. Of these, the most important are proximity to a motorable road and accessibility from either the land or water to part, or the whole, of the lochside. While all the lochs under consideration are proximal, the proportion, actual length and distribution of the proximal part (or parts) of the lochside vary greatly. The types of proximity exhibited by 42 lochs studied in the field (5 per cent of all proximal lochs) is illustrated in Figure 38. The size-distribution of this sample approximates to that of all the proximal lochs and hence is heavily weighted in favour of the small loch. Category 1 contains a number of small reservoirs where access is by an untarred road to a point beside the dam. A high proportion of the lochs less than 25 ha are proximal to dead-end roads constructed to give access to the water for supply, or for boat or bank-fishing. The number of water bodies in

1. POINT PROXIMITY

SINGLE (1S)   MULTIPLE (1M)

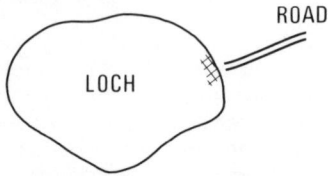
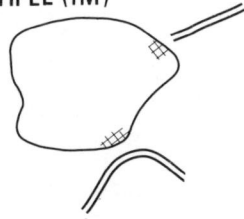

(1S) A single road comes within 100m of the lochside at one point only and occcupies <5% of the shoreline

(1M) As above, but more than one road approaches the lochside

2. LESS THAN 25% PROXIMITY

CONTINUOUS (2C)   DISCONTINUOUS (2D)

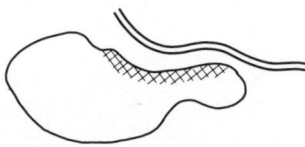
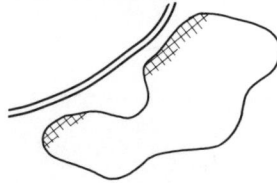

6 - 25% of the lochside is continuously within 100m of a road

3. 25-50% PROXIMITY

CONTINUOUS (3C)

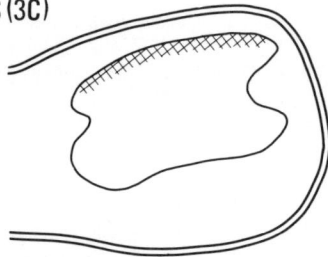
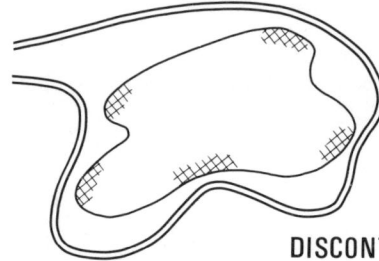

DISCONTINUOUS (3D)

4. 50-75% PROXIMITY

CONTINUOUS (4C)

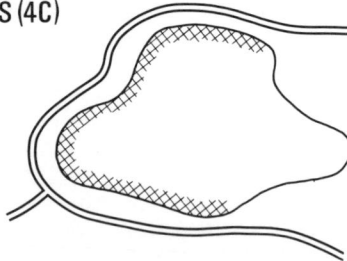
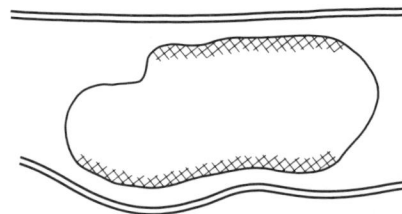

DISCONTINUOUS (4D)

5. MORE THAN 75% PROXIMITY

CONTINUOUS (5C)

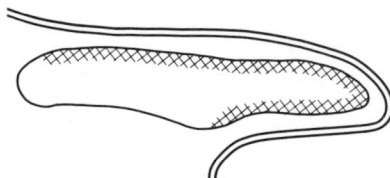
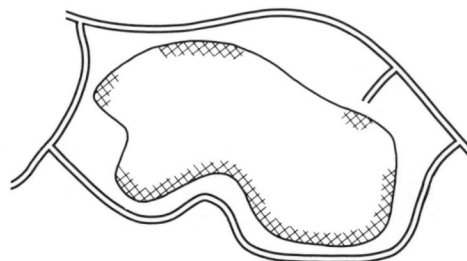

Scale ⌴ 100 m   ▨ Proximal part of lochside

DISSECTED (5D)

Fig. 38   Categories of lochside road proximity

Table 57

Comparison of dominant site-type (S) of entire lochside (L) with that of the proximal part of the lochside (P)

(a) Lochs where the dominant site-type of lochside and proximal lochside coincide

Per cent lochside occupied by dominant site-type (S) for entire lochside (L)	Per cent lochside occupied by dominant site-type (S) for proximal lochside (P)	Per cent lochside proximal	Category of proximity (see Fig. 38)	Loch
31	28	71	4D	L. Rannoch
44	44	65	4D	L. Ness
41	18	59	4D	L. Morlich
50	50	58	4C	L. Glenboig
26	23	51	4C	Talla R.
42	42	45	3C	Beecraigs R.
69	35	45	3C	Ryat Linn R.
37	37	43	3C	Loch of the Lowes
86	15	29	3D	L. Garten
53	20	20	2C	Holl R.
51	13	20	2D	Linlithgow L.
30	8	11	2D	Castle L.
31	10	10	2D	L. Insh
46	3	3	1S	Craigluscar R. (2)
90	3	3	1S	Lochend L.
43	2	2	1S	Balla R.
82	1	1	1S	Roscobie R.

(b) Lochs where the dominant site-type of lochside and proximal lochside differ

Per cent lochside occupied by dominant site-type (S) for entire lochside (L)	Per cent lochside occupied by dominant site-type (S) for proximal lochside (P)	Per cent lochside proximal	Category of proximity (see Fig. 38)	Loch
37	32	56	4D	St Mary's L.
19	11	40	3C	L. Tummel
66	29	34	2D	L. Tarff
36	22	28	2D	Johnston L.
26	18	25	2D	L. Lomond
42	12	22	2C	Kirk L.
69	13	20	2D	Mill L.
24	12	19	2D	L. Duntelchaig
52	16	18	2C	Craigluscar R. (1)
31	17	16	2D	Black L.
36	8	15	2D	L. Ken
25	6	10	2C	Arnot R.
44	10	10	2C	Woodend L.
49	6	6	2C	Glanderston Dam
44	5	5	1S	Stenhouse R.
33	5	5	1S	Lochcote R.
47	3	5	1S	Harperleas R.
41	3	3	1S	Harelaw Dam
32	2	2	1S	L. Glow
54	2	2	1M	L. Fitty
45	2	2	1S	L. Gelly
45	2	2	1S	Cullaloe R.
14	1	1	1S	L. Ruthven
82	1	1	1S	Lochgoin R.
47	1	1	1S	Bennan R.

Category 2 is slightly smaller and includes a wider range of sizes. The reservoirs in this category are larger than in the preceding one and access is usually by metalled road.

On lochs used for formal or informal recreation, where attractive sites are available, single or multiple-point proximity encourages a high concentration of visitors, which increases the site vulnerability and often results in a very high degree of localised impact. As a result, many of these lochs are characterised by a rapidly eroding cliff and a small beach at the proximal point. Extreme damage to the backshore can occur rapidly. These features do not occur elsewhere round the lochside but a recreational site, once established, will attract increasing use. Although point-proximity may not affect the intrinsic vulnerability of the whole lochside, it can have a visual impact which will be the greater the smaller the loch-size, and the greater the inter-visibility of one part of the lochside from another. In addition, sites with point access suitable for car-parking, camping and caravanning are particularly attractive for development.

The number of sample lochs in Categories 3 and 4 is small, while there are none in Category 5 (more than 75 per cent proximal). The number of proximal lochs in Category 5 is probably much smaller than was estimated from maps (particularly on a scale of 1:50,000) because of the practical difficulties of measurement to the nearest 100 m. Also, the selection of any threshold-distance is very difficult because, although most of a lochside may (as in the case of Linlithgow Loch) be over 100 m from the road, it is never very far, so that the whole of the lochside is well-used.

While point-access concentrates land-based activities, it makes the whole loch available to water-based recreation. Vulnerability of the entire lochside may, therefore, be increased. This will depend on the volume of water-use generated and, more importantly, the type of use. In the case of sailing, water-skiing and power-boating the use of one and the same starting and finishing point, based on formal facilities, is usual, though by no means invariable, and beaching of boats elsewhere is exceptional. As a result, these water-based activities may have little or virtually no impact on the rest of the lochside. This is particularly well exemplified in lochs such as the Lake of Menteith, Loch Insh and Castle Semple Loch. However, for reasons already analysed (see Chapter 6) the impact of boat-beaching by fishers, casual-boaters and canoeists can be greater than might be expected given the amount of use.

As the proportion of linear proximity increases, so too does the potential vulnerability of the lochside. Also, dependent on the nature of the adjacent lochside, discontinuity of the proximal road could either reduce or greatly increase the lochside vulnerability to physical or visual impact by recreational activities and developments. Indeed, the relationship between the site-types and the proximal part of the lochside in this respect is all important, but it can only be considered in the context of the individual loch. For each loch in the sample, a comparison was made between the nature of the dominant site-type (S) of the entire lochside (L) and the dominant site-type (S) of the proximal part of the lochside (P). The results are given in Table 57 (a) and (b). Of the 42 lochs sampled, 17 showed a coincidence between the dominant site-type for the loch and that for the proximal part of the loch. It is interesting to note that these are fairly evenly distributed between proximal Categories 1 to 4. In contrast, lack of coincidence between (L) and (P) is more characteristic of lochs with point-access where the chances of a coincidence are very much less, and will be smaller the greater the variety of the lochside.

The extent, however, to which people will move from the proximal to more remote parts of the lochside depends on the following factors:

(a) Physical (*de facto*) accessibility. This influences the ease with which it is possible to walk along the backshore and/or shore. The presence of visible footpaths plays a large part in determining the distance a given number of visitors will walk from the proximal part of the loch;

(b) Visibility (the amount and nature of the lochside that can be seen from the road). Small lochs, with open or low-growing vegetation, all of whose perimeter can be seen from the proximal lochside, usually have a well-developed footpath all around them;

(c) The types of recreational activities pursued. Generally, bank-fishermen tend to be the most enterprising, and they are often prepared to walk further and through site-types which would be considered unattractive to the usual recreational visitor.

In the final analysis, however, the relationship between proximal road, lochside vulnerability and intensity of recreational impact will depend on the availability of space for either informal or formal car parking.

A major factor determining the vulnerability of the lochside to recreational impact is the type and level of management for recreational and other activities. The aims of this chapter are:

(a) to illustrate, with reference to specific lochs or types of lochs, the range of development and management found within the sample studied;

(b) to assess as far as is possible, the effectiveness of such lochside recreational development and management observed within the sample studied;

(c) to assess how far the lochsides studied reveal a need, and provide guidelines, for the future management of lochsides for recreation in Scotland.

The survey indicated that development and management of lochs and lochsides is relatively limited in extent and intensity (in terms of capital and/or labour). Although the sample studied contains a high proportion of the most highly developed lochs for recreation in the country, they exhibit a very wide range of types and intensities of development and management. Around the majority of lochs, however, not only is development relatively embryonic, but management of the lochside *per se* is incidental to other land- or water-based objectives. Where management does occur, it is loch, resource, or activity-specific, and there are insufficient examples to provide case-studies which illustrate more general or more widely applicable management principles or techniques. For these reasons, it has been decided to analyse the existing range of lochside management in terms of the objectives and intensity of recreation. These are discussed under the following headings: suburban and urban-fringe park lochs; country park lochs; other local authority and public lochside sites; Water Department reservoirs and Forestry Commission lochs; commercial lochside camping and caravan sites; private fishing lochs; and water users association lochs.

8.1 Suburban and Urban-fringe Park Lochs

Suburban and urban-fringe park lochs (see Fig. 39) of which there are a number in Central Scotland (with a marked concentration in Strathclyde Region), are the most intensively developed and managed within the size range under consideration. The survey sample included three which serve to illustrate the function, management objectives and formal characteristics of these types of lochs particularly well. Hogganfield Loch (NS 632674), managed by Glasgow District Council, is a very intensively developed and heavily used suburban park on the east side of the city, within view from the Cumbernauld and Stirling trunk road (A80). It is an impounded loch, constructed in 1922, on the site of a former puddling-clay subsidence hollow. Lochend Loch (NS 705663), managed by Monklands District Council, is an urban-fringe loch in open land between Coatbridge and Glasgow. Formerly a compensation reservoir feeding the Monkland Canal, it was gifted,

together with the adjacent Drumpellier Park, to Coatbridge Town Council in 1919. Linlithgow Loch (NS 027760) falls into an intermediate category of a rural/urban-fringe loch, and it differs further from the two preceding lochs in being Crown Property managed, together with the adjacent Linlithgow Palace, by the Department of the Environment. It has the most scenically attractive setting of the three.

The type of management of these three park lochs varies in relation to the managing authority, their budgets and the nature of the facilities provided. Hogganfield Loch serves a large accessible urban population, and is one of the very popular, heavily used, city parks which provides more user-oriented types of activities than the others. High intensity boating (including a pleasure steamer) with associated facilities, a large car park and tearoom pavilion concentrate activities near the entrance. Three-quarters of the lochside is embanked and paralleled by hard-surface paths; the grass and tree parkland is carefully maintained.

Lochend Loch, in contrast, is managed in the interests of free, public bank-fishing and boating. Hard-surfaced paths, car parks and a boat-house, together with the reinforcement of soft lochside cliffs, are concentrated on the access side of the loch. One of the important management objectives is to attract major spectator events (particularly fly-casting competitions at national and, it is hoped, at international levels) which will provide entertainment and a source of revenue that can be used in the continuing management and development of the loch in the interests of the local community. However, unlike the two other lochs its use is not constrained by bye-laws, and access is open and hence uncontrolled.

Linlithgow Loch has an even larger area of formal grass parkland than Hogganfield Loch located between the Palace and the lochside. It is intensively managed to support a wide range of land- and water-based uses, while maintaining the amenity setting which is the hallmark of the ancient monument. Increasing demands on the water-space and the ensuing problems of managing and reconciling the conflicting pressures from fishers and other boat-users has resulted, only recently, in the leasing and general control of sailing and fishing to clubs which issue permits to the public. Linlithgow Loch is not only much larger than Hogganfield Loch, but its management is also more resource-oriented and the development and concentration of facilities on the lochside is minimal. Despite variations in the level of recreational provision, the three lochs have one significant management objective in common. That is to maintain, around at least part of the lochside and on the islands, a natural appearance, and to protect and conserve their varying wildlife resources in the interests of amenity and education. The particular method used is to keep overt development to a minimum so that a large part of the lochside remains fringed by semi-natural vegetation such as willows, reeds, rushes, and often backed by damp grassland. These areas pose the most difficult management problems, particularly on heavily used lochsides subjected to a high intensity of bank-fishing. Indeed, on Lochend Loch, these are the

118

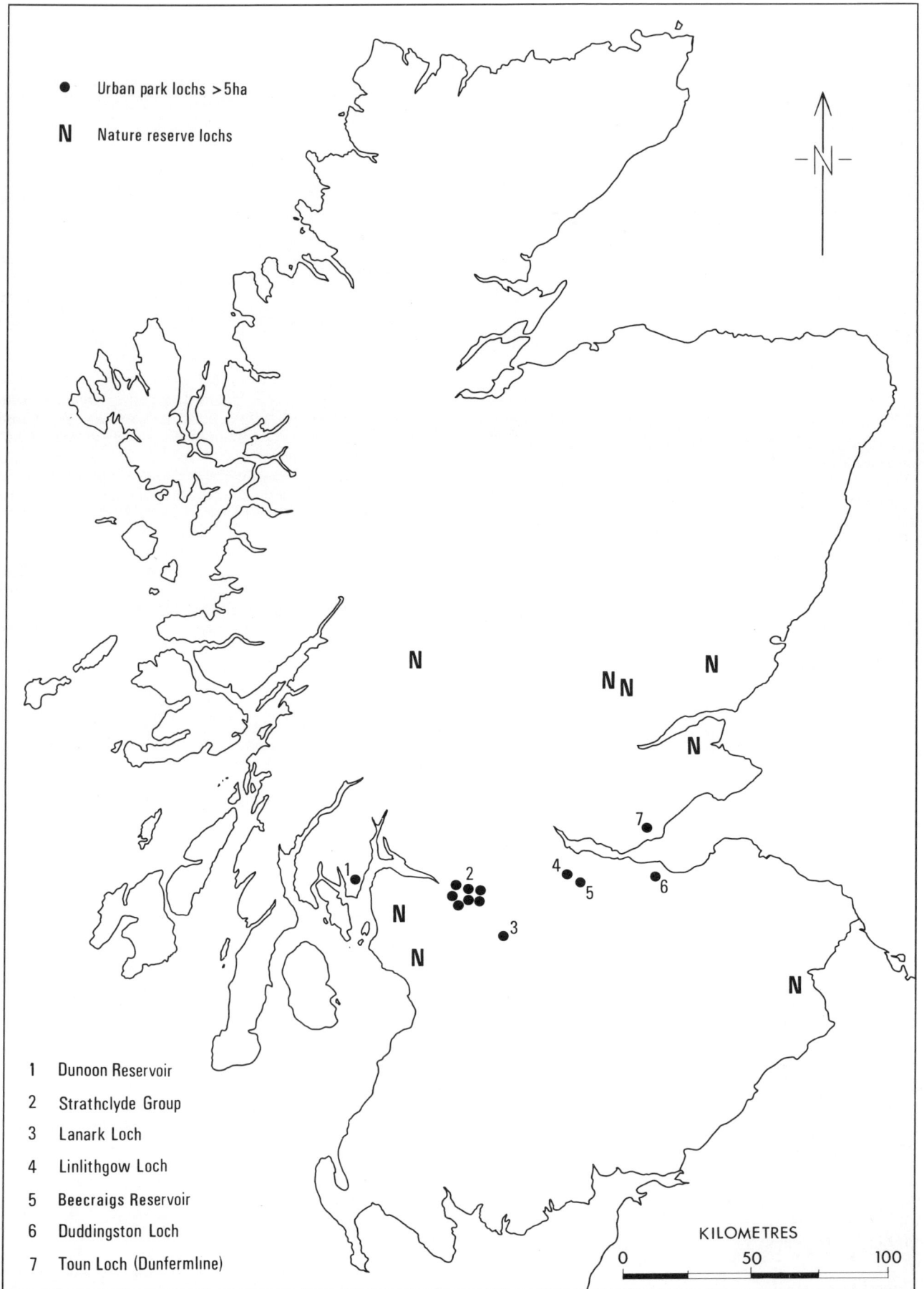

● Urban park lochs >5ha

N Nature reserve lochs

1 Dunoon Reservoir
2 Strathclyde Group
3 Lanark Loch
4 Linlithgow Loch
5 Beecraigs Reservoir
6 Duddingston Loch
7 Toun Loch (Dunfermline)

KILOMETRES

0   50   100

Fig. 39   Urban park and nature reserve lochs

shores which are most favoured by the bank-fishers, and on Linlithgow Loch it is these same shores which are zoned for bank-fishing.

Around Linlithgow Loch, much care has been devoted to maintaining the footpath fringing the less-developed side of the loch, as well as planting trees on the backshore and replacing old or fallen trees along the lochside. However, because of the scenic and tourist value of the loch, the problems of reconciling conflicting demands from public and private local bodies, particularly interested in fishing or wildlife or sailing, are greater on this loch than on the other two. Around Hogganfield Loch, the semi-natural lochside vegetation is not so extensive, and for most of its length the damp fringe is wide enough to permit a degree of self-protection which, if not subject to bank-fishing, can minimise management input. On the south-west side of Lochend Loch, however, the immediate lochside is subjected to exceptionally heavy bank-fishing pressure, and the low water-side peat cliffs and the backshore have suffered considerable erosion and fire damage.

8.2 Country Park Lochs

Apart from the suburban and urban-fringe park lochs, and those managed for conservation and nature study (see Fig. 39), few other lochs have been deliberately developed and managed for outdoor recreation. Among the best examples are those which have been developed by local authorities within the concept of a Country Park. Three lochs of this type in Scotland are: Castle Semple Loch (Renfrew District), Beecraigs Reservoir (West Lothian District) and Strathclyde Loch (Motherwell and Hamilton Districts). The last is a newly created loch which, because of its fairly recent development, was not included in the survey. The first two provide interesting and marked contrasts in site, origin, development and management. Together they exemplify many of the problems associated with the use and management of lochs and lochsides for recreational use.

8.2.1 Castle Semple Water Park

Castle Semple Loch (NS 365590) which was acquired in 1962 and opened as a Water Park in 1971 is managed by Strathclyde Regional Council. Development and subsequent management have been deliberately directed towards the provision of facilities for either participating in or watching a wide range of sailing, canoeing and rowing activities, whether competitive or casual. The maximum number of boats allowed on the loch is 65, though it rarely exceeds 50. The only other land-based activities are bank-fishing, a traditional sport inherited with the loch, and bird-watching. Facilities for the latter are provided by the Royal Society for the Protection of Birds which leases part of the southern shore of the adjacent wetland. Fishing club members can use both shores: day-permit users, with no restriction on numbers, are confined to the Lochwinnoch shore.

A·consistent management policy has been to concentrate water-sports and spectator facilities (the marina and car park) on the Lochwinnoch shore, and to manage the remaining shore in as natural a state as possible. Use of the extensive fringing marshes is now discouraged. A former sleeper-track across the marshes increased the use of the area through which a network of informal paths had developed. However, the track was washed away each winter and was costly to repair, and therefore, it has not been replaced since 1974. In the interests of water-sports, weeds which include floating plants (water-lilies, and floating pondweed), and emergent reeds and sedges are cleared twice a year by a mowing boat.

The main management problems are: the very rapid increase in the number of water sport spectators as compared with participants (1971 - 1,600; 1972 - 48,000; and 1975 - 98,000); the reconciliation of the conflicting demands of the fishers and the boat-users; the conflict of interest between different types of boats; and the policing of the park. The first has led to attempts to increase the size of the picnic area which was small in relation to the car park capacity. This is being achieved by the dumping of demolition spoil on adjacent wetland and by sowing it down to a hard-wearing grass sward. A marina extension is planned which will include an additional 100 car-parking and 200 dinghy-spaces, a new rowing club, sailing ramp and increased number of moorings. An access path to Parkhill Wood (a new inland extension to the park) will follow the water's edge and will necessitate raising the ground surface sufficiently to clear high winter water-levels.

Ongoing management of the lochside resource base is limited and subordinated to that of the water-based recreational facilities and activities. It is carried out by a small staff which includes three full-time rangers, one summer ranger, three to four part-time rangers and one estate worker.

Bank-fishers, who maintain that water-sports disturb the fish, would like to restrict boats to within 50-60 m of the bank; this would effectively limit boats to the southern, wide end of the loch, and hence zoning on this basis is proposed. There is no zoning of boating activities at present by either area or time – except for five days each year when the loch is booked for the exclusive use of rowing and/or sailing regattas. The more serious sailors and rowers complain about amateurs, particularly canoe hirers, who tend to hug the jetty and make mooring hazardous. However, the new sailing jetty should solve this problem. Finally, there is the problem of overseeing the loch; the ranger's office is not sited to provide the necessary overview, and to this end it is proposed to add another storey to the building.

8.2.2 Beecraigs Reservoir

Beecraigs Reservoir (NT 010744) has been developed more recently than Castle Semple Loch. When it was abandoned as a water supply at the end of the nineteen-sixties, the reservoir and adjoining land was transferred from the Water Board to West Lothian District Council. The original idea, that it should be one of the units in a

120

future Bathgate Hills Country Park, however, was shelved until regionalisation was completed. Eventually, in 1975, it was leased for a nominal sum to West Lothian District Council on the understanding that any profits accruing from its recreational use should be ploughed back into the scheme. It is run by the District's Department of Leisure and Recreation. The management plan, written in conjunction with the Countryside Commission for Scotland over the intervening period, was approved in principle in 1975, and following the recent development of facilities, Beecraigs has now been registered as a Country Park by the Commission.

The management for recreational purposes is to be based on the development of a combination of special (or unusual) and visitor-attractive land-use elements. These are intended to serve as foci for countryside interpretation which will facilitate the distribution and movement of people in such a way as to minimise impact on the secluded woodland fringes of the loch, and thereby to conserve its scenic value.

Two land-use elements already in operation are a fish and deer farm. The former, on the site of the old filter beds (adjacent to the dam) was started in 1974 and is run as a commercial-educational enterprise. It sells fresh fish (trout) to the public, as well as to other lochs, reservoirs and fish farms, and tours are provided for all visitors free of charge. It already draws a daily maximum of between 2,000 - 3,000 visitors at weekends. Entrance to, and exit from, the fish farm is by a circular path which also provides an attractive short walk around the loch. Sawdust is used to maintain the path-surface, and although the grass-verge between the road and the lochside is cut, a two-metre waterside verge is left to grow naturally. It is deliberate policy not to provide facilities for car-parking or picnicking around Beecraigs Reservoir, instead these are sited on other less-vulnerable sites away from the reservoir. These facilities include a car-parking loop providing space for 60 cars on the opposite site of the main road, a picnic-barbecue area and interpretive centre at Balvormie, and further car-parking at Whitebaulk near the deer farm. This is also a commercial educational unit and is linked by controlled access with Beecraigs Reservoir.

Direct water-based recreational activities are kept to a minimum in the interests of the fish farming and of the scenic value of the loch and its surroundings. The necessity to maintain water quality and correct nutrient status inevitably puts a constraint on the recreational use of the water. To this end, the water is recycled in order to maintain biological productivity, and hence to realise the full recreational potential of the fish farm. Boat-fishing, for which there is a high demand, is restricted to two boats. This facility is open to the public, and daily permits are issued by the District Council. Bank-fishing and boat-beaching, although difficult to control, are prohibited around the loch or the central island in an attempt to protect the wildfowl which nest in the fringing reed and sedge beds. In addition, wildfowl nesting is encouraged by controlling predatory crows and squirrels from the sur-

rounding coniferous plantation. The only other water-based activity is competitive canoeing, with one or two events a year permitted in the closed fishing season. Another management problem is reservoir silting which is very costly to remedy. The development of the park had been restrained by financial cut-backs, and had not proceeded in the preferred order of car-parking and toilets in advance of recreational attractions. Furthermore, the number of visitors to the Country Park had increased more rapidly than originally envisaged and before the completion of all ancillary services.

8.3 Other Local Authority and Public Lochside Sites

The most common and widespread type of lochside recreational site is that associated with off-road parking facilities for which local authorities at Regional and/or District level are responsible. A distinction can be made between two groups of local authority sites.

The first group includes lay-bys which are related primarily to the road and only incidentally to the lochside. They are designed to provide short-stay, off-road parking facilities rather than a base for recreational activities specifically related to the lochside. They are usually of small capacity (6-10 cars), have a hard surface, and are provided with litter bins and, occasionally, toilets. Lay-bys on the lochward side of the road may provide a viewing point and are often situated at the crest of a steep bank or cliff. The majority of lochside lay-bys have an inner kerb which prevents vehicles running on to the soft verge; a considerable number, however, do not. Such officially signposted lay-bys inevitably attract visitors to often vulnerable sites and on to unprotected surfaces.

Another characteristic feature of lochsides is informal verge parking where space permits either a view of or access to the loch. Many of these pull-offs have become formalised over time either by the provision of a conspicuous litter receptacle and/or a tarmac surface. The number of such informal pull-offs is particularly high where classified, particularly A, roads run alongside large scenically attractive and well-known lochs, such as Ness, Rannoch and Tummel. Apart from the obvious traffic hazards, they attract and concentrate impact at specific points. Both the formal lay-by and the informal pull-off provide, and are used as, lochside recreation sites by default, since they have not been designed nor are they managed for this purpose.

The second group of local authority sites include car parks and picnic sites specifically designed to provide lochside recreational opportunities. These are less common in the sample studied, since, in the first instance, their provision is dependent on the availability of suitable land between the road and the lochside. Among the most highly developed are those located on old road loops made available by recent road improvements. The most notable examples are at Duck Bay on the south-west shore of Loch Lomond and along the north side of Loch Earn, and their standards of development and management are usually high. However, since they tend to be located on the landward side of the loop road, the lochside cliff or bank and lochshore are not, from the management point of view, an integral part of the site.

Examples of less highly developed local authority car parks and picnic sites can be found on Lochs Venachar (north side), and Lomond (Milarrochy) and Lubnaig (east side) where sufficient space is combined with easy access to the lochshore. Development and management are variable though aimed primarily at providing hard-surface access points, tracks, and parking areas, together with litter bins and, in some cases, picnic tables. One of the main aims of management at present is to prevent car access to the shore by means of vehicle barriers such as logs, or wooden or stone bollards. On Lochs Lomond and Lubnaig, plans are presently being formulated and implemented to manage the lochside in order to minimise the impacts of recreational activities.

8.4 Water Department Reservoirs and Forestry Commission Lochs

The development of lochs and lochsides has recently been associated with the growing use of publicly controlled forest and water resources for outdoor recreational purposes. National and local government, as well as other statutory bodies have, particularly within the past three decades, given increasing attention to, and funds for, the provision of public recreational opportunities and facilities within a framework of multi-purpose resource use. Management of lochs and lochsides for recreation has generally had to be secondary to, if not merely incidental to, that designed to maintain the economic value or quality standards of the primary resource, be it for wood products, pure water or water-power generation. In addition, the major constraints on recreational development and management under which the bodies concerned have to operate are those of land tenure and water rights, already discussed in Chapter 2.

8.4.1 Public water-supply reservoirs

Public water-supply reservoirs, under the jurisdiction of the new Regional Water Departments, account for a large proportion of the proximal freshwater bodies in Scotland. As has been previously noted (see Chapter 2), recreational use and development is relatively limited in relation to the resource potential, the conditions of the primary use having constrained recreational development. Where management has been effected it has tended to be directed towards minimising the detrimental impact of land- or water-oriented recreational use on water quality. The recreation potential of public supply reservoirs is much less developed in Scotland than in England and Wales. Nevertheless, there would appear to be a growing awareness, within all the Water Departments, that although there has been no spectacular increase in demand (such as was experienced south of the Border) a gradual increase in pressure for more recreational water-space must be anticipated and taken into consideration in the formulation of current management policies.

Concern about how to maximise the recreational potential of reservoirs is focussed in those areas of Central Scotland where local and/or tourist demand is highest.

The main recreational use to date is fishing (trout and coarse fish). The potential for land-based activities around many public supply reservoirs is limited either by private land ownership and/or land uses still considered incompatible with recreation, or by spatially restricted unsuitable sites around those reservoirs where the Water Department is the riparian owner. Even with respect to fishing, management policies and practices can be effected only where the Department holds the fishing rights (see Table 13). All Water Departments would prefer to acquire these rights. Once acquired, however, management may vary depending on the differences in Regional policies and practices as well as the particular circumstance at each reservoir.

The majority of Water Departments, with the exception of Lothian and Tayside Regions, favour the leasing of fishing rights to a responsible club on the basis of short (3-5 years) but easily renewable lease, provided the arrangement has proved mutually satisfactory. Generally, the Department agrees the terms of the lease in relation to the number of boats, the number and types of permit, and fishing levels. Some clubs have to keep and submit records of fish catches, and some clubs are responsible for maintaining fish stocks, and it is thus in their own interests to protect the loch and its periphery from poachers, vandals and other unauthorised users. Leasing of fishing rights to clubs in this way is regarded as the most efficient and least expensive (in terms of capital and labour) means of managing the activity and at the same time protecting the resource. The Water Department and the Club then exist in a symbiotic relationship which effectively maintains a low level of recreational use and development of the water body in question.

Strathclyde and Fife Regional Water Departments favour non-exclusive clubs which allow daily permits for an approved percentage of rods to be made available to non-club members. In Central Region, however, only some leases have clauses allowing public access, and these are mainly for those reservoirs in the more scenic and tourist-attractive parts of the Region; elsewhere non-club members can only fish by invitation. Also, in these Regions the control of fishing is usually only retained by the Water Department where its employees are resident at the reservoir; permits are issued to the public on a daily basis and fishing is normally by boat only.

Tayside and Lothian Regions are the exceptions. The former prefers to retain control and management of recreation on reservoirs, and clubs are organised where it is thought necessary by the Water Department. In the Lothian Region, the Water Department handed the responsibility for the recreational use of reservoirs over to the Department of Leisure and Recreation. Where possible the latter's policy is to retain all sporting and fishing rights, and to arrange the provision of recreation from within the

Department rather than through clubs. Exceptions to this policy might be made in the case of small, remote reservoirs where it would be an advantage to lease the fishing rights to a club.

The Countryside (Scotland) Act 1967 gave the existing Water Boards powers to frame bye-laws for the purpose of either prohibiting, or regulating, the recreational use of land or water in the interests of water purity and in order to help minimise conflicts which might arise as a result of recreational use. There are, however, very few reservoirs for which bye-laws have been instituted, though some are in preparation for particular reservoirs or groups of reservoirs in Strathclyde and Central Regions.

8.4.2 Forestry Commission lochs and lochsides

Land owned and managed by the Forestry Commission accounts for nearly 10 per cent of Scotland's land area. The Commission is the main or sole riparian landowner of a considerable number of lochs, particularly in the Highlands and, to a lesser extent, in Galloway (see Fig. 40). However, the proportion of proximal lochs where it owns both the surrounding land, together with the sporting- and water-rights, are surprisingly few. The Commission's policy towards countryside recreation is inevitably land oriented. It is "to develop unique recreational features and the potential of its forests particularly where they are readily accessible to visitors from major cities and holiday centres" (Mithen, 1975). A closely related aim is to protect the forest environment from damage and to ensure its conservation. Management, therefore, is directed towards providing facilities for as wide a range of recreational opportunities for the visitor as possible, which will help reconcile these aims. Within budgetary limits, the level of development and management is dependent upon the type of visitor to be catered for, whether it be day, overnight, specialist or educationalist.

Two types of visitor have made a greater impact on the lochside than the others: the day-visitor and the specialist. Facilities for the first are most frequently located on the backshore of lochsides owned by the Forestry Commission, and include car-parks, with or without picnic-sites, and way-marked walks. These range from highly developed, well-designed sites at scenically attractive and durable locations as at Rowardennan, on Loch Lomond (NN 359996), to limited facilities and off-road parking on unprotected sites as at the head of Loch Chon (NN 419064). They may be under high pressure, usually in popular locations; or low pressure in more remote, less frequented spots.

Resources and/or facilities provided for the specialist user include those which allow such activities as shooting, fishing, orienteering, pony-trekking, sailing and nature study to take place. The most common is fishing, but, as in the case of the Water Departments, management by the Forestry Commission is confined to those lochs on which they have acquired the water-rights. The majority of fishings are organised through angling clubs, but day

permits are often available from local Forestry Commission caravan sites or offices. Of the total overnight and educational facilities provided by the Forestry Commission, very few are located on lochsides. There is a high concentration of the lochside recreational facilities in a few areas such as Loch Awe (Dalavich village), Lochs Rannoch and Tummel, and more particularly in the Argyll Forest Park (Loch Eck); the Queen Elizabeth Forest Park (Lochs Lomond, Chon, Ard and Achray), Strathyre (Loch Lubnaig), and Glen More Forest Park in which Loch Morlich is the focus of one of the major and most highly concentrated areas of outdoor recreational activities in Scotland.

8.4.3 Loch Morlich

Loch Morlich is in the centre of a major tourist area and, with the exception of a small area belonging to the Rothiemurchus Estate, the Forestry Commission are the sole owners. Developments on or within easy access of the shore are considerable and highly concentrated. Intensity of use over the short summer period must be of the highest order in Scotland. An indication of the level of use is a Forestry Commission camping and caravan site with a capacity of 230 units, a wide sandy/gravel beach on which the peak numbers recorded on one day were 1,600 people, and a water surface on which more than 100 boats of varying types may be out at one time. A high level of development and management, particularly of the camping and caravan site, car parks, and Sailing Club site, combined with a wide variety of relatively durable lochside sites, has allowed a high carrying capacity without marked deterioration around all but one part of the loch.

The most vulnerable and badly damaged area is that of the backshore sand-dunes; a type of lochside rare outside the machair and other coastal lochs of north and north-west Scotland. Passage of people from the road to the shore has, over time, resulted in a high degree of instability, and loss of sand by wind transport, to such an extent that the dunes have all but disappeared and the remaining trees which once occupied them are being rapidly undermined. Efforts are being made to trap and stabilise sand, to plant trees and to protect areas that are being rehabilitated from use by visitors. The other main management problem is the maintenance of the cleanliness of a heavily used marine-type beach which is not subjected to natural tidal flushing.

8.5 Commercial Lochside Camping and Caravan Sites

The commercial development and management of lochs and lochsides for outdoor recreational activities by private individuals and organisations are also limited; the most common being camping and caravan sites. The number of such sites on lochsides is relatively small when compared to the total number of sites in Scotland, and this is related to a number of constraints. Of these, the most significant are physical and legal access to a developable site, space sufficient to accommodate an economically viable enterprise between a good road and the lochshore, and

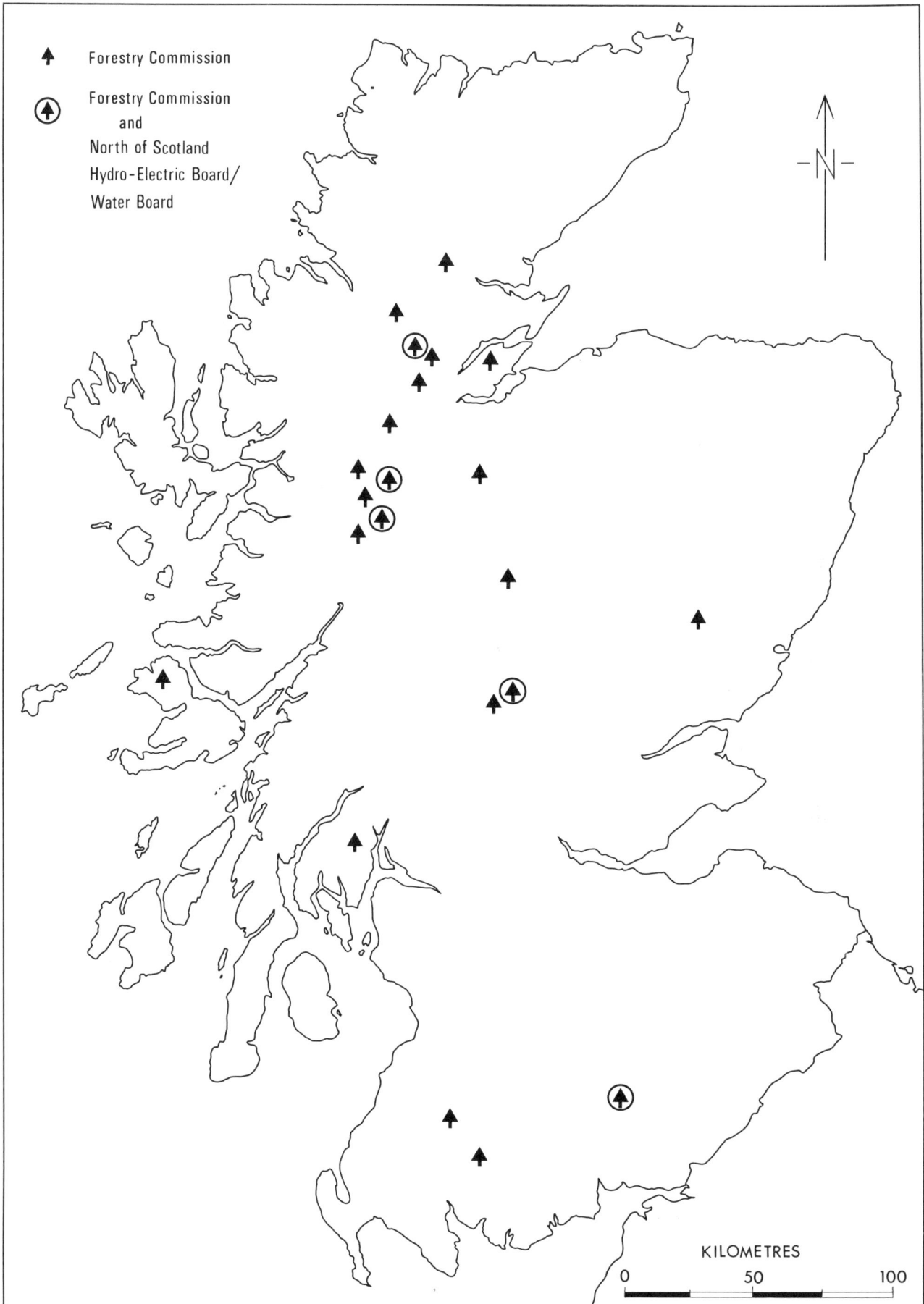

Fig. 40   Forestry Commission and North of Scotland Hydro-Electric Board/Water Board lochs

planning controls. Management on existing private sites is primarily directed towards the provision of facilities and to the general upkeep of the area. The extent to which amenity is taken into consideration varies widely and depends partly on the natural attributes of the site and the policy of the owners.

Intensity of use, in terms of the numbers and density of people and length of stay, is such that camping and caravan sites on the lochside generate the greatest concentration of informal water-based, boat-using activities along the loch fronts on which they are situated. The cliff between the backshore and the shore on every camping and caravan site visited was subject to the highest levels of erosion and bank retreat observed, with or without excessive tree damage. Little is being done to protect cliff-edges from such impact; there is either a lack of awareness of the problem, or a refusal to admit that recreational or related activities are the principal factors involved.

8.6 Private Fishing Lochs

Apart from camping and caravanning, the main activity on all privately controlled lochs is fishing for personal pleasure or, particularly in the case of salmon, for private lease. However, many more privately owned, less valuable trout and/or coarse fishings are leased to angling clubs, local hotels and/or angling associations (which involve several lochs and are found more particularly in the Highlands) or larger organisations such as the Forth Federation of Anglers, one of the largest in Central Scotland.

Within the sample of lochs investigated in the field, the Lake of Menteith, one of the most popular and attractive of the Trossachs Group, provides an example of another type of management. In this case, the five principal riparian owners, including the Lake Hotel, came together in the late nineteen-sixties to form a limited fishing company. No bank-fishing is permitted, the number of boats and rods is regulated and permits are sold to the public from the hotel. Launching and beaching of fishing boats is concentrated at one point which has a natural and very durable shingle beach, beside the hotel. A manager is employed to control fish-stocking and to maintain correct nutrient-levels which, together with Linlithgow Loch and Loch Leven, are among the highest in Scotland. The Company has proved a most effective means of conserving the fish resources while protecting the lochside which has a combination of good agricultural and forest land, and the associated scenic and wildlife resources, from the very considerable and mounting pressures from day visitors and tourists.

8.7 Water Users Association Lochs

In the private sector, a wider development of the fishing association, federation or company, are the water users associations. The rapid development of water-sports in the late nineteen-fifties and early nineteen-sixties created a large number of conflicting interests, particularly on those lochs which for reasons of accessibility, space or other favourable natural conditions were most attractive for sailing, canoeing, water-skiing and power-boating. The conflicts resulting from the recreational use of lochs involve those between incompatible water-uses, between water- and land-based activities, between land- and/or water-based recreational activities and nature conservation, and not least between riparian land owners or holders who wish to promote recreational development of a given loch and those who are not in favour or would actively oppose such moves.

Recognition of the need to try to reconcile these conflicts resulted in the setting up of water users associations on a group of lochs in the Highlands which included Lochs Tay, Lomond, Earn, and Venachar. That on Loch Lomond was abortive largely because of the numbers of riparian owners involved, the variety of vested interests, and the fairly rapid turnover of many commercial operators. Associations normally comprise representatives of riparian owners and occupiers, clubs and organisations who use the loch in question and other interested bodies, such as the Nature Conservancy Council and local authorities. They operate in accordance with formal, legally ratified constitutions. While the basic concept and organisation is common to all water users associations, their objectives, and their powers, vary from one loch to another and reflect differences in the nature of the conflicts.

The earliest Association was set up in 1960 on Loch Earn to reconcile the growing conflict between sailing, which started in 1950, and water-skiing, initiated by a local hotelier and riparian owner in 1955, and to control unauthorised access by non-riparian land owners or holders. The primary objectives of the Association as stated in their Constitution are "(a) to regulate the use of the waters of Loch Earn for the enjoyment, convenience and safety of those using it for sport or recreational activities; and (b) the conservation of fishing and wildlife in the vicinity of and on Loch Earn". Its management powers in terms of the Constitution are wide and varied and include, among others, power to purchase, rent or lease land adjoining the loch, to construct roads, paths and erect fences, build sheds, jetties and piers or any other works required for the furtherance of the Association; own, hire or charter boats, authorise surveys, publish plans and lay marks and moorings.

Management controls, however, were not implemented and enforced in time to keep pace with the rapid and continuing growth of water-based sports and recreational activities on Loch Earn. Zoning to concentrate sailing activities based at the east end, and water-skiing at the west end of the loch was not enforced, and unauthorised access is still a problem. Since 1960, the increase in second-home and/or boat-owners in St Fillans, together with other commercial lochshore developments, have resulted not only in an increase in the number of the boat users but also in the often inconsiderate and dangerous power-boaters; a situation which is exacerbated by the fact that some riparian developers are putting more than their quota of boats on the loch. The combination of the loch's great depth, of very rapidly changing weather conditions and of the large numbers and variety of craft using the loch at any one time increase the hazards for those participating in boating activities. The main impact on the lochside is from the increasing number of car-based spectators taking advantage of recently provided car park and picnic facilities on old road loops (see Section 8.3).

In contrast to Loch Earn, the Water Users Association on Loch Venachar, founded in 1969 with a constitution based on that of Loch Earn, appears to be a more successful management tool. This is partly because its water use is not so highly developed as that of Loch Earn, and partly because its objectives are more specific. The latter are concerned primarily with the protection of boat-fishing, and wildlife resources, from conflict with other forms of boating. To this end, the use of motorised boats is permitted for control and rescue only, and the number of boats is controlled. In addition, only a limited number of launching points is permitted and the loch is zoned with sailing prohibited in an area of considerable wildlife interest, and within fifty yards of the shore. In addition, more of the constitutional powers (which are comparable to those on Loch Earn), are being used. There is active liaison with the local authorities, the British Trust for Conservation Volunteers and other interested bodies over the landward use of the north shore which is more accessible and vulnerable to road traffic. Attempts have been made to prevent or minimise informal car-parking, particularly along verges in hazardous places, and management of cliff edges is carried out in such a way as to prevent car access to the lochshore. Vulnerable points, such as gates and broken fences are monitored and where necessary repaired or replaced. An important element in the relative success of the Venachar Water Users Association is the stronger local interest and control than on Loch Earn, where a high proportion of the users and the small riparian owners in the village of St Fillans are recent second-home owners.

The management of Loch Tay, one of the most famous salmon lochs in Scotland, involves the ecologically, economically and politically sensitive problems associated with salmon-fishing. Salmon was, in the past, the main resource, and income from fishing-rights still provides a considerable source of private and public revenue. Changes in land-ownership have complicated an originally complex system of salmon fishing-rights, and land tenure differs from the preceding lochs discussed in that, in most cases, the titles to riparian land do not include the loch salmon. The latter is vested with salmon fishing-rights as conveyed when the Breadalbane Estates were sold. For the purpose of salmon-rights, the loch is divided into three areas: those in the east are owned by an hotel company and subject to one individual right; those in the centre are owned by the Central Loch Tay Salmon Fishing Company, limited shares in which are owned by various riparian owners around the whole loch, and many of these rights are in turn let to hotels; while those in the western area are owned by an hotel, and subject to six individual

rights, not all of whom are riparian owners. Salmon fishing-rights are heritable and are rated, and they bring valuable business, mainly to hotels, in the off-season period between January and May.

Although the recreational development of Loch Tay has not reached the level of that on Loch Earn, increasing use of the former for sailing, power-boating and water-skiing had alerted those concerned about a possible replication of the situation on Loch Earn. The resulting Loch Tay Water Users Association, set up in 1970, has as its main objective the control and regulation – subject to existing rights of owners of riparian land and/or of salmon fishing – of the various uses of the waters of the loch for the "enjoyment, convenience and safety of all users with minimum disturbance to those people residing beside and close to the loch; to the amenity of the area; and to bird, fish and plant life". The constitution was designed to be consistent with the designation of Loch Tay by the previous local planning authorities as a multiple-use loch. Under the terms of its constitution, rules were prepared covering swimming, water-skiing and sub-aqua sports, boating, trout and salmon fishing, and the loch was zoned according to use. Zoning included the designation of specific areas for particular uses, for example, a central area not less than 200 yards from each bank was zoned for use by high-speed boats only, a peripheral area 200 yards from the bank for rowing and small boats, and an area for swimming on the Kenmore shore. Sailing boats are not zoned, but are advised to avoid certain areas where winds are unpredictable.

Although conflicts on Loch Tay are not as acute as on Loch Earn, the Water Users Association's problems are similar and as difficult to solve. Problems result particularly from the unregulated use of the loch by speed-boats, unauthorised launching of private boats, and the inability of the Association to enforce rules and finance management projects. As well as highlighting problems associated with salmon-fishing, Loch Tay illustrates other problems particular to the recreational development of the long, narrow and often very deep Highland loch. These include limited accessibility and steeply shelving lochsides with very deep water immediately off shore which together may, as on Loch Tay, result in the over-concentration of activities at one or two points along the lochside that are suitable for such activities as boat-launching, boat-beaching, and swimming. Noise-levels, particularly from power-boats, also increase as a consequence of the deep, narrow troughs in which the loch is situated.

Finally, the need for, and problems involved in, setting up water users associations is exemplified by Loch Ken where the actual and potential conflicts on a relatively small loch are more numerous and probably more involved than on the previous examples. Loch Ken is an impounded loch owned by the South of Scotland Electricity Board, though the lower half is the flooded River Dee, and hence is subject to somewhat different riparian and water-use rights than in the case of the upper loch. The Board owns all but 275m of the total shoreline of the loch up to the 45m contour and there are, in addition, thirty-five riparian owners including the Forestry Commission. The upper part of the basin is an area designated by local planning authorities as an Area of Great Landscape Value, and three wetland Sites of Special Scientific Interest, one of which is a Grade 1 wildfowl site, fringe its shores.

The loch has experienced a rapid increase in recreational use particularly for coarse fishing, water-skiing, sailing and power-boating. Potential conflicts are great and are increased by the differences in legal rights covering the upper and lower part of the loch, and also by the fact that the public have an historic right to launch boats at the site of an old ford. The central conflict is undoubtedly that between those who wish to promote recreational activities and those who wish, for one reason or another, to conserve the loch and its perimeter in as unmodified a state as possible. This conflict came into the open in the early nineteen-seventies when planning permission for a caravan-site was granted in a conservation-sensitive area, and when commercial water-skiing was developed on another part of the lochside. The only body with legal powers to frame bye-laws to control the use of the loch-waters in the interests of amenity is the South of Scotland Electricity Board which, naturally, does not wish to be involved as an arbitrator. Eventually, in 1974, it was proposed that some form of water users or riparian owners association be set up. That this is still being debated is a measure of deeply entrenched and conflicting interests, which are common to many other lochs in Scotland.

8.8 Effectiveness of Lochside Management

Any attempt to assess the effectiveness of existing types and levels of lochside management which have been described in this report must inevitably be extremely limited and very tentative, for the reasons detailed below:

(a) Direct management of the lochside *per se* in Scotland is limited and piecemeal in extent and *ad hoc* in type and execution. Also, it is rarely, with a few outstanding examples such as Castle Semple Loch, co-ordinated with or related to other types of recreational development or management objectives on and around any one loch or reservoir;

(b) Management specifically concerned with the maintenance and/or protection of the lochside resource-base is minimal in comparison with either the direct or indirect management of lochs or lochside users. As already indicated, the most direct and clearly defined management objectives are concerned with water-use and water-users;

(c) There is no clear-cut distinction at either the policy-making or technique-implementation levels between development and management (see Glossary), and hence between direct and indirect management of a resource. Development which involves provision of a recreational facility such as a car-park, picnic or

camping site is rarely deliberately designed to manage the recreational use of the lochside so as to maintain, conserve and protect the values of the lochside resource-base;

(d) Effectiveness of management can only properly be assessed in relation to management policies and objectives. In many cases, it was not possible to identify these within the terms of this project; they had either not been clearly formulated, or, in some cases, they did not exist. In many other instances, direct management merely involved *ad hoc* repair techniques rather than the implementation of on-going, long-term policies;

(e) Finally, effectiveness of management can only be assessed realistically in the knowledge of site conditions before management commenced, and of the time during which known types and levels of management have been in operation. Assessment is thus dependent on a type and level of monitoring that was far beyond the scope of this survey.

It is an accepted truism that the more vulnerable a site to impact the greater will be the need to protect it from physical and/or biological damage at given levels of use. Recreational development and management may, however, have one or more objectives of which the principal noted in the sample of lochs studied are: the installation of facilities specific to a particular activity, such as jetties, marinas, way-marked trails; or for the general use of visitors, such as toilets, car-parks, litter-disposal units; the management or organisation of people so as to allow efficient and enjoyable use of the resources and associated facilities; and the management of the resource-base, in part or in whole, in the interests of safety, nature conservation, and amenity. However, unless development and management are specifically designed to prevent or mitigate environmental damage, they may, within the context of any of these objectives, increase or decrease the physical-biological vulnerability of a given lochside or lochside site. Also, the level of management associated with any of these objectives will depend on whether the financial investment is seen in terms of economic and/or social returns and whether it is controlled by the public or private sector.

8.8.1 Provision of recreational facilities

Facilities for water-based activities such as marinas, jetties, board-walks, and boat-houses can combine user-management and resource-protection. To a degree dependent on size and sophistication of the constructions involved, as well as on the level and type of on-going management, they concentrate both water-users and spectators (the latter usually out-numbering the former) at one point on the lochside. Although the resulting impact on the lochside will, inevitably, be intensified at this point, it can be more easily controlled and contained while the remainder of the lochside can be the more effectively protected from impact. This is illustrated in two outstanding examples, the marina at Castle Semple Loch and the Sailing Club at Loch Insh.

Jetties and board-walks go some way to protecting lochshores and lochside cliffs and banks from abrasion as a result of boat-launching and beaching, as well as from the impact of trampling. These facilities appear to be particularly effective where the wooden jetties or board-walks are parallel to the lochside bank or cliff, as at Balmaha (Loch Lomond) and Loch Earn (north-west shore). However, jetties or piers at right angles to the lochside facilitate berthing, but fail to protect the lochside from the worst abrasive effects of boat-beaching and launching. This is well demonstrated at Loch Earn and Loch Lomond where a concentration of jetties at lochside camping and caravan sites have intensified and accelerated bank or cliff erosion and lochshore scour.

Facilities for general land-based recreational activities, such as formal car parks (including those associated with camping and caravan sites), informal car parks, lay-bys, picnic sites, toilets and litter receptacles, also have the effect of attracting and concentrating visitors at particular lochside sites. Except on specialised sites, such as Water Department reservoirs and suburban and urban-fringe park lochs, management of the lochside tends to be confined to the provision of one or more of these facilities with varying degrees of hard-surface protection. However, careful siting of facilities is necessary to prevent local impact from recreation activities, such as that which has occurred on the south-east side of Loch Ness, where steep slopes cleared of woodland adjacent to recently developed picnic sites, have been subject to footpath and gully erosion in only a short time despite relatively low levels of use. In contrast, the provision of picnic fire-places and barbecue pits, particularly by the Forestry Commission within the Queen Elizabeth Forest Park along the east side of Loch Lomond, appears to have been very effective in protecting trees from fire damage, the evidence of which was more obvious at these sites five years ago.

Barriers may be installed to protect individual trees or adjacent areas of unprotected verge from the impact of vehicles, or to prevent car parking on the loch-shore. However, the bank or cliff, the most vulnerable part of the lochside site, tends to remain unprotected and, inevitably, these features of the lochside have suffered most from the concentrated impact of recreationists. This is a characteristic feature of many otherwise well-managed camping and caravan sites and lochside car parks as for example around Loch Lomond, Loch Venachar, Loch Lubnaig, Loch Ken and the Speyside lochs.

8.8.2   Types of user-management
The three main types of lochside user-management which were noted during this survey are use-zoning, recreational clubs and user associations, and prohibitory regulations, particularly bye-laws.

(a)   Use-zoning
The zoning of water-based activities aims to reduce conflicts between different types of water-users; between water-users and riparian land-holders; and between private and commercial water-users. The main limitations

of zoning are the possible build-up of use above the physical capacity of the loch to accommodate the potential conflicting uses, and the sharp conflict in interests between private and commercial recreational development, as on Loch Earn. Effective zoning is dependent on mutual co-operation.

Water-use zoning is only occasionally linked with resource protection; a case in point is the protection of fringing reed beds on the west and south side of the Lake of Menteith. Zoning may indirectly protect the lochside resource by concentrating water-based recreation activities at particular points, especially where appropriate facilities are provided. Use-zoning alone, however, rarely protects the lochside generally from the impact of boat-launching or beaching, or from the impact of trampling.

Water-zoning can, however, be most effective in the prohibition of motorised craft, particularly power-boats. It can also help to reduce noise, water pollution and user conflict, especially adjacent to popular lochside sites where swimmers, paddlers, sailors, canoeists and motor boaters compete for the same water space.

(b)   Recreational clubs and water-users associations
Basically, these are designed to promote and protect the interests of water and land-users (see Section 8.7). At best, they tend to reduce competition from other users and may encourage lower levels of water and/or lochside use than might otherwise be the case. To some extent they indirectly aid site protection by reducing the intensity of recreational impact.

(c)   Prohibitory regulations
The use of statutory bye-laws designed to control undesirable water or lochside activities (recreational or otherwise) are, with one or two exceptions, confined to suburban and urban-fringe lochs and, more particularly, to urban park lochs. In all cases noted, the lochs are small enough and have the necessary financial and labour resources to ensure that regulations are not infringed. Within the Water Departments, however, there is general agreement that the bye-law is not a very effective management tool, because of the difficulties of enforcing laws with inadequate staff and slender budgets, particularly on more remote water bodies. Certainly, the wide variety of prohibitive notices found around Water Department reservoirs do not appear to deter visitors from using sites otherwise accessible and attractive, for off-road parking, picnicking, general land-based recreational pursuits and other undesirable water-side pursuits, such as paddling, swimming and bank-fishing.

Prohibition of activities in the interests of safety, noise, water pollution, and nature conservation is most successful in the case of power-boating and water-skiing, and on small lochs where infringement of regulations can more readily be detected. Prohibition is probably least effective in restricting undesirable impacts from both bank and boat-fishing. Both tend to be more elusive recreational pursuits, less amenable than others to control and management, particularly on medium to large size lochs with

128

irregular shorelines and varied lochside sites which provide considerable natural cover.

8.8.3   Types of resource management

In contrast to user-orientated development and/or management, that of the lochside resource base, for whatever purpose, is even more limited. From the sample of lochs studied, the following objectives related to direct resource management have emerged: bank or cliff protection; lochside amenity; nature conservation; and recreational use. Bank or cliff protection includes a broad spectrum of management techniques of varying level of effectiveness, which are set out in Table 58. The hard bank (and edge) is the most effective in erosion control; it is not, however, possible on the basis of data available to assess the efficiency of the other measures.

Table 58

Examples of types of cliff or bank protection

Type of cliff or bank protection	Type of loch or reservoir	Examples of lochs
Hard embankment (stone, brick, concrete)	Reservoirs Park lochs	
	Lochs with A/B roads running near loch side	L. Lomond St Mary's L.
Reinforced banks, complete or partial edging with vertical or parallel wood or metal strips	Park lochs Small urban-fringe lochs; heavily used sites on other medium/large lochs	Linlithgow L. Lochend L. L. Lubnaig
Embankment constructed of loose boulders and/or coarse gravel	Heavily used sites on large lochs	L. Lomond
Maintenance of tall grass and/or wetland lochside fringe of varying widths	Urban park lochs Country park lochs	Hogganfield L. Beecraigs R.

In one or two cases, lochside cliff-edge trimming has been observed which involves the regrading of the cliff to increase or reduce its height and steepness. Measures to reduce cliff steepness and undercutting have been undertaken in the interests of safety, as at Luss on Loch Lomond, rather than for the protection of the site value. They can, at most, be temporary short-term measures which tend to exacerbate the effect of impacts, because they make an already vulnerable point more attractive to visitors or less durable.

Tidying-up operations, which may also include clearing shrub, bracken and long grass vegetation on lochside slopes as on Loch Ken, more often than not have the effect of making sites more vulnerable to impact.

Management in the interests of lochside amenity around water bodies other than the urban or country park type of loch is also exceedingly limited in extent. Apart from the high intensity park-type management, the only techniques which could legitimately be considered under this heading are tree planting and maintenance of a natural appearance. The only example noted of tree planting and replacement to this end was at Linlithgow Loch. Examples of the deliberate maintenance of a natural appearance around part of the lochside were the woodland fringe at Balloch Park by Loch Lomond; the long grass and wetland fringe at Hogganfield Loch and Beecraigs Reservoir; and the grass strip backed by birch scrub at Lochend Loch. The success of these techniques would appear to be directly related to the intensity and level of management (determined by existing financial resources) in reducing impact on what were usually vulnerable sites, or in the constant repair of the damage resulting from impact.

Finally, there were within the sample surveyed, a number of lochs, other than those, the whole, or part of, which had been designated as nature reserves or Sites of Special or Scientific Interest, where part of the lochside had been informally zoned or posted as a wildlife reserve, usually by private riparian owners. All were areas of emergent reed/sedge beds sometimes associated with floating-leaved vegetation. Other than notices, and occasionally fencing, there were no signs of management. It was not possible to assess the extent to which the conservation aims were being realised.

8.9   Management Principles and Guidelines

There appears to be a low level of concern about the recreational value, use and vulnerability of the lochside resource-base in Scotland. This may be due partly to a lack of awareness about the nature and effects of recreational impact, and partly to the resulting lack of management principles and techniques available to minimise or prevent the deterioration of the value of this particular type of site. There is a need, then, for applied research on which management principles can be formulated and management techniques developed. The main research areas requiring investigation are identified in the next chapter. Meanwhile, it is possible, at this stage, to indicate some broad guidelines appropriate to the management of lochsides which have emerged in the course of this study.

General principles of lochside recreational management should take into consideration the following points:

(a) The need to decide which resources should be protected or conserved in the interests of aesthetic, ecological or educational values;

(b) The need to guide the development of activities and

facilities in such a way that the relative vulnerability of the lochside site can be assessed at an early stage, in order that containment of or protection from known impacts can be built into on-going management plans. This emphasises the need for lochside resource zoning to be undertaken on the basis of vulnerability to impact, and resource values other than those of recreation, in conjunction with use-zoning. This would allow priority to be given to activities and associated facilities that are compatible with the maintenance of all resource values;

(c) The need to consider the relationship between the type and total number of boats on the water surface and the possible resulting impact on the lochside;

(d) The need to consider the relationship between the type and amount of lochside access and water-use. Uncontrolled access, either from one point or all round the lochside, can rapidly lead to the use of the water surface in excess of its capacity.

A basic consideration in the management of a lochside must be an understanding of the inter-relationships between land and water-based activities and facilities, their various impacts, as well as the direct and indirect effects they may have on the vulnerability of not only particular sites used for recreation, but also of the whole lochside resource. Facilities for land-based activities do not necessarily generate an increase in water-use, and the land-based activities and sites where they are pursued are more easily managed than most types of recreational water-use. In contrast, facilities for water-based activities can increase lochside impact not only directly, but also indirectly by attracting spectators to sites that might not otherwise have been used. This is particularly relevant for local authority planning departments since the development of such facilities are frequently subject to planning permission.

Given this situation, it is probably best to accept and allow for a high intensity of use on the most extensive and popular lochside recreation sites available, and to manage them for as high a capacity as possible. It should be remembered that maintenance of environmental quality requires a level of resource management commensurate with the expected type and level of use. More effort, therefore, needs to be made to concentrate land-based recreational facilities on highly developed hard-surface sites which, when properly managed, can survive heavy use and remain attractive.

Given the limited extent of most lochside sites and the variety of recreational activities associated with them, car parking should be more strictly controlled than at present, and consideration should be given to the possibility of providing a wider range of types of car park around accessible lochsides. At present, parking takes up a disproportionate amount of the naturally restricted lochside sites, and therefore the intensity of damage tends to be out of proportion to the amount of use by people themselves. If parking was managed at higher densities

than at present, many sites could probably carry a higher intensity of recreational use. Also, the capacity of car parking facilities should be as closely related as possible to the extent of the lochside recreation site, its durability and the planned or expected type and intensity of use or uses. Where access and parking are provided, management of the whole site to minimise the effect of concentrated impact should be taken into consideration, particularly on cliffs and banks.

Parking of cars and other vehicles on unprotected lochsides and lochshores or on other unofficial parking places should be prohibited or discouraged. Also, more attention needs to be given to the location of parking places on major and minor roads proximal to the same loch in relation to the present, or envisaged, recreational activities, particularly associated with long-duration visits with picnicking, and short-duration viewing with or without picnicking.

More consideration should also be given to the aesthetic and ecological value of lochside woodland fringes and the need to protect, maintain and renew them in face of fire, erosion damage and road improvement schemes. Lighting of picnic (or other) fires except at authorised places should be prohibited.

Boat-launching and beaching, particularly on shores with emergent reed and sedge beds, should, where possible, be concentrated at sites where impact on the shore and the backshore can be contained and minimised. Camping and caravan sites give rise to the highest concentrations of boat-launching and beaching and the most intensive cliff erosion and associated tree-fringe damage. This is largely because the backshore and backshore/lochshore cliff are not protected and proper access for boat trailers is not provided. In view of this, and the limited scale and small number of available sites, serious consideration should be given to the appropriateness of locating large caravan sites, particularly those with a high proportion of static caravans, on what are often most suitable and accessible lochside sites for a wide variety of recreational activities.

Management is easier to implement and has been shown to be more effective and successful when fewer riparian owners are involved and where water rights and/or use are similarly concentrated. The main problem however (on which an effective solution of those related to either resource or user-management is contingent) is that for many of the large, recreationally attractive lochs and lochsides in Scotland, use of the freshwater surface, once access has been obtained, is free and statutorily unrestricted, except in the case of water-supply reservoirs. It has already been demonstrated that this limits the effectiveness of either user or resource management on some of the most heavily used lochs. Hence, there is a real need for the various public and private bodies concerned to consider how the rights of freshwater surface-use could be rationalised in the interest of more effective resource management.

The final task of this Report is to highlight the main points which have emerged from the preceding analysis and which are also the conclusions derived from it, and also to draw some further final conclusions, mainly pointing to the need for further study or experiment. In the interests of clarity, the conclusions will be tabulated under the main aims of the study as set out in the introduction to this Report. It should be noted that, in so far as the use of the lochside cannot realistically be divorced from that of the water body itself, relevant recreational characteristics of the loch itself have had to be considered.

9.1 The Recreational Use of Lochsides

The level of lochside recreational use in Scotland is, in relation to the actual and potential resource, very low. Eighty per cent of lochs proximal to a motorable road have one or no recorded recreational activity, and fishing accounts for most one-activity lochs.

Relatively few proximal lochs have over four recreational activities; most of these are located in areas of high scenic and/or tourist value, as in the South-West and Central Highlands and in the Dumfries and Galloway Region. Only ten proximal lochs have been ranked as high to very highly developed for land and/or water based recreation.

The extent to which the effects of recreational activity constitute a problem on lochsides in Scotland is, therefore, very limited at present. Recreational uses and impacts are concentrated on a few lochs and lochsides, most of which are of high scenic and/or tourist value, and this in itself may constitute a problem since it would appear, on the basis of the evidence available, that the recreational use of these lochsides is increasing at a rate greater than the less-used and less-developed lochs and lochsides.

Some of the main reasons for this situation at a national scale are given below:

(a) The smaller demand for water-space and associated recreational activities in Scotland in comparison to that in England and Wales;

(b) The relationship between potential and actual resources:

(i) The proportion of all lochs over 5 ha proximal to a motorable road is small (approximately 25 per cent), and of these three-quarters are water bodies of less than 55 ha, half of which are situated in Central Scotland.

(ii) *De facto* or *de jure* accessibility further limits the actual use of proximal lochs. Only 25 per cent of their lochsides are within 100m of a classified road; of these, about half have less than a quarter of their total lochside proximal, while half have a proximal lochside of less than 5 km in length. In the case of 50 per cent of the proximal lochs, riparian ownership and fishing rights are in private hands; and approximately 18 per cent of the proximal lochs are under public control; 80 per cent of these being Water Department reservoirs.

9.2 The Nature and Effects of Recreation

The lochside is a recreational interface which attracts a wide range of both land- and water-based activities none of which, however, are exclusive to or characteristic of this particular type of site. The variety and combination of possible activities, rather than the activities themselves, tend to distinguish lochside recreational pursuits from those which take place elsewhere.

In terms of numbers of activities pursued, water-based equals or exceeds land-based recreational activities on proximal lochs in Scotland. First-ranking group activities on the lochs investigated in the field were the informal, passive to only moderately active, car-based pursuits. With the exception of a relatively small number of lochs, water-based activities accounted for less than 20 per cent of all group activities, with bank-fishing and casual boating the most popular. In contrast to land-based activities, water-based recreation is more formally organised.

Following from this, formal lochside recreation facilities are mainly associated with the water-based activities and are related to the problems of launching, mooring and berthing a wide variety of crafts.

On the basis of the processes involved, five main types of direct impact on the lochside have been identified: trampling by human feet and rolling by vehicle wheels; digging; building; abrasion (or scouring) during the launching and beaching of boats; and fire. Other types of recreational impact noted but not investigated in detail include: deliberate mechanical breakage of vegetation; plant-picking; water pollution; and litter.

Recreational facilities occurring most frequently on lochsides are those which tend to be used prior to and/or immediately after the particular activity. They tend, therefore, to focus and hence concentrate both formal and informal activities in their vicinity.

Evidence of picnic fires is probably greater on lochsides than on other recreational sites. Remains of picnic fires and burning of vegetation are a characteristic feature of the recreationally used lochsides in Scotland. A lochside without this evidence is either not used for recreational purposes or is subject to a very low level, usually of private, use.

Effects of these recreational impacts identified vary not only according to the type of activity and nature of the impact, but also with the physical characteristics of the lochside and the distribution of activities on this site.

The sphere of influence of land and water-based recreational activities are in such close proximity that, particularly on the lochshore, they overlap and the lochshore is, therefore, a potential zone of conflict between these two groups of activities. In addition, there are three conflicting directions of movement which, on heavily used lochshores, exacerbate the mutual hazards for users, as well as intensifying the physical impacts on the site.

The most widespread non-recreational impacts on lochsides include natural wave action, livestock movements and grazing, wildfowl feeding and/or nesting, and sand and gravel winning.

Natural wave action is universal, but the loch wave does not appear to be a very effective agent of erosion and the minimum loch size dimensions necessary for the formation of true beaches are: a surface area of 120 ha – a fetch of 1,500m, and a shore-slope less than 0.05 degrees. On the basis of available evidence, boating activity on Scottish lochs is such that it does not appear to increase the effect of the natural loch wave.

Impact of livestock (mainly cattle) and wildfowl is distinctive and their effects can readily be distinguished from those of recreational activities. The former tend to be highly localised on those parts of lochsides either not used by recreationists (because of *de jure* or *de facto* constraints), or on wetland shores where recreational impact is not only limited but is confined to narrow Indian-file fishermen's tracks, which can be easily distinguished from the heavy grazing and flattening of vegetation by wildfowl.

Other non-recreational impacts, including water pollution, appear to be minimal.

9.3 Ability of Selected Lochsides to Withstand Recreational Impact

Two of the most important factors affecting the lochside's ability to withstand the impacts outlined above are the narrowness of the lochside recreational site and its close nature.

Ability of the lochside to withstand recreational impact is, all other things being equal, partly a function of the relative durability (or resistance) of its physical and biological components to the impacts identified in Chapter 6.

Durability of the lochside is dependent on the nature of the slope, the vegetation, and the substratum. It can vary around any one lochside and also across any one lochside site dependent on the nature and combination of the three main components: the backshore; the main backshore /lochshore discontinuity (cliff or bank); and the lochshore. The most durable components are the bare, dry shore composed of mixed sand and gravel to cobbles and boulders, and the backshore (with or without a cliff) formed of hard rock *in situ*. The least durable are steep to very steep backshores, and banks with some depth of organic or unconsolidated mineral substratum and cliffs made of similar material.

The main backshore/lochshore cliff is the principal weak point on which the vulnerability of the whole lochside site frequently depends. Its durability varies with height, with the material of which it is composed, and its location, either at the water's edge, or at the back of the summer shore.

Because of the complexity, particularly the multi-faceted and variable nature of the lochside, some method of categorisation was required. The one presented in the Report is based on a selection of those features deemed to be significant for recreational use, these being the dominant slope and vegetation-form, together with the particular shore features recorded on the lochs studied in the field. The method allows for different levels of categorisation, and also for the addition of further large- or small-scale features not represented in the sample.

The ability of lochsides to withstand the impacts identified is dependent not only on the durability of the lochside site as a whole, or on that of one or more of the individual site components, but on the type (or types) and intensity of recreational use. Hence the concept of intrinsic vulnerability, which combines attractiveness for recreation and durability, has been used to rank the ability of the lochside (or particular parts or components) to withstand impact.

Two factors which influence the attractiveness for use of the lochside are: the physical character of the shore, and more particularly the presence of a beach; and the composition and form of the shore vegetation, and more particularly the combination of shore and backshore vegetation on a particular site.

Very steep and/or densely wooded backshores and wetland shores tend to deter all but specialist users.

Vulnerable parts of the lochside are sites which combine short grass or open woodland, through which the loch is visible, and with a gentle to moderate slope on the backshore and a shore with little or no vegetation cover. They combine a high degree of attractiveness for almost all land-based and (dependent on the size of the shore material and off-shore conditions) the majority of the water-based activities. Also vulnerable is the major cliff discontinuity; when it occurs at the back of the summer shore its vulnerability is increased by its attractiveness for very concentrated and destructive activities. Also, cliff erosion takes place at the expense of a relatively small backshore area and an aesthetically valuable tree fringe.

Much less vulnerable because of low attractiveness and spatially restricted use are wetland shores with emergent reed and sedge beds. However, although their direct recreational value may be low, their scenic and ecological values are often high and thereby such shores make a significant contribution to the recreational value of the site in question.

Emergent reed and sedge beds stabilise bottom sediments, provide habitats for a variety of fish and wildfowl, and are effective wave dampeners. Their durability to recreational impact varies with the type of species and the nature of the substratum.

Tree damage and cliff erosion are the two most important diagnostic features by which intensity of recreational impact on lochsides may be assessed.

The vulnerability of one lochside component may increase or reduce that of another with which it is associated on a given site. The vulnerability of a lochside site-type must therefore be ranked on the basis of the comparative durability and/or attractiveness of the main backshore components together with those of the lochshore and the main cliff-discontinuity.

The vulnerability of an entire lochside will then be dependent on: the scale, variety and pattern of the site-types of which it is composed; the proportion of the lochside proximal to a motorable road; the continuity or discontinuity of the dominant site-type (or types); the scale and durability of the non-proximal parts of the lochside, and the proportion of the lochside occupied by various shore features.

Vulnerability of a lochside will also be a function of the amount and type of access from either the land or the water; the continuity or discontinuity of the proximal road; the volume and type of use generated; and the visibility of the loch and lochshore from the proximal part of the road.

Ultimately, however, the relationship between sites which are proximal to a road and lochside vulnerability will depend on the availability of suitable space for either informal or formal car parking.

9.4 Resource and Recreation Management Techniques
The development and/or management of lochsides *per se* for recreational uses in Scotland is very limited in extent, and is unco-ordinated and piecemeal in execution. The emphasis at present is on the management of water resources and of water or lochside users rather than on the lochside resources. Management of the latter tends to be *ad hoc* and generally incidental to other management objectives and techniques. The following three paragraphs summarise the management problems that have been identified on the sample of lochs studied; the management objectives or policies that have been formulated; and the management techniques that have been or are employed which have a direct or indirect bearing on the lochside in Scotland.

Management problems:

(a) Increasing demand for water space
(b) Conflicting recreational uses of water space:
  (i) fishers and other boaters
  (ii) bank-fishers and other users
  (iii) water-skiers and power-boaters and other boaters
(c) Conflicting recreational uses of backshore:
  (i) informal and formal activities, particularly for camping and caravan sites
  (ii) bank-fishers and other users

(d) Conflicting recreational use and other non-recreational use:
  (i) recreation and conservation
  (ii) recreation and domestic/industrial water-use
(e) Unauthorised access and use of lochside for bank-fishing, boat-launching and beaching and other informal recreational uses, particularly car parking
(f) Informal car parking on the lochside and the lochshore
(g) Increasing noise levels on and round lochsides
(h) Accumulation of hard-litter (particularly glass) on lochshores
(i) Vandalism and poaching.

Management objectives/policies:

(a) Multiple water use
(b) Minimising conflicting water uses
(c) Protection of fishing resources and interests
(d) Wildlife conservation
(e) Provision of facilities for specialised land or water use
(f) Provision of facilities for informal lochside recreation
(g) Maintenance of a natural appearance around part or whole of lochside.

Management techniques:

(a) Management of users:
  (i) zoning of water-use
  (ii) private or public sailing and fishing clubs
  (iii) bye-laws controlling water-use
  (iv) notices about water and lochshore use or restriction
  (v) prohibition of motorised boats, boat-beaching, boat-launching and bank-fishing, around part or whole loch
(b) Management of users and resources:
  (i) provision of formal lochside sites with facilities such as boat-houses, marinas, cat-walks, jetties and piers
  (ii) zoning of lochsides
  (iii) provision of hard-surfaced car parks and footpaths
  (iv) provision of fire-places
(c) Management of resource:
  (i) bank or cliff reinforcement
  (ii) tree-cutting, clearing and planting
  (iii) footpath maintenance
  (iv) planting emergents
  (v) clearing floating and emergent plants
  (vi) trimming cliff edges.

Since most of the present problems are concerned with the water-surface rather than land-use, it is not surprising that management policies and techniques are similarly biased.

Management on and around lochsides is primarily concerned with the users rather than the resource. Most effective in this respect are the organisation of the main activities by clubs, or the statutory prohibition of incompatible activities where, as in suburban or urban-fringe park lochs, resources are available to enforce bye-laws. Except in the latter cases, zoning, prohibitive notices and bye-laws are of limited effectiveness.

Where the management of users involves the provision of facilities the latter may or may not protect the lochside resources. In many cases, they concentrate and increase use without commensurate protective management of the lochside.

Management of the resource tends to be specific to a given loch, lochside site or, more usually, a particular component of the lochside-site.

Lochside resource management is minimal and is often a low-cost, one-off cosmetic treatment which in some instances, such as cliff-trimming, water-weed cutting, and tree and shrub cutting or clearing, may make the site more, rather than less, vulnerable.

Management of land and water-based activities is more easily implemented and more effective on smaller lochs where there is only one riparian land holder.

9.5 Further Areas for Research and Field Experiment

The low level of concern about resource management is probably partly due to a lack of awareness of the nature and implications of recreational impact, and partly to a lack of available management principles and techniques that could be employed to minimise and/or prevent resource deterioration.

Apart from the need for more basic ecological research on the impact of land- and water-based recreational activities on lochshore swamps and other freshwater biota, there is a need and scope for investigation into management techniques applicable to lochsides.

This study suggests the following areas in which field experiments and trials should, if possible, be conducted:

(a) Methods of preventing or minimising bank or cliff erosion taking into consideration:

   (i) height and composition of bank or cliff

   (ii) location at or back from the water's edge

   (iii) type and level of recreational use envisaged on the site in question and on the rest of the loch and lochside

   (iv) capital versus management costs

(b) Methods of lochside tree-fringe maintenance and, where necessary, rehabilitation:

   (i) identification of areas where this is necessary for amenity and/or physical stability of cliff or bank and of the backshore

   (ii) experimental planting of characteristic lochside trees such as alder and willow at the foot of backshore, bank or cliff

(c) Methods of lochshore wetland management in relation to a variety of uses and combination of uses with particular attention to the ecological effects of:

   (i) clearing, and

   (ii) planting emergent and floating vegetation

(d) Design of lochside car parks and picnic sites to minimise erosion damage on both backshore and banks or cliffs as well as on paths giving access to the lochshore;

(e) Demonstration of the relative effectiveness of zoning lochsides on the basis of their suitability for particular types and combinations of land- and water-based recreational activities, compared with zoning to minimise user conflict alone.

In all these areas there is a need to monitor the effectiveness of different types and levels of management.

9.6 Promoting an Awareness of the Effects of Recreation

Many of the problems discussed in this Report had been previously identified in general terms in the report on *Recreation at Reservoirs* (Countryside Commission, 1973) and also, with more specific reference to Loch Lomond, in the present writer's unpublished report to the Countryside Commission for Scotland (1974).

This study indicated that there is less appreciation of, and concern about, the effects of recreation on lochsides among fishermen, all other boat users and commercial managers of recreational sites, than among those normally involved with some aspect of either private or public land or water use or management. Those most directly affected by impacts on the lochside resources are conservationists, wildfowlers and, paradoxically, fishermen.

However, the nature and scale of lochside recreational impact is not fully realised by all managers or users. In comparison to problems of user-management, those of resource-management are generally regarded as of minor importance. In all but a few instances, this is reasonable in the context of a single loch or reservoir.

In addition, the differences between direct and indirect impacts resulting from the provision of facilities for land-based as compared with water-based recreational activities, and their implications for the whole of any lochside, are not always fully appreciated.

More consideration needs to be given, particularly by local authorities, to the control of road-verge and informal car parking, and to the provision and continuing management of sites for land-based recreation, so that they will support heavy use and remain scenically attractive.

It would appear that insufficient attention is given to the effect of the development of one lochside component, for example the backshore car parks and caravan sites, on the other components such as the lochside cliff or bank and/or lochshore. This applies similarly to the possible effects of the development of one lochside site on others around the same loch.

There is a lack of concern, particularly on commercially developed lochsides, and on many local authority lochsides, for the amenity value of the narrow lochside woodland fringe and of its susceptibility to damage and loss as a result of recreational impact.

It is not fully realised, however, that the number of accessible lochside sites along proximal lochsides is relatively few or that the majority are of very small scale. They are, therefore, extremely vulnerable to impact which, at its worst, may cause rapid deterioration of the physical and biological quality of the site, effect a serious reduction in its area, and hence its ability to maintain a given level of recreational activity.

This report could not have been written without the help of numerous public and private bodies, and private individuals. The author would like to record her indebtedness to the following:

1. Public Bodies

1.1 *Water Boards and Water Departments*

Argyll Regional Water Board
Ayrshire and Bute Water Board
East of Scotland Water Board
Fife and Kinross Water Board
Highland Regional Water Board (Central Division)
Inverness-shire Water Board
Lanarkshire Water Board
Lower Clyde Water Board
Mid-Scotland Water Board
North-east of Scotland Regional Water Board
North of Scotland Regional Water Board
Ross and Cromarty Water Board
South-east of Scotland Water Board
South-west of Scotland Water Board
Central Regional Council Water Department:
Mr J. T. Robertson
Central Scotland Water Development Board:
Mr Fraser, Mr Bamber
Fife Regional Council Water Department: Mr Crawford
Lothian Regional Council Water Supply Services:
Mr J. P. Williamson
Strathclyde Regional Council Water Department of
Resources and Planning: Mr Devenney, Mr Cranston

1.2 *Other local authority departments*

Central Regional Council Department of Physical Planning:
Mr F. Bracewell, Mr R. Ferguson, Mr S. Clark
City of Glasgow District Council Parks Department:
Mr K. Fraser
Dumfries and Galloway Regional Council Planning
Department: Mr Dobbie, Mr D. G. Thomson, Mr. T. E. Girgan
Lothian Regional Council, Department of Recreation and
Leisure: Mr E. Langmuir
Monklands District Council Department of Recreation and
Leisure: Mr Barron, Mr O'Hara
Stirling District Council:
Mr M. Dobson, Mr. Kinneard, Mr. K. Graham
Strathclyde Regional Council, Department of Recreation
and Leisure: Mr D. Skelley, Mr T. Robinson
West Lothian District Council, Department of Recreation
and Leisure: Mr P. Sutherland

1.3 *Forestry Commission (Scotland)*

The Conservators for the North, South, East and West of
Scotland: Mr W. Blake (Forester), Rowardennan; Mr Forbes
(Recreation Officer), South of Scotland Conservancy;
Mr W. Henman (District Officer), Queen Elizabeth Forest;
Mr Howells (Forester), Loch Ard; Mr R. Hurst (Conservation
and Recreation Officer, Scotland); Mr D. McCaskill
(Forester), Strathyre; Mr W. McGivern (Forester), Loch Ken;
Mr B. Martin (Recreation Forester), Queen Elizabeth Forest;
Mr Morison (Forester), Glenmore; Mr T. Polwarth (Tourist
and Recreation Forester), Stirling; Miss H. Pottinger
(Information Officer), Edinburgh; Mr Reynard (Park Warden),
Glenmore; Mr Robertson (Forester), Achray.

1.4 *Other public bodies*

British Waterways Board:
Mr R. B. Davenport, Glasgow; Mr C. E. Wall, Clachnaharry
Clyde River Purification Board
Department of Agriculture and Fisheries for Scotland:
Mr J. L. Murray
Department of the Environment, Edinburgh:
Mr Geary and Mr Mitchell
Game Fisheries Association: Mr Huish
Institute of Terrestrial Ecology:
Dr R. Britton, Dr P. Maitland, Dr I. Smith
Lanchester Polytechnic, Coventry:
Wave Power Research Group
National Trust for Scotland
Nature Conservancy Council: Mr A. Allison, Miss N. Gordon,
Mr E. Idle, Mr A. Kerr, Miss J. Martin, Mr J. Mitchell
North of Scotland Hydro-Electricity Board: Mr I. Smith
Scottish Sports Council: Dr I. Davies and Mr D. McCallum
Scottish Tourist Board:
Mr Humphries, Mr Hughes, Mr Carter, Lt.-Col. H. C. Paterson
South of Scotland Electricity Board: Mr S. Fraser
Sports Council
Water Space Amenity Commission, London

2. Private Bodies

2.1 *Sports organisations*

Angling Associations and Clubs (numerous)
British Canal Union
British Caravanners Club
British Water Ski Federation and affiliated clubs
Cairdsport, Loch Morlich
Camping Club of Great Britain and Ireland Ltd:
Mr M. Smith
Caravan Club
Glasgow University Athletics Club
Glenmore Lodge: Mr J. King
Holiday Fellowship Ltd
Loch Insh Sailing School: Mr C. Freshwater
National Caravan Council Ltd
Pony Trekking Centres (several)
Royal Yachting Association, Scottish Council: Mr A. Mitchell
Sailing Schools (several)
Scottish Amateur Rowing Association
Scottish Anglers Association and affiliated clubs
Scottish Canoe Association and affiliated clubs
Scottish Dinghy Association and affiliated clubs
Scottish Schools Sailing Association
Scottish Sub-aqua Club and affiliated clubs
Strathyre Riding Centre: Miss K. Hutton
Trossachs Canoe and Boat Club: Miss M. McLure
Wildfowlers Association of Great Britain and Ireland
Youth Hostels Association

2.2 *Other private bodies*

Callander District Tourist Association: Mr A. Samuel
Dartington Amenity Research Trust, Edinburgh:
Mr R. Aitken
Scottish Landowners Federation: Mr A. F. Roney-Dougal

Killin and District Tourist Association

Royal Society for the Protection of Birds:
Mr P. Bowyer, Lochwinnoch; Mr S. Taylor, Loch Garten

Scottish Recreational Land Association:
Capt. M. Collins

Scottish Water Association Ltd

Scottish Wildlife Trust: Dr J. A. Gibson, Mr I. Grant

Tourist Operators on the Caledonian Canal:
Mr R. D. T. Black, Mr J. Dodds, Mr P. Harreschau,
Mr J. Hagan, Mr C. Hughes

Water Users Association:
Mr E. Cameron (Loch Earn); Mr J. McNab of McNab
(Loch Tay); Mr P. Tennant (Loch Venachar)

3. The Offices of the following Estates

Ardgour; Atholl; Auchleeks; Balmoral; Barbreck; Benmore; Bradford (Dell Estate); Brin; Conchra; Coulin; Dalhousie; Dalnaspidal; Elderslie; Fairburn; Glen Cannich; Glencarron; Haddo House Estate Trust; Islay Estates Co.; Keir & Cawdor Estates Ltd; Kinlochlaggan; Latheronwheel; Leverhulme; Lochdochart; Lochluichart; Lovat; McLeod; Managed Estates, Cambusbarron; Meggernie; Melfort Farming Co.; Monar & Pait; Moray Estates Development Co.; Mullandoch Hotel & Estate Co.; Novar; Pitlochry; Seafield; Strathraich; Strathspey; Westminster (Lochmore Estate).

4. Members of Staff of Glasgow University

Professor A. S. G. Curtis (Cell Biology); Dr H. Duncan (Chemistry); Mr A. M. Ferguson (Superintendent, Hydrodynamics Laboratory); Mr M. Moore (Computing Services); Mr J. Fulton (Fluid Dynamics); Mr N. S. Miller (Naval Architecture and Ocean Engineering); Mr D. A. Pirie (Aeronautics & Fluid Mechanics); Mr T. J. Scott (University Garage); Dr R. Tippet (Zoology Department Field Station, Rowardennan); Dr G. Wyllie (Physics).

5. Helpers in Surveys and Experimental Work

5.1 *User surveys*
Loch Ken: Mr G. Anderson, Principal Teacher of Geography, Kirkcudbright Academy, and pupils; Mr Greenwood and members of the Kirkcudbrightshire Sea Scouts
Trossachs: Mr Dunn, Principal Teacher of Geography,

McLaren High School and pupils; Mrs Allison and Falkirk Ranger Guides; Mrs Hird and Killermont Ranger Guides; Mrs Sproul and Bearsden Ranger Guides; Dr J. Sime and Falkirk Venture Scouts; Dr D. Vass and Dunblane Venture Scouts; Undergraduate and Postgraduate students, members of staff of Glasgow University who together with wives and families acted as recorders

5.2 *Trampling experiments*
Scottish Association of Geography Teachers:
Mr Gilbert Gray

Mr H. Aitchison, Principal Teacher of Geography, John Neilson High School, Paisley

Mr P. Cortopassi, Principal Teacher of Geography, St Ninian's High School, Kirkintilloch

Mr J. Davis, Principal Teacher of Geography, Bearsden Academy

Mr J. Graham, Adviser in Social Studies, Renfrewshire

Mr P. Knowles, Principal Teacher of Geography, Johnstone High School

Mr A. Leitch, Principal Teacher of Geography, Graeme High School, Falkirk

Mr R. McAllister, Principal Teacher of Geography, Braidfield High School, Clydebank, and Mrs McAllister

Mr & Mrs I. McLean, Teachers of Geography, Graeme High School, Falkirk, and numerous pupils

6. Other Individuals

Mr M. Bryson, Loch Ken Holiday Park; Mr Carruthers, Manager of Rossdhu House, Loch Lomond; Major D. Crichton-Maitland, Elderslie Estate; Mr E. Finch, formerly of Stirling University Outdoor Centre; Col. P. H. Gough, Barbreck; Mr Groome, Inversnaid; Mr J. D. Hamilton, Paisley College of Technology; Admiral Sir Nigel Henderson, Hensal, Loch Ken; Mr W. Heron, Balmaclellan; Mr Kerr, Heriot Watt Former Pupils Angling Association; Mrs McCabe, Achray Farm; Mr & Mrs R. McClymont, Kirkstead Farm, Selkirk; Mr H. Macgregor, Factor, Elderslie Estate; Mr E. Millar, Miller Systems, Barrhead; Mrs Paullin, Loch Ken; Mr Scott, Milton Farm, Loch Venachar; Col. G. R. Simpson, Port of Menteith; Mr G. Smee, Loch Ken; Dr Weir, Department of Mining, Heriot Watt University (Loch Rusky).

1. Coverage

Data is available for all accessible (a minor road within 100m) freshwater lochs and reservoirs in Scotland, including the Inner Hebrides but excluding the Outer Hebrides, Orkney and Shetland.

2. Variable List

No.	Name	Type	Description
1	CODENUMBER	Ident. No.	
2	GRIDREF	Grid Ref	Mid-point in loch or reservoir
3	COUNTY	Region	
4	LAREGION	Region	
5	LADISTRICT	Region	
6	TYPE	Logical	Natural, impounded, artificial
7	SIZE	Real	Area in hectares
8	SHAPE	Real	Ratio of length to breadth (km)
9	ORIENTATION	Logical	N-S, E-W, NW-SE, NE-SW
10	SHORE	Real	Length of shore in km
11	DEPTH	Real	Maximum depth ⎫ in metres
12	MEANDEPTH	Real	Average depth ⎭
13	MAXLEVEL	Real	
14	MINLEVEL	Real	
15	MEANLEVEL	Real	
16	MEANRANGE	Real	
17	LANDUSECNT	Value Count (2)	Number of types of land use
18	LANDUSETYPE(a)	Logical	⎫ Repeated
19	LNTHSH	Real	Length of shore occupied by type ⎭
20	TOPFEATCNT	Value Count (3)	Number of topographical features
21	TYPETOP	Logical	Type of feature ⎫ Repeated
22	TOPGR	Grid Ref	Grid ref location ⎭
23	TENURECNT	Value Count (2)	Number of types
24	TENURETYPE(b)	Logical	Type of holder ⎫ Repeated
25	TENURECODE	Logical	Code of holder ⎭
26	WATERCNT	Value Count (2)	Number of types
27	WATERTYPE(b)	Logical	Type of holder ⎫ Repeated
28	WATERCODE	Logical	Code of holder ⎭
29	RECACTCNT	Value Count (5)	Number of recreational activities
30	RECACTTYPE	Logical	Type of activity
31	RECACTGR	Grid Ref	Location of activity
32	RECACTNUM	Integer	Number of users ⎬ Repeated
33	RECACTUSER	Real	Average daily users
34	RECACTMAX	Integer	Maximum demand
35	RECFACCNT	Value Count (6)	Number of recreational activities
36	RECFACTYPE	Logical	Type of facility
37	RECFACGR	Grid Ref	Location of facility ⎫ Repeated
38	RECFACCAP	Integer	Capacity ⎭
39	RECFACSTAT(c)	Logical	Status (built, under const., planned)
40	RECFACDM(c)	Integer	Month ⎫
41	RECFACCY(c)	Integer	Year ⎭
42	MIDISSTA	Real	Minimum dist. to any road (km)
43	MINDISTB	Real	Minimum dist. to A/B road (km)
44	PERCENTCAR	Integer	Percentage accessible to car
45	PERCENTPTT	Integer	Percentage accessible to path/track/trail
46	ROADTRACKCNT	Value Count (2)	Number of roads and tracks
47	RTTYPE	Logical	Type of road or track ⎫ Repeated
48	RTFUNCTION	Logical	Function of road or track ⎭
49	WATER(c)	Logical	Water quality
50	OTHERCNT	Value Count (2)	Number of other lochshore activities
51	OTHERACT	Logical	Type of activity
52	OTHERGR	Grid Ref	Location of activity

2.1 *Sub-divisions of variables*

(a) Types of surrounding land use (or vegetation cover)
woodland and shelter belts (deciduous)
plantations (young, mature)
scrub
rough grazing
wetlands
farmland
parkland

(b) Tenure and water rights
Private individual
(a) single
(b) multiple
Private organisation (number and names)
Public body (number and names)

(c) No data collected.

3. Sources

Most recent editions of Ordnance Survey 1:63360 (or 1:50,000 where available), 1:25,000 and 1:10,000 maps; Bathymetrical Survey of Scotland; Department of Agriculture and Fisheries for Scotland; Nature Conservancy Council; Scottish Sports Council; Scottish Tourist Board; Forestry Commission (Scotland); South of Scotland Electricity Board; North of Scotland Hydro-Electricity Board; Regional Water Boards; National Trust for Scotland; Scottish Wildlife Trust; Royal Society for the Protection of Birds; Local Authorities; Scottish and National Bodies associated with recreational activities; individual Angling Clubs; Estate Offices; Scottish Youth Hostels Association; Countryside Commission for Scotland; and miscellaneous publications dealing with Scotland.

4. Collection Procedure

Location
Type
Size   for all accessible lochs from maps
Form   (a) 1:63360 for all accessible lochs
Land-use   (b) 1:25,000 and 1:10,000 for lochs in special area
Other topographical features
Accessibility
Roads and paths
Other lochshore activities
Depth/water levels – Bathymetrical Survey of Scotland, Water Departments (Reservoirs); riparian owners and/or water users where available.
Recreational Activities/facilities – for each activity a search was made of relevant literature; correspondence with public bodies; private organisations (local, regional, national representatives); and with private clubs. In the special area any lochs for which data was not complete, was checked out in the field.

5. Comprehensiveness of Data

Location ⎫
Type   ⎬ complete
Size   ⎪
Form ⎭
Depth and water level: circa 50 per cent complete, particularly for small lochs/reservoirs (e.g. less than 50 ha).
Land use complete: in respect of most recent Ordnance Survey maps
Topographical features: complete, as above
Loch tenure ⎫ circa 80-90 per cent complete
Water rights ⎭
Recreational facilities: presence, as complete as possible;
Capacity: incomplete or estimates, dependent on data available.
Recreational activities: presence as complete as possible;
32, 33, 34 incomplete
Accessibility: complete
Roads/tracks: complete
Other lochshore activities: as complete as possible.

6. Reliability of Data

Data collection commenced 1974 and continued, with up-dating as new data became available, until December, 1976.

1. Method of Vegetation Survey

For each of the 39 lochs studies (see 1.1 and Figure 41), the main types of vegetation, on the basis of form and dominant species, were mapped for both shore and backshore, at a scale of 6 inches to the mile.

Detailed sampling was carried out on selected stands from as wide a range of communities as possible, including sites subject to light, moderate and heavy recreational use, as well as sites not used for recreation. Systematic point sampling with a 1 metre, 10 pin frame was undertaken every 0.5m or 1m (depending on the environmental gradient) along transects at right angles to the shore. The number of transects per stand varied with the area of the stand. Transects were selected subjectively, and whenever possible were placed so as to include parts of the community used for recreation and parts not used by visitors. The transects

extended from at least 2m inside the first backshore community to a water depth of 28cm (which includes the zone most used for paddling and boat launching). The number of short hits for each species, and the presence of any other species was recorded and expressed as a percentage of the total hits. Given the high degree of species dominance or co-dominance characteristic of communities on aquatic margins, this percentage frequency of occurrence of species can also be used as a measure of relative abundance and hence of physical dominance.

The following species lists (Tables 2 - 3) indicate the variety and frequency of species occurring in the various shore and backshore communities identified. Further background to the method is given in Greig-Smith (1964) and Rees and Tivy (1977 and 1978).

1.1 *Lochs and reservoirs on which vegetation transects were recorded*

Loch (L) or Reservoir (R)	Area (ha)	Grid Ref.	Loch (L) or Reservoir (R)	Area (ha)	Grid Ref.
L. Ruthven	149	NH 620276	Johnston L.	11	NS 697688
L. Tarff	69	NH 425100	Queen's (Glenboig) L.	8	NS 717687
L. Garten	80	NH 973180	Woodend L.	24	NS 705667
L. Morlich	122	NH 965095	Lochend L.	11	NS 705663
L. Insh	170	NH 830045	Castle Semple L.	78	NS 365590
L. Laggan	699	NN 490873	Barr L.	114	NS 353575
L. Rannoch	1902	NN 600580	Kilbirnie L.	79	NS 330545
L. Lomond	7125	NS 375975	Ryat Linn R.	7	NS 521573
L. Lubnaig	216	NN 575137	Harelaw Dam	33	NS 475537
L. Venachar	394	NN 575055	Black L.	10	NS 497513
L. Achray	79	NN 515064	Lochgoin R.	52	NS 537477
L. Chon	106	NN 420053	Gladhouse R.	152	NT 300535
L. Ard	235	NN 465019	Portmore L.	41	NT 261502
Lake of Menteith	264	NN 578006	St Mary's L.	244	NT 250228
Balla R.	78	NO 225050	Loch of the Lowes	40	NT 237197
L. Gelly	60	NT 201935	Castle L.	78	NY 088815
Stenhouse R.	14	NT 211877	Kirk L.	15	NY 079823
Linlithgow L.	42	NT 002776	Mill L.	12	NY 077833
Beecraigs R.	8	NT 010744	L. Ken	464	NX 690698
Black (Slamannan) L.	44	NS 860700			

Surveyed for present survey

+ Surveyed by Spence, 1964

✦ Surveyed for present survey
and by Spence, 1964

R River sites surveyed
by Spence, 1964

S Swamp sites surveyed
by Spence, 1964

KILOMETRES

0   50   100

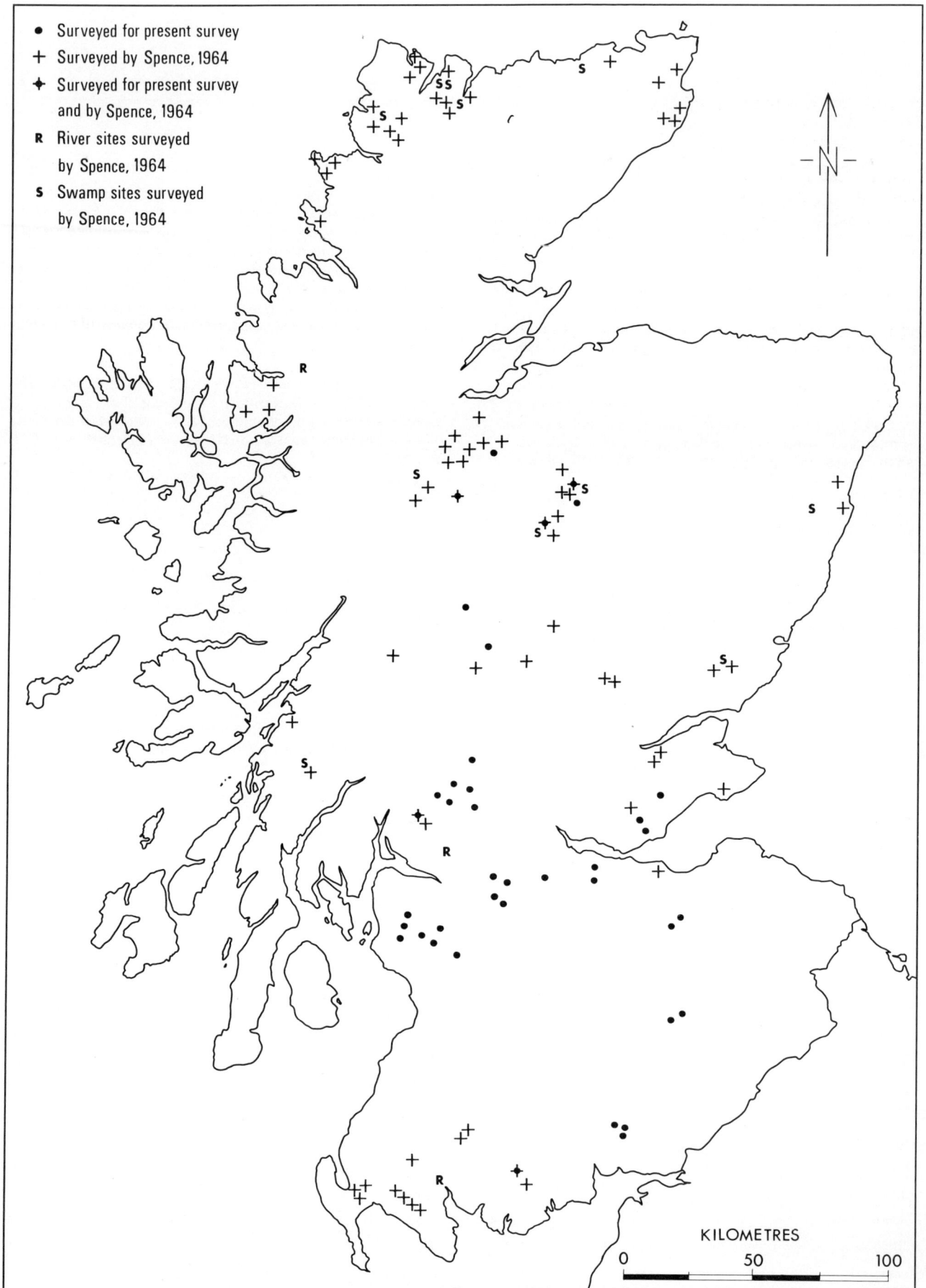

Fig. 41   Lochs included in vegetation survey

2. Species Lists for Shore Vegetation

Constant species are recorded in terms of percentage frequency; + = species occasionally present at frequencies less than 10 per cent.

2.1 *Shore vegetation: emergent communities I*

	Eleocharis palustris/ Littorella uniflora	Dense E. palustris	Mixed E. palustris	Typha spp.	Dense Phragmites communis	Open P. communis	Transitional P. communis
Number of stands	18	12	20	6	11	4	7
Soil	Sand/Stones	Sand	Sand/Stones	Mud	Mud	Mud/Stones	Mud/Loam
Water depth	0 - 30cm	0 - 30cm	0 - 30cm	1m	0 - 1m	0 - 1m	0
Bare ground	20 - 60	0 - 30	20 - 60	0	0 - 60	50 - 80	0 - 10
Calluna vulgaris							0 - 20
Myrica gale				0 - 10	60 - 100	20 - 40	10 - 25
Equisetum fluviatile	+	+	+		+		0 - 15
Agrostis spp.			0 - 20				0 - 30
Glyceria maxima					0 - 10		
Phalaris arundinacea		+	0 - 15	0 - 15	0 - 10		
Phragmites communis				0 - 10	60 - 100	20 - 40	10 - 25
Poa annua			+				
Carex curta						0 - 10	
Carex echinata							+
Carex nigra			0 - 10				0 - 25
Carex serotina	+		+				+
Carex rostrata	0 - 10		+		0 - 20	0 - 30	+
Carex vesicaria		+	0 - 10				+
Eriophorum spp.							+
Eleocharis palustris	15 - 65	65 - 100	15 - 65		+		+
Alisma plantago-aquatica		+		+			
Iris pseudacorus				0 - 20			+
Juncus articulatus	+	+	0 - 10		0 - 40		+
Juncus bulbosus	0 - 10	+				0 - 20	+
Juncus effusus				0 - 10			
Lemna minor				+			
Nuphar lutea						0 - 10	
Nymphaea alba						0 - 10	
Scirpus lacustris						+	
Typha spp.				80 - 100			
Achillea ptarmica			+				
Caltha palustris			+				+
Circuta virosa				0 - 10			
Epilobium angustifolium		+					
Epilobium palustre					0 - 10		+
Gallium palustre			0 - 10	+		+	0 - 15
Littorella uniflora	20 - 55						
Lobelia dortmanna	0 - 15				+		
Lythrum salicaria		+	0 - 15				
Mentha aquatica		+	0 - 10				
Myosotis scorpioides		+	+				+
Polygonum amphibium		0 - 30	0 -30				+
Potentilla palustris			+		+		+
Ranunculus flammula	0 - 10		0 - 10				+
Rorippa islandica			+				
Rumex acetosa			+				
Rumex obtusifolius			+				
Senecio jacobea			+				
Solanum dulcamara				0 - 20			
Succisa pratensis			+				+
Trifolium repens			+				+
Viola palustris			+				+
Polytrichum spp.						0 - 20	
Sphagnum spp.						0 - 20	

2.2 Shore vegetation: emergent communities II

	Glyceria maxima	Dense Phalaris arundinacea	Mixed P. arundinacea	Open P. arundinacea	Sparganium erectum	Scirpus lacustris	Equisetum fluviatile
Number of stands	12	47	6	10	8	4	6
Soil	Loam/Silt	Loam	Loam/Sand	Loam/Sand	Loam/Silt/Sand	Clay/Sand	Sand
Water depth	0.5cm	0	0	0	0 - 35cm	30 - 1m	0 - 31cm
Bare ground	0 -20	0 - 20	10 - 25	25 - 75	40 - 60	25 - 75	0 - 50
Rubus fruticosus				+			
Equisetum fluviatile	+					0 - 10	50 - 100
Agrostis tenuis	+	0 -10	0 - 40	0 - 20			
Anthoxanthum odoratum			+				
Dactylis glomerata			0 - 30				
Deschampsia caespitosa		+	0 - 20	0 - 20			
Festuca rubra	+			+			
Glyceria maxima	75 - 100	0 - 20	0 - 30				
Holcus lanatus				0 - 20			
Phalaris arundinacea	0 - 35	60 - 100	35 - 65	0 - 30	0 - 30		
Phragmites communis	0 - 10						
Poa annua		+	0 - 30	0 - 30			
Carex aquatilis			0 - 30		0 - 30		
Carex nigra			+				
Carex rostrata			+		0 - 10	+	0 - 50
Carex vesicaria		0 - 10	0 - 40	0 - 10			
Eleocharis palustris					0 - 15	+	
Elodea canadensis					0 - 10		
Alisma plantago-aquatica		+					0 - 10
Iris pseudacorus	+	+	+				
Juncus articulatus		+	0 - 30	0 - 30			
Juncus bulbosus				0 - 10		0 - 10	
Juncus effusus	+	+	0 - 15	0 - 20			
Lemna minor	+				+		
Scirpus lacustris						25 - 50	
Sparganum erectum	+		+		50 - 100		
Achillea ptarmica		+	+				
Angelica sylvestris		+	+				
Caltha palustris		+	+	+			
Carum verticillatum			+	+			
Cardamine amara			+				
Centaurea nigra			+	+			
Circuta virosa	+						
Epilobium angustifolium			0 - 30				
Epilobium palustre		+	+				
Filipendula ulmaria	+	0 - 30	10 - 45	0 - 10			
Galium palustre		+	+	+			
Littorella uniflora						0 - 50	
Lychnis flos-cuculi			+				
Lythrum salicaria		+	+				
Mentha aquatica	+	+	+				
Myosotis scorpioides	+	+	+				0 - 30
Myriophyllum spp.					0 - 10		
Plantago lanceolata				0 - 15			
Potentilla palustris	0 - 10	+					+
Ranunculus flammula		+	+	+			
Ranunculus repens	+	+	+				
Rumex acetosa	+						
Silene spp.			+				
Succisa pratensis			0 - 30				
Symphytum offininale	+						
Viola palustris				+			

2.3 Shore vegetation: emergent communities III

	Iris pseudacorus	Dense Carex rostrata	Mixed C. rostrata	Open C. rostrata	Carex vesicaria	Carex lasiocarpa	Carex aquatilis
Number of stands	6	11	17	26	10	1	5
Soil	Loam	Sand/Mud	Sand/Mud	Sand	Loam	Peat	Sand/Mud
Water depth	0 - 10cm	0 - 30cm	0 - 25cm	0 - 50cm	0 - 25cm	0	0 - 50cm
Bare ground	0	0 - 20	0 - 40	40 - 80	0 - 20	+	+
Calluna vulgaris						+	
Equisetum fluviatile	0 - 10	0 - 20	10 - 30	0 - 10	0 - 20	+	+
Agrostis stolonifera var. palustris			+	+	+		
Agrostis tenuis					+	+	+
Brizia media					+		
Deschampsia caespitosa	+			+	+		+
Festuca rubra						0 - 15	
Glyceria maxima			+				
Holcus lanatus					+		+
Phalaris arundinacea	0 - 20		+	0 - 15	0 - 40		0 - 15
Phragmites communis						0 - 10	
Carex aquatilis			+		+		40 - 100
Carex curta						+	
Carex echinata						+	
Carex lasiocarpa						50 - 100	
Carex nigra		+	10 - 20	+	+		+
Carex panicea						+	
Carex rostrata	0 - 30	85 - 100	50 - 80	10 - 50	0 - 30		0 - 40
Carex serotina				+			
Carex vesicaria			0 - 20	0 - 10	60 - 100		0 - 10
Eleocharis palustris		+	0 - 35	0 - 15	0 - 15		
Iris pseudacorus	80 - 100						
Juncus articulatus		0 - 15	0 - 20	0 - 20	0 - 15		0 -15
Juncus bulbosus			10 - 40	0 - 20		0 - 10	
Juncus effusus			+		0 - 10	+	+
Lemna minor	0 - 20						
Scirpus lacustris				+			
Sparganium erectum	0 - 20						
Caltha palustris		+	+		+		+
Epilobium hirsutum	+						
Epilobium palustre					+		
Filipendula ulmaria			+		0 - 15		0 - 10
Galium palustre			+	0 - 10			0 - 20
Hydrocotyle vulgaris			+		+	+	
Littorella uniflora		+	0 - 30				
Lobelia dortmanna				+			
Lotus comiculatus							+
Lythrum salicaria			+	+			
Mentha aquatica		0 - 15			0 - 30		+
Menyanthes trifoliata		+	0 - 10	+	+		
Myosotis scorpioides			+	+			
Pedicularis pediculare			+	+			
Potentilla palustris		0 -10	+		0 - 10	+	0 -10
Ranunculus flammula		0 - 20	0 - 20	+	+		
Ranunculus repens					+		
Rumex acetosa						0 - 15	
Succisa pratensis				+			
Viola palustris					+	+	
Acrocladium spp.			0 - 10				
Sphagnum spp.		0 - 20	0 - 50		0 - 30	0 - 10	0 - 20

2.4 Shore vegetation: floating-leaved communities

	Nuphar lutea	Nymphaea alba	Menyanthes trifoliata	Polygonum amphibium	Potamogeton natans	Sparganium angustifolium
Number of stands	7	5	9	9	7	4
Soil	Mud	Mud	Mud	Sand/Mud	Sand/Mud	Sand/Mud
Water depth	1m	1m	0 - 1m	0 - 1m	15cm - 1m	1m
Bare ground	0 - 10	0 - 10	0 - 50	20 - 50	15 - 45	0 - 90
Equisetum fluviatile	0 - 20	+			0 - 10	
Alisma plantago-aquatica					+	
Deschampsia caespitosa				+		
Holcus lanatus				+		
Carex rostrata	10 - 60	+	0 - 30		0 - 30	
Eleocharis palustris	+	+				
Elodea canadensis		+		+	0 - 20	
Iris pseudacorus					0 - 20	
Juncus acutiformis			+			
Juncus articulatus				+		
Juncus bulbosus	+	+		0 - 10	0 - 50	
Juncus effusus				0 - 10		
Sparganium angustifolium					+	20 - 40
Sparganium erectum						0 - 20
Potamogeton natans			5 - 10		40 - 60	
Callitriche spp.			0 - 25			
Littorella uniflora	+			+	+	
Lobelia dortmanna					0 - 15	
Menyanthes trifoliata			50 - 100		+	
Myosotis scorpiodes				+		
Myriophyllum alterniflorum					+	+
Nuphar lutea	90 - 100					
Nymphaea alba		90 - 100				
Polygonum amphibium				50 - 100	0 - 25	
Ranunculus aquatilis				+	+	
Fontinalis spp.					+	0 - 10
Sphagnum spp.			0 - 50			

2.5 Shore vegetation: submerged/low-shore communities

	Fontinalis spp.	Myriophyllum alterniflorum	Littorella uniflora/ Lobelia dortmanna	Dense L. uniflora	Discontinuous L. uniflora	Open L. uniflora/ Juncus bulbosus	Sparse L. uniflora/ J. bulbosus
Number of stands	6	5	10	8	22	11	36
Soil	Stones	Sand	Sand/Stones	Sand	Sand/Stones	Stones	Stones
Water depth	0 - 1m	0 - 50cm	0 - 50cm	0 - 1m	0 - 1m	0 - 1m	0 - 1m
Bare ground	50 - 100	80 - 100	20 - 80	0 - 25	10 - 40	47 - 75	75 - 100
Equisetum fluviatile			+				
Agrostis stolonifera var. palustris				+	+	+	+
Deschampsia caespitosa					+	+	
Poa annua						+	
Carex curta						+	
Carex nigra				+	+	+	
Carex serotina					+	+	
Carex rostrata					+		
Eleocharis palustris			0 - 20	+	0 - 15	+	
Juncus articulatus					+	+	+
Juncus effusus				+			
Juncus bulbosus			+			0 - 30	0 - 30
Epilobium palustre					+	+	
Gnaphilium uliginosum						+	+
Littorella uniflora			35 - 75	75 - 100	35 - 75	20 - 30	0 - 30
Lobelia dortmanna			5 - 30				
Mentha aquatica				+	+		
Myosotis scorpioides						+	
Myriophyllum alterniflorum	+	10 - 20			+		
Polygonum amphibium							+
Ranunculus flammula			0 - 10		0 - 15	+	
Rorippa islandica					+		
Fontinalis spp.	10 - 70	+		+	+		

2.6 Shore vegetation: sparsely vegetated stony shores

	Open Carex nigra	Sparse C. nigra	Open Juncus articulatus	Stony, Weedy, Grazed	Shores Ungrazed
Number of stands	6	8	11	6	21
Soil	Stones	Stones	Stones	Stones	Stones
Water depth	—	—	—	—	—
Bare ground	15 - 50	50 - 70	25 - 80	50 - 100	50 - 100
Equisetum fluviatile	+	+			
Equisetum palustris					+
Agrostis stolonifera var. palustris	+	+			+
Agrostis tenuis		+	0 - 20		0 - 20
Anthoxanthum odoratum		+	+		0 - 30
Cynosurus cristatus					0 - 15
Deschampsia caespitosa			+	+	+
Rumex obtusifolius		+		+	+
Festuca rubra		+	0 - 30		0 - 20
Holcus lanatus			+	+	+
Molinia caerulea		+	+		
Phalaris arundinacea	+	+	+		+
Poa annua		+		0 - 10	
Prunella vulgaris					+
Carex nigra	10 - 70	10 - 30	0 - 15		
Carex pulicaris					0 - 10
Carex serotina	0 - 10	+	+		
Eleocharis palustris	0 - 20	0 - 10			
Juncus articulatus			20 - 80		
Juncus bulbosus	0 - 20	0 - 10	0 -15		0 - 10
Juncus effusus	+	+	0 - 20		+
Luzula campestris					0 - 10
Alchemilla vulgaris					+
Caltha palustris			+		
Centaurea nigra	+				+
Chenopodium spp.					0 - 30
Cirsium palustre				+	+
Cruciata laevipes					+
Filipendula ulmaria			+		0 - 20
Galium saxatile			0 - 10		0 - 10
Leontodon spp.					+
Littorella uniflora	0 - 20	+	+		
Lotus corniculatus					+
Matricaria matricoides				0 - 20	
Plantago lanceolata			+	+	+
Plantago major				+	
Potentilla anserina	+				
Potentilla erecta			+		+
Prunella vulgaris					+
Ranunculus flammula	0 - 10	+	0 - 20		+
Ranunculus repens					+
Rumex acetosa			+		+
Rumex acetosella					+
Rumex obtusifolius		+		+	+
Sagina procumbens					+
Senecio jacobea				0 - 20	+
Senecio viscosus					0 - 20
Succisa pratensis					+
Thymus drucei					+
Trifolium pratense					+
Trifolium repens	+				+
Tussilage tartara					0 - 20
Urtica dioica				+	+

3. Species Lists for Backshore Vegetation

3.1 *Backshore vegetation: wetland and short grass/forbs*

	Bog	Short Grass (grazed)	Short Grass (recreated)	Short Grass Rush Fringe	Short Grass + Sedges
Number of stands	6	48	83	15	31
Bare ground			0 - 75		
Calluna vulgaris	0 - 30			+	+
Equisetum fluviatile					+
Agrostis stolonifera				0 - 10	
Agrostis tenuis		30 - 70	10 - 60	0 - 40	0 - 30
Anthoxanthum odoratum		0 - 80	0 - 50		+
Cynosurus cristatus		0 - 10	0 - 30		
Dactylis glomerata		+			+
Deschampsia caespitosa	0 - 10	0 - 20	0 - 10	0 - 20	+
Festuca pratensis		0 - 20			
Festuca rubra	10 - 40	0 - 60	0 - 60	0 - 40	0 - 30
Holcus lanatus		0 - 30	0 - 50	0 - 20	0 - 10
Lolium perenne			10 - 50		+
Molinia caerulea	20 - 40				0 - 20
Nardus stricta	+		+		+
Phalaris arundinacea		+	+	0 - 20	0 - 20
Phleum pratense				0 - 20	
Poa annua		+	1 - 90		
Carex curta	+				
Carex demissa					+
Carex echinata	+				+
Carex flacca					+
Carex nigra	+	0 - 20	0 - 10	0 - 10	20 - 60
Carex serotina					+
Carex vesicaria				0 - 10	+
Eriophorum spp.	50 - 90				
Eleocharis palustris					+
Juncus articulatus		0 - 20	0 - 20	0 - 30	10 - 40
Juncus bulbosus					+
Juncus effusus	0 - 10	0 - 30	0 - 30	30 - 90	+
Juncus squarrosus	+				+
Luzula campestris		+	+	+	+
Achillea millefolium		0 - 20	0 - 15	+	0 - 20
Achillea ptarmica	+			+	+
Alchemilla vulgaris					+
Bellis perennis		+	0 - 10		
Caltha palustris				+	+
Centaurea nigra		0 - 30	+		+
Cerastium arvense			0 - 10		
Cirsium palustre		+			
Cruciata laevipes		+			
Conopodium majus		0 - 20	0 - 25		
Endymion non-scriptus		0 - 10	+		
Filipendula ulmaria		0 - 10	+	0 - 20	0 - 20
Galium palustre					0 - 15
Galium saxatile	0 - 10	+	+	+	
Galium verum		0 - 20			
Leontodon spp.			+		+
Lotus corniculatus		0 - 30	0 - 10	0 - 10	+
Matricaria matricoides		+	+		
Mentha aquatica		+	+	+	0 - 15
Plantago lanceolata		+	+	+	0 - 10
Potentilla erecta	+	0 - 10	+	+	+
Potentilla palustris				+	0 - 20
Prunella vulgaris		+	+	+	+
Ranunculus ficaria		+	+		
Ranunculus flammula		0 - 20	0 - 10	+	0 - 20
Ranunculus repens		0 - 15	0 - 30	+	0 - 10
Rumex acetosa		0 - 20	+	0 - 10	+
Rumex obtusifolius		+	0 - 10	+	
Senecio jacobea			+		
Stellaria graminea			+		
Stellaria media		0 - 20	+		+
Succisa pratensis		+	+	+	+
Trifolium repens		0 - 20	0 - 30		0 - 10
Trollius cressida			+		+
Urtica dioica			+		
Acrocladium spp.		+	+	+	0 - 30
Climacium spp.			+	+	0 - 30
Polytrichum spp.	0 - 10			+	+
Sphagnum spp.	50 - 100	+	0 - 20	0 - 20	0 - 40

3.2 Backshore vegetation: Long grass/forbs and moor grass

	Weedy Grass	Deschampsia caespitosa Grassland	Juncus articulatus	Filipendula ulmaria Communities	Molinia caerulea Grassland
Number of Stands	21	15	12	20	15
Calluna vulgaris					0 - 50
Erica tetralix					0 - 30
Myrica gale		+			0 - 30
Rubus fruticosus	0 - 20	+		+	
Rubus idaeus	0 - 20			0 - 30	
Vaccinium myrtillus					0 - 15
Agrostis tenuis	0 - 50	0 - 40	0 - 20	0 - 40	0 - 50
Anthoxanthum odoratum	0 - 20		+	+	+
Arrhenatherum elatius	0 - 70	+	+	+	
Brizia media	+		+		
Cynosurus cristatus	0 - 40				+
Dactylis glomerate	0 - 70	0 - 15	+	+	
Deschampsia caespitosa	0 - 20	35 - 70	0 - 30	0 - 20	0 - 30
Festuca rubra	0 - 50	0 - 10	0 - 10	+	+
Holcus lanatus	0 - 40	0 - 20	0 - 30	0 - 15	+
Lolium perenne	0 - 15				
Molinia caerulea	+	0 - 10	0 - 60	+	75 - 100
Nardus stricta			0 - 40		0 - 40
Phalaris arundinacea	0 - 15	0 - 50	0 - 20	0 - 50	+
Phleum pratense	0 - 30	+			
Poa annua		+			
Carex aquatilis				+	
Carex curta					+
Carex echinata					+
Carex lasiocarpa					0 - 30
Carex nigra	+	0 - 30	+		0 - 20
Carex panicea		+			
Carex rostrata		+			+
Carex serotina					+
Carex vesicaria		0 - 30			0 - 20
Eriophorum spp.					0 - 30
Juncus articulatus	+	0 - 20	20 - 80	0 - 40	+
Juncus bulbosus			+		
Juncus effusus	+	0 - 30	0 - 30	0 - 25	0 - 30
Juncus squarrosus					+
Scirpus caespitosa					0 - 15
Achillea millefolium	+	+			
Achillea ptarmica		0 - 15	+	+	+
Alchemilla vulgaris		+		+	
Aster novi-belgii	0 - 20			0 - 40	
Angelica sylvestris	0 - 10	+	+	+	+
Bellis perennis					+
Caltha palustris		+	+		
Centaurea nigra	0 - 20	0 - 15		+	
Drosera rotundifolia					+
Epilobium angustifolium	0 - 20	0 - 30	+		
Epilobium palustre	+	+	+		
Filipendula ulmaria	+	0 - 30	0 - 10	40 - 100	+
Galium aparine	+	0 - 10		0 - 20	
Heracleum sphonylium	+		+		
Leontodon spp.	+				+
Lotus corniculatus	+	0 - 15	0 - 10	+	+
Mentha aquatica		+			
Pedicularis pediculare					+
Plantago lanceolata	0 - 20	+	+		+
Prunella vulgaris		+	+		
Potentilla anserina		0 - 15			
Potentilla erecta		+		+	+
Potentilla palustris		+			
Ranunculus flammula			+		+
Ranunculus repens	0 - 20	0 - 15	0 - 35	+	+
Rumex acetosa	0 - 15	+	+	+	
Silene alba	+				
Silene dioica	+		+		
Succisa pratensis	0 - 15	+	+	+	0 - 15
Taraxacum officinale	+				
Trifolium repens	0 - 20	+	+		+
Senecio jacobea	0 - 20	+			
Stachys palustris				0 - 20	
Urtica dioica	+			+	
Valerian officinalis		+	+	0 - 10	
Vicia cracca	+	+		+	+
Vicia sepium	+	+	+		+
Viola palustris					+
Acrocladium spp.					+
Polytrichum spp.		+			0 - 20
Sphagnum spp.					0 - 65

3.3 Backshore vegetation: dwarf shrub and shrub (understorey)

	Calluna vulgaris	C. vulgaris/ Grass	Myrica gale/ Grass	M. gale	Pteridium aquilinum	Ulex europaeus	Alnus glutinosa/ Salix spp.
Number of stands	10	10	2	13	5	7	21
Bare ground						0 - 40	0 - 75
Alnus glutinosa							50 -100
Salix spp.		+	+	+			50 -100
Ulex europaeus					50 -100		
Calluna vulgaris	45 - 100	15 - 45	20 - 40	0 - 30			
Erica tetralix	+	0 - 20	+				
Myrica gale		+	20 - 45	45 - 100			
Rubus idaeus					+		+
Vaccinium myrtillus	0 - 15	0 - 15					
Pteridium aquilinum					50 - 100		
Agrostis tenuis	0 - 15	0 - 20			0 - 25	20 - 70	0 - 20
Anthoxanthum odoratum		0 - 10					+
Deschampsia caespitosa		+	50 - 60	0 - 50	0 - 20	+	0 - 15
Deschampsia flexuosa			+				
Festuca rubra	0 - 55	0 - 25	+		+	10 - 40	0 - 30
Holcus lanatus			+	+	0 - 25	0 - 20	+
Molinia caerulea	0 - 55	20 - 40	50 - 65	0 - 50	0 - 10	0 - 10	0 - 20
Nardus stricta	0 - 55	10 - 30	+				
Phalaris arundinacea		+	+	+			+
Phragmites communis				+			
Carex curta		+	+	+			
Carex echinata		+					
Carex nigra	+	+	+	+			+
Carex rostrata		+					+
Carex serotina	+		+	+			
Carex vesicaria		+					+
Juncus articulatus		0 - 20	+	+			0 - 10
Juncus bulbosus			+				+
Juncus effusus		+	+	+	+		0 - 15
Juncus squarrosus		0 - 15	+				
Achillea ptarmica		+					
Angelica sylvestris					+		
Drosera rotundifolia		+					
Filipendula ulmaria					+		+
Galium saxatile			+				
Geum rivale					+		
Leontodon spp.		+				+	
Oxalis acetosella					+		
Pedicularis pediculare		+					
Plantago lanceolata		+					
Potentilla erecta		+	+	+		+	
Ranunculus flammula			+				
Ranunculus repens				+		+	
Rumex acetosella					+		
Succisa pratensis	+	0 - 15				+	
Urtica dioica					+		+
Viola palustris		+					
Sphagnum spp.	0 - 10	0 - 60	0 - 30				

3.4   Backshore vegetation: Woodland (understorey)

Number of Stands	Deciduous (undisturbed) 44	Deciduous (used for recreation) 38	Coniferous (undisturbed) 5	Coniferous (used for recreation) 9
Bare ground	0 - 75	10 - 100	0 - 20	10 - 100
Calluna vulgaris			30 - 100	0 - 40
Erica tetralix			+	0 - 20
Myrica gale				+
Rubus fruticosus	0 - 15	+		
Vaccinium myrtillus			0 - 50	0 - 20
Pteridium aquilinum	+	0 - 10	+	+
Agrostis tenuis	0 - 25	0 - 65	0 - 20	0 - 40
Anthoxanthum odoratum	+			+
Dactylis glomerata	0 - 15	0 - 15		
Deschampsia caespitosa	0 - 15	+		+
Festuca rubra	+	0 - 25		0 - 20
Holcus lanatus	0 - 30	0 - 30	+	0 - 10
Nardus stricta			+	0 - 20
Phalaris arundinacea	0 - 50	0 - 50		
Poa annua		0 - 60		0 - 20
Carex nigra	+	0 - 10	+	0 - 20
Juncus articulatus		+		+
Juncus effusus	0 - 10	+		0 - 20
Luzula sylvatica	0 - 50	0 - 20		
Achillea millefolium	+	+		
Achillea ptarmica	+	+		
Anemone nemorosa	0 - 40			
Angelica sylvestris	0 - 20	+		
Bellis perennis		+		+
Centaurea nigra		+		
Chrysosplenium oppositifolium	0 - 15			
Conopodium majus	0 - 15	0 - 20		
Endymion non-scriptus	0 - 45	0 - 20		
Galium aparine	+	+		
Geum rivale	+			
Geum urbanum	0 - 40	+		
Lamium album		+		
Leontodon spp.		+		+
Oxalis acetosella	0 - 20	+		
Plantago lanceolata		0 - 30		+
Potentilla erecta			+	+
Primula vulgaris	0 - 15	+		
Prunella vulgaris		+		
Ranunculus ficaria	0 - 60	+		
Ranunculus repens	0 - 14	+		0 - 10
Rumex acetosa	+	+		
Rumex obtusifolius		+		
Silene alba	+	+		
Silene dioica	+	+		
Stachys sylvestris	0 - 50			
Stellaria graminea	+	+		
Succisa pratense			+	+
Taraxacum officinale		+		
Trifolium repens		+		0 - 20
Valerian officinalis	0 - 15	+		
Veronica chamaedrys		+		
Viola palustris	+	+		+

152

4. Species Readily Eliminated by Trampling

Wetland: Bog moss (*Sphagnum* species)

Grassland: Yorkshire fog (*Holcus lanatus*)
False oat-grass (*Arrhenatherum elatius*)
Timothy grass (*Phleum pratense*)

Heathland: Heather (*Calluna vulgaris*)
Bog myrtle (*Myrica gale*)
Cross-leaved heath (*Erica tetralix*)
Heath tormentil (*Potentilla erecta*)
Heath bedstraw (*Galium saxatile*)

Woodland: Wood sorrel (*Oxalis acetosella*)
Wood anemone (*Anemone nemorosa*)
Wood rush (*Luzula sylvatica*)
Pignut (*Conopodium majus*)

5. Trample Resistant Species Characteristic of Sites Visited

5.1 *Very resistant species: tolerate a wide range of conditions*

Annual meadow-grass (*Poa annua*)
Meadow-grass (*Poa pratensis*)
Red fescue (*Festuca rubra*)
Common bent (*Agrostis tenuis*)
Couch-grass (*Agropyron repens*)
Rye-grass (*Lolium perenne*)
Greater plantain (*Plantago major*)
Pineapple weed (*Matricaria matricoides*)

5.2 *Moderately resistant species: many of these species occur naturally in grassy swards, but become increasingly important under recreation.*

(a) Dry grassland
Crested dog's-tail (*Cynosurus cristatus*)
Birdsfoot trefoil (*Lotus corniculatus*)
White clover (*Trifolium repens*)
Daisy (*Bellis perennis*)
Dandelion (*Taraxacum officinalis*)
Hawkweed (*Leontodon* species)
Self heal (*Prunella vulgaris*)
Pearlwort (*Sagina procumbens*)
Yarrow (*Achillea millefolium*)
Mouse-ear Chickweed (*Cerastium arvense*)
Ragwort (*Senecio jacobea*)
Knapweed (*Centaurea nigra*)
Heath grass (*Sielingia decumbens*)
Field woodrush (*Luzula compestris*)
Ribwort plantain (*Plantago lanceolata*)

(b) Damp grassland
Common sedge (*Carex nigra*)
Carnation sedge (*Carex panicea*)
Glaucous sedge (*Carex flacca*)
Common yellow-sedge (*Carex demissa*)
Creeping buttercup (*Ranunculus repens*)
Jointed rush (*Juncus articulatus*)

(c) Moorland grass/heathland
Purple moor-grass (*Molinia caerulea*)
Mat-grass (*Nardus stricta*)
Sheep's fescue (*Festuca ovina*)
Wild thyme (*Thymus drucei*)
Cotton-grass (*Eriophorum* species)
Heath rush (*Juncus squarrosus*)

6. Pathways through Emergents: species composition

6.1 *Pathways through emergents: species frequency*

Community	Species	Main Vegetation	Path	Path Margin
Phragmites communis (Mill Loch)	Bare	—	—	—
	Litter	—	100	—
	Phragmites communis	100	—	100
(Loch Ard)	Bare	—	—	—
	Litter	—	75.0	—
	Phragmites communis	72.0	5.0	50.0
	Juncus articulatus	8.0	25.0	50.0
	Potentilla palustris	4.0	—	50.0
Glyceria maxima (Linlithgow Loch)	Bare	—	—	—
	Litter	—	100	—
	Glyceria maxima	100	—	100

Phalaris arundinacea

Community	Species	Main Vegetation			Path			Path Margin		
		Light use	Mod. use	Heavy use	Light use	Mod. use	Heavy use	Light use	Mod. use	Heavy use
(Castle Semple Loch)	Bare	—	—	—	—	20.0	83.0	—	—	12.0
	Litter	—	—	—	100	40.0	—	—	—	—
	Phalaris arundinacea	100	100	78.0	—	—	—	85.0	70.0	82.0
	Agrostis tenuis	—	—	—	—	40.0	17.0	—	—	—
	Deschampsia caespitosa	—	—	—	—	—	—	4.0	25.0	—
	Iris pseudacorus	—	—	22.0	—	—	—	—	—	—
	Filipendula ulmaria	—	—	—	—	—	—	9.0	—	—
	Myosotis scorpioides	—	—	—	—	—	—	—	3.0	2.0
	Polygonum amphibium	—	—	—	—	—	—	—	2.0	4.0
	Urtica dioica	—	—	—	—	—	—	2.0	—	—
	Mentha aquatica	—	—	—	—	—	—	—	—	+
	Cardamine amara	—	—	—	—	—	—	—	—	+
	Lychnis flos-cuculi	—	—	—	—	—	—	—	—	+
	Epilobium palustre	—	—	—	—	—	—	—	—	+

Community	Species	Main Vegetation	Path	Path Margin
Carex rostrata (Loch Achray)	Bare	—	40.0	—
	Carex rostrata	97.8	60.0 (flat)	60.0
	Equisetum fluviatile	7.0	—	—
	Potentilla palustris	1.0	—	10.0
Low/moderate use	*Filipendula ulmaria*	—	—	25.0
Carex rostrata (Loch Lubnaig) Heavy use	Bare	20.0	100	—
	Carex rostrata	70.0	—	70.0
	Potentilla palustris	1.0	—	10.0
	Filipendula ulmaria	10.0	—	—
	Ranunculus flammula	1.0	—	23.3
Carex vesicaria (Loch Achray)	Bare	—	40.0	—
	Carex vesicaria	60.0	—	40.0
	Sphagnum spp.	70.0	30.0	50.0
	Mentha aquatica	12.5	—	25.0
	Potentilla palustris	2.7	—	55.0
	Carex nigra	20.0	40.0	40.0
	Viola palustris	—	—	25.0
Carex lasisocarpa (Loch Garten)	Bare	4.2	20.0	4.2
	Carex lasisocarpa	57.8	50.0 (flat)	62.4
	Juncus bulbosus	8.5	10.0	10.0
	Carex curta	2.1	—	2.0
	Calluna vulgaris	2.1	—	0.7
	Sphagnum spp.	7.8	—	7.8
	Hydrocotyle vulgaris	2.8	10.0	10.0
	Potentilla palustris	4.2	30.0	10.0
	Festuca rubra	11.4	—	2.0
	Agrostis tenuis	5.7	10.0	8.5

6.2 *Recovery of unused pathways on Barr Loch: species frequency 1974 and 1975*

Community	Transect No.	Main Community 1974	1975	Pathway 1974	1975
Glyceria maxima	1	36.0	42.0	13.0	55.0
	2	60.0	66.0	12.0	52.0
	3	86.0	82.0	10.0	72.0
	4	74.0	82.0	No path	
	5	69.0	76.0	45.0	82.0
	6	30.0	30.0	No path	
Phalaris arundinacea	1	69.0	71.0	40.0	75.0
	2	17.0	25.0	—	—
	3	58.0	65.0	45.0	62.0
	4	20.0	23.0	No path	
	5	55.0	63.0	47.0	62.0
	6	87.0	89.0	No path	

6.3 *Path vegetation changes induced by experimental impact. The data shows the vegetation type (as listed on Table 4) before impact (left hand figure) and after impact (right hand figure) for each intensity of impact; 1 = 16; 2 = 32; 3 = 64; 4 = 128 passages each month*

Site	Running 1	2	3	4	Walking 1	2	3	4	Control 1	2	3	4
Littorella uniflora/Lobelia dortmanna	3 1	1 1	3 3	3 (7)	3 (7)	*3* 2	*3* 2	*3* 2	3 1	3 1	3 3	3 3
Open *Littorella uniflora*	*2* 4	*2* 5	*2* 5	*2* 4	*2* 4	*2* 4	*2* 4	*2* 4	2 2	2 2	2 2	2 2
Carex rostrata	19 19	19 19	19 19	*19* 20	19 19	19 19	*19* 20	*19* 20	19 18	19 18	19 18	19 18
Carex rostrata/Sphagnum spp.	20 (19)	20 20	20 20	20 20	20 (19)	20 20	20 20	20 20	20 20	20 20	20 20	20 20
Phalaris arundinacea	*15* 16	*15* 16	*14* 16	*14* 16	*15* 16	*15* 16	*14* 16	*14* 16	15 15	15 15	15 15	14 14
Glyceria maxima (Barr Loch)	*13* *	*13* *	*13* *	*13* *	*13* *	*13* *	*13* *	*13* *	13 13	13 13	13 13	13 13
Glyceria maxima (Lake of Menteith)					*13* *	*13* *	*13* *	*13* *	13 13	13 13	13 13	13 13
Phragmites communis					11 11	*11* *	*11* *	*11* *	11 11	11 11	11 11	11 11

Figures in italics indicate a site where modification towards a vegetation type characteristic of recreation has taken place; * indicates a site where modification has been towards path-vegetation as identified on other lochside sites; ( ) indicates a site where natural changes have resulted in modification, e.g. in the *Littorella/Lobelia* site, the seasonal growth of *Eleocharis palustris* has altered the vegetation type.

7. Boat-nests in Emergents: Species Frequency

Community	Species	Main Community	Boat-nest	Boat-nest Margin
Eleocharis palustris (Loch Lubnaig)	Bare	32.0	94.0	40.0
	Eleocharis palustris	60.0	+	60.0
	Littorella uniflora	4.0	2.0	—
	Ranunculus flammula	2.0	—	2.0
	Achillea ptarmica	4.0	—	—
Phragmites communis (Castle Loch)	Bare	—	100	—
	Phragmites communis	95.0	—	15.0
	Glyceria maxima	10.0	—	85.0
	Poa annua	—	+	—
	Phalaris arundinacea	—	—	7.5
	Potentilla palustris	—	—	5.0
Phalaris arundinacea (Loch Ken)	Bare	—	38.7	5.7
	Litter	—	18.7	—
	Phalaris arundinacea	96.0	20.0	52.8
	Filipendula ulmaria	1.4	1.2	11.4
	Galium palustre	6.0	—	8.5
	Centaurea nigra	—	6.2	—
	Deschampsia caespitosa	—	6.2	7.1
	Carex vesicaria	—	—	12.8
	Carum verticillatum	—	—	4.2
	Ranunculus flammula	—	—	2.8
(Loch Gelly)	Bare	—	100	—
	Litter	—	—	—
	Phalaris arundinacea	100	—	96.0
	Agrostis tenuis	—	—	4.0
Carex rostrata (Loch Lubnaig)	Bare	—	100	20.0
	Carex rostrata	97.0	—	80.0
	Equisetum fluviatile	7.0	—	2.0
	Potentilla palustris	1.0	—	—
	Agrostis stolonifera (var. *palustris*)	—	+	—

+ = present at percent frequency less than 0.5

K

156

8. Lochside Vegetation Communities of Relatively Uncommon Occurrence in Scotland, which are probably worthy of conservation

8.1 *Those shore communities dominated by:*
(a) Reedmace (*Typha* spp.) – e.g. Johnston Loch (and other small lowland lochs).
(b) Bulrush (*Scirpus lacustris*) – e.g. Lochs Garten, and Ard.
(c) Branched bur-reed (*Sparganium erectum*) – e.g. Johnston Loch and Barr Loch.
(d) Water Lobelia (*Lobelia dortmanna*) – e.g. Loch Ard and Highland Lochs.
(e) Yellow flag (*Iris pseudacorus*) and water horsetail (*Equisetum fluviatile*) e.g. Queen's Loch, Glenboig.

8.2 *Shore/shore margin communities with relatively rare associates:*
(a) Reed canary-grass (*Phalaris arundinacea*) with whorled caraway (*Carum verticillatum*) – e.g. Loch Ken.
(b) Reed canary-grass (*P. arundinacea*) with cowbane (*Circuta virosa*) – Lochmaben lochs.
(c) Woody nightshade (*Solanum dulcamara*) at inner margin of reed canary-grass – Kirk Loch.
(d) Reed (*Phragmites communis*) with trifid bur-marigold (*Bidens tripartita*) and marsh ragwort (*Senecio aquaticus*) – Castle Loch.
(e) Common spike-rush (*Eleocharis palustris*) with water plantain (*Alisma plantago-aquatico*) – Rossdhu Bay (Loch Lomond); Johnston Loch.

8.3 *Areas of considerable variety as yet little disturbed by recreation:*
(a) Loch Ard (south west bay): shoreweed/water lobelia (*Littorella uniflora*/*Lobelia dortmanna*) and bulrush (*Scirpus lacustris*).
(b) Johnston Loch (west end): reedmace (*Typha* spp.) water plantain (*Alisma plantago-aquatica*) branched bur-reed (*Sparganium erectum*) bistort (*Polygonum bistorta*) bogbean (*Menyanthes trifoliata*) yellow flag (*Iris pseudacorus*) woody nightshade (*Solanum dulcamara*) and wetland behind.
(c) Loch Garten (south end): reed (*Phragmites communis*) Bulrush (*Scirpus lacustris*) water-lily (*Nuphar lutea*) and wetland behind including slender sedge (*Carex lasiocarpa*).
(d) Loch Ken (south-west side): extensive marshes with very mixed communities, unusual succession presumably resulting from the flooding of an irregular ground surface: includes communities dominated by marsh Cinquefoil (*Potentilla palustris*).
(e) Loch Ruthven (north-east side): water-milfoil (*Myriophyllum* spp.)/water lobelia (*Lobelia dortmanna*)/shoreweed (*Littorella uniflora*) and bottle sedge (*Carex rostrata*).
(f) Loch Tarff (south-west corner): bogbean (*Menyanthes trifoliata*) and bottle sedge (*Carex rostrata*).
(g) Queen's Loch (Glenboig): remarkable beds of yellow flag (*Iris pseudacorus*) and water horsetail (*Equisetum fluviatile*).

1. Recording Procedures

Recreation user surveys were undertaken on four presumed peak days in the summer of 1975 around the Trossachs Lochs, and in 1976 around Loch Ken. Recording of selected data by questionnaire (see Section 2) was undertaken simultaneously between 11.00 and 17.00 hours on all the main informal (car parks, lay-bys, road verges) and a few formal recreational sites.

Recorders worked in pairs, and one questionnaire was given to each group of visitors (for example a car-load) and to individual cyclists and walkers, on arrival at the site. Visitors were asked if they would be willing to accept a questionnaire, fill it in and return it as they left the site. In the case of refusals the recorder, noted the rejection and recorded the number of people in the group and time of arrival.

A record was kept of the number of questionnaires distributed, completed, returned but not completed, rejected, and not returned. An analysis of the returns is given in Section 3, and a qualitative description of the weather conditions on the particular date is tabulated in Section 4. The same questionnaire was used in both areas, with the exception of questions 7-10 which were included only for Loch Ken for purposes other than those specifically required for the final report.

Analysis of the Trossachs User Survey, 1975, was undertaken as a separate report for Central Region and Stirling District Planning Departments and the Scottish Sports Council. Analysis of the Loch Ken data was incorporated in an undergraduate thesis written by Miss Sheila Morgan (Middlesex Polytechnic at Enfield).

2. Recreation User Questionnaire

The exact wording of the questionnaire is given below.

This project is being undertaken by the Department of Geography of the University of Glasgow as part of a programme in which types of recreation and their effect on different types of lochshore are being examined. This work is being done for the Countryside Commission for Scotland, a Government organisation concerned with recreation and with the conservation of the natural beauty of the countryside.

We would be most grateful if you would assist us by completing this questionnaire *just before you leave* and handing it to our representative. You are *not* required to give your name.

1. How did you get here today?   (Please tick)

          Car   . . . .

          Coach   . . . .

          Minibus   . . . .

          Dormobile   . . . .

          Motorbike   . . . .

          Bicycle   . . . .

          Walked   . . . .

          Other   . . . .

2. How long did you spend here today?   (Please tick)

          ¼ hour or less   . . . .

          ¼ – ½ hour   . . . .

          ½ – 1 hour   . . . .

          1 – 2 hours   . . . .

          2 – 4 hours   . . . .

          4 – 6 hours   . . . .

          Over 6 hours   . . . .

          Overnight   . . . .

3. Approximate time of arrival   . . . .

   Approximate time of departure   . . . .

4. Did you do any of the following while you were here?
   (Please tick box)

   1. Sat in car

   2. Picnicked in car

   3. Picnicked near car

   4. Picnicked over 50 yards from car

   5. Played games

   6. Relaxed out of car (e.g. sitting, reading, chatting, viewing, sunbathing, etc.)

   7. Took photographs

   8. Strolled around parking area

   9. Walked some distance along lochside

   10. Hill walking

   11. Competitive bank fishing

   12. Horseriding/pony trekking

   13. Casual bank-fishing

   14. Bathed or paddled

   15. Boat-fishing (specify type of boat)   . . . .

   16. Casual boating (specify type of boat)   . . . .

   17. Competitive boating (specify type of boat). . . . .

   18. Water-skiing

   19. Sub-aqua

   20. Bird-watching

   21. Other activities (specify)

          . . . .

5. Did you bring and/or use any of the following personal equipment? (Please put a tick in box(es).)

                                   Brought   Used

1. Boat

2. Boat trailer

3. Tent

4. Touring caravan

5. Other (specify)

                          . . . .

6. How many people are in your group? . . . .

     How many are under 18 years? . . . .

     How many are over 18 years? . . . .

7.   Please tick appropriate box

1. Are you on holiday?

2. Are you on a weekend trip?

3. Are you on a day trip?

8. 1. Which is your home town? . . . .

    2. Where did you travel from today? . . . .

    3. Where are you heading for tonight? . . . .

9. Why did you visit Loch Ken today? . . . .

. . . .

. . . .

. . . .

10. How many times have you visited Loch Ken before?

                               Please tick appropriate box

1. Never (this is first visit)

2. Once

3. 3 – 5 times

4. 6 – 10 times

5. 10 times

3. Analysis of Questionnaire Returns

3.1 *Trossachs user survey, 1975*

Number of questionnaires (D) distributed; (A) returned; (NR) not returned; (R) rejected; (NU) not used; + not recorded.
* Site used for Trossachs Water Festival (Ardchullaire).

Loch Site No.	Sun 6th July (D)	(A)	(NR/R)	Mon 21st July (D)	(A)	(NR/R)	Sun 3rd August (D)	(A)	(NR/R)	Sun 24th August (D)	(A)	(NR/R)
Chon 1	35	29	6/0	3	3	0/0	23	18	5/0	16	15	1/0
2	9	5	1/2	20	17	3/1	7	4	0/3	16	14	2/0
3	33	28	5/0	19	16	3/0	37	36	0/0	12	12	0/0
Total	77	62	12/2	42	36	6/1	67	58	5/3	44	41	3/0
Ard 1	15	14	1/0	9	7	1/1	24	17	6/1		} 20	0/1
2	13	10	3/0	0	0	0/0	19	11	8/0			
3	24	13	11/0	0	0	0/0	17	13	3/1		30	0/2
Total	52	37	15/0	9	7	1/1	60	41	17/2		50	0/5
Menteith 1	33	20	10/3	35	35	0/0	34	34	0/0	47	40	3/4
Achray 1	+	+	+	+	+	+	+	+	+	+	+	+
2	17	14	1/2	0	0	0/0	7	7	0/0	1	1	0/0
3	20	11	7/2	22	22	0/0	26	26	0/0	15	15	0/0
4	35	29	6/0	12	12	0/0	27	14	9/4	12	10	0/2
5	60	50	10/0	36	30	6/0	60	44	6/10	35	35	0/0
6	+	+	+	+	+	+	+	+	+	+	+	+
7	+	+	+	24	20	2/2	30	21	9/0	28	17	9/0
8	+	+	+	18	18	0/0	NU	NU	NU	NU	NU	NU
Total	132	104	24/4	112	102	2/2	150	112	24/14	91	78	9/2
Venachar 1	122	90	32/0	84	70	13/1	127	91	28/8	81	71	10/0
2	46	42	0/4	30	23	6/0	32	24	4/4	32	29	3/0
3	+	+	+	9	9	0/0	54	29	25/0	39	33	5/1
Total	168	132	32.4	123	102	19/1	207	144	57/12	152	133	18/1
Lubnaig 1	31	28	1/2	25	23	0/2	36	29	2/5	29	29	0/0
2*	22	20	0/1	1	1	0/0	NU	NU	NU	NU	NU	NU
3	19	10	8/1	10	10	0/0	17	13	0/4	2	2	0/0
4	58	50	9/1	44	39	3/2	80	40	40/0	24	24	0/0
5	47	45	2/0	18	16	1/2	130	80	46/4	25	22	0/3
6	151	111	40/1	84	60	14/10	175	121	57/3	109	105	4/0
7	9	5	4/0	+	+	+	+	+	+	+	+	+
8	1	0	1/0	0	0	0/0	7	7	0/0	0	0	0/0
Total	338	269	65/6	182	149	18/16	445	290	145/16	189	182	4/3

3.2 Loch Ken user survey, 1976.

Numbers of questionnaires (D) distributed; (A) returned; (R) rejected; (NR) not returned; + not recorded.

Site No.	(D)	Sun 4th July (A)	(R)	(NR)	(D)	Sun 18th July (A)	(R)	(NR)	(D)	Mon 2nd August (A)	(R)	(NR)	(D)	Sun 22nd August (A)	(R)	(NR)
West 1	24	21	0	3	16	16	0	0	10	10	0	0	19	14	2	5
2	32	32	0	0	30	28	0	2	16	16	0	0	25	17	2	8
3	10	8	1	2	11	11	1	0	22	21	0	1	18	18	1	0
East 4	22	21	2	1	27	23	0	4	3	3	0	0	39	24	0	15
5	21	20	0	1	11	9	0	2	38	37	1	1	45	42	1	3
6	19	16	0	3	22	21	1	1	27	26	2	1	17	10	2	7
7	22	22	0	0	14	14	0	0	12	6	0	6	40	40	0	0
8	22	10	0	12	36	33	1	3	31	29	2	2	23	23	0	0
9	+	+	+	+	28	28	5	0	+	+	+	+	+	+	+	+
Total	172	150	3	22	195	183	8	12	159	148	5	11	226	188	8	38

4. Types of Recreational Sites Monitored

4.1 Trossachs Lochs

Loch	Site Number	Type of Site
L. Chon	1	Forestry Commission car park/picnic site
	2	Forestry Commission car park/picnic site
	3	Off-road parking
L. Ard	1	Off-road parking
	2	Off-road parking
	3	Off-road parking
Lake of Menteith	1	Off-road parking
L. Achray	2	Off-road parking
	3	Off-road parking
	4	Off-road parking
	5	Forestry Commission car park
	6	Water-ski Club
	7	Walled lay-by
	8	Scout club camping site
L. Venachar	1	Local Authority car park
	2	Local authority car park
	3	Off-road parking
L. Lubnaig	1	Lay-by
	2	Off-road and shore parking
	3	Off-road and shore parking
	4	Lay-by
	5	Off-road parking
	6	Off-road parking
	7	Forestry Commission log cabins
	8	Private camping site

4.2 Loch Ken

Loch	Site Number	Type of site
West Loch Ken	1	Off-road parking
	2	Off-road parking
	3	Off-road parking
East Loch Ken	4	Car park
	5	Roadside verge
	6	Lay-by
	7	Water-ski Club
	8	Caravan site
	9	Rally-site (one day)

5. Weather Conditions During Recreation User Surveys

	July 6 Sunday	July 21 Monday	August 3 Sunday	August 24 Sunday
Trossachs Lochs (1975)	Very warm Clear skies Good visibility Light breeze	Cool Continuous rain Low persistent cloud Poor visibility Windy	Very hot Clear skies Slight heat haze No wind	Cool Sunny Occasional showers Fresh wind
	July 4 Sunday	July 18 Sunday	August 2* Monday	August 22 Sunday
Loch Ken (1976)	Dry, sunny Very warm No wind	Dry, overcast Cool Fresh SW wind	Cloudy Occasional showers Cool NW wind	Dry, sunny Warm Light E/SE wind

* English Bank Holiday

1. Introduction to the Methodology

A method, relevant to the needs of site planners and managers, whereby the physical suitability of lochside sites may be assessed for recreational activities is outlined below. It must be emphasised that:

(a) only the physical character of the lochside is being assessed; it is assumed that other constraints such as demand, land ownership, and economic feasibility are evaluated at some other stage of the planning process;

(b) suitability refers to the ability of an area in its present state to accommodate a particular recreational activity, and should not be confused either with carrying capacity, which gives an indication of the level of use the area can withstand before specified changes occur in the landscape, or with potentiality, which refers to the suitability of the area allowing for a future management input.

1.1 Methodological procedure

The procedure falls into three stages:

(a) Division of the lochside into site-types. The physical nature of lochsides varies considerably from swamp conditions to a vertical rock-face. It is, therefore, useful to be able to categorise the lochside into site-types which are reasonably homogenous for recreation, particularly where a length of lochside is being evaluated for a number of possible activities. The site-type is described in terms of the dominant slope category and dominant vegetation form. Since participants of all activities have to pass through the backshore, this forms the basis of the classification with shore features forming the second stage of the classification.

(b) General assessment of suitability of site-types. For each site-type, information is available relating to the dominant vegetation form, slope category and any shore features present (see Chapter 4). This is used to give a preliminary suitability ranking, in order to eliminate sites which do not warrant further consideration.

(c) Detailed assessment of suitability of site-types. The remaining sites, which are potentially suitable for a particular activity, are then assessed in more detail using field information which is analysed using the suitability matrices. Each site is finally allocated to a Suitability Class.

1.2 Use of the method

It is envisaged that the method may be used in a number of situations:

(a) in assessing the suitability of a particular site for a specific activity;

(b) in determining the most suitable site for a particular activity;

(c) in assessing which activities a lochside is most suited for when developing a previously unused loch for recreation;

(d) in assessing which activities a lochside is most suited for when there is a situation of heavy and conflicting use.

2. Division of the Lochside into Site-Types

This constitutes the reconnaissance stage where the whole lochside is inspected on the ground and, if necessary, viewed from the water and simultaneously divided into site-types using visually-detectable characteristics of vegetation-form and land-form.

2.1 Explanation of standardised site-types

Site-types (see Fig. 21 and Glossary) are defined in terms of the assemblage of vegetation-form and land-form characteristics present on the lochside.

Types 1 - 6. The number of the site-type is determined by the combination of slope category and dominant vegetation form of the primary facet. Shore features such as rocky shore, wetland fringe, beach, cliff, bank, strip, and embankment, which may or may not be present, form sub-classes within the major site-type, for example 4c is a steep, wooded slope extending to the water's edge into the water; BC 4c is a steep wooded slope with a beach and cliff present. The assemblage of characteristics refers to the combination occurring along a line extending in a perpendicular direction from the water's edge to the landward limit of the lochside, in other words features occurring along a profile of the lochside.

Types 7 - 10. These are based on slightly different characteristics than 1 - 6.

Type 7   is characterised by strongly undulating terrain.

Type 8   commonly occurs where the road is close to the shore and is raised slightly above it. The two characteristic features of this site-type are the steep slope, often an artificial embankment, and the narrow nature of the lochside, usually less than 20m.

Type 9   is similar to Type 8 except that the steep slope is of greater height, making access to the shore more difficult. However, if the height is greater than 5m, the site belongs to Type 11.

Type 10   can be described as a rocky promontory in which more than 50 per cent of the surface is outcropping rock.

Type 11   consists of a steep to extremely steep slope, which is dominantly rock outcrop, and is greater than 5m in height.

2.2 Procedure

Equipment required includes a clinometer, 1:10,000 or six-inch to the mile base map of the loch and a field-notebook. The standard procedure is to walk the length of the lochside recording the vegetation and land-form characteristics in the form of annotated profiles in a notebook. The profile extends from the water's edge to the landward limit of the lochside. The definition of lochside given in the Glossary is intended to be used in a flexible manner, for example, in the case of many upland reservoirs the fence surrounding the loch may be taken as the landward limit of the lochside. When a significant change occurs in the assemblage of profile features, a new profile, to represent a new site-type, is drawn and the boundary between the two marked on the base map (preferably at a scale of 1:10,560). The annotated profile consists of a profile of the landform, showing the slope category, dominant vegetation form and such features on rocky shore, wetland fringe, beach, cliff, bank, strip and embankment if present.

The slope category is measured using the clinometer; with experience it can be judged by eye. Another useful feature which was noted is the location of a trash-line which indicates the extent of flooding. The annotated profile is then designated to a site-type by comparison with the types outlined in Figure 28. This may be carried out in the field or, if all the relevant information is contained on the annotated profile, it may equally well be a desk operation.

Figure 42 gives examples of annotated profiles and the site-types into which they fit and Figure 43 shows this information after it has been transferred onto the base map.

2.3 Complex profiles

Where the lochside is wide, there may be a number of subsidiary site-types behind primary site-type. These are defined when the slope category or dominant vegetation form changes in a direction perpendicular to the shoreline (see Figs. 42 and 43).

162

a. BASIC PROFILE WITH ONE SITE-TYPE AND NO SHORE FEATURES

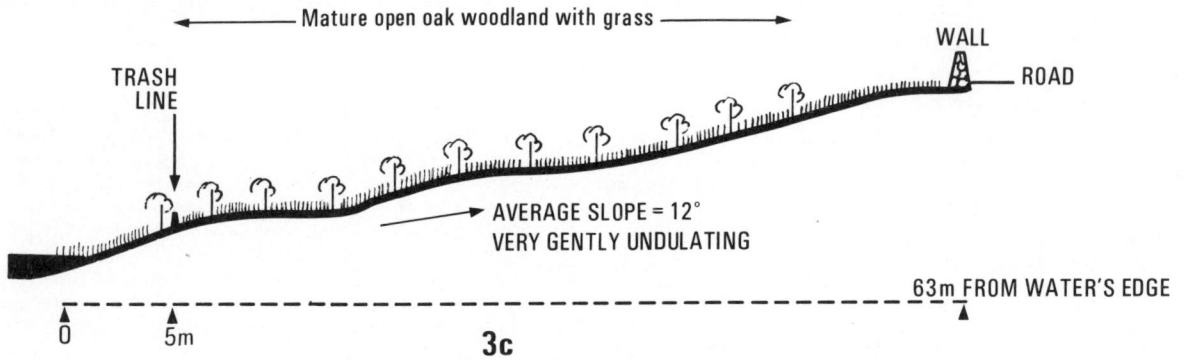

b. PROFILE WITH SHORE FEATURES

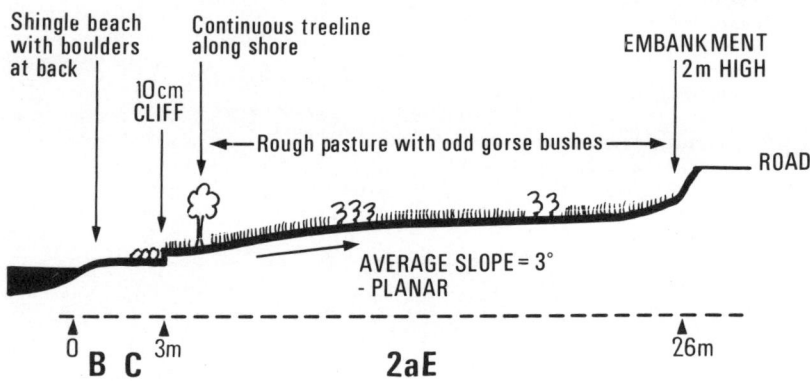

c. COMPLEX PROFILE WITH SUBSIDIARY SITE-TYPES

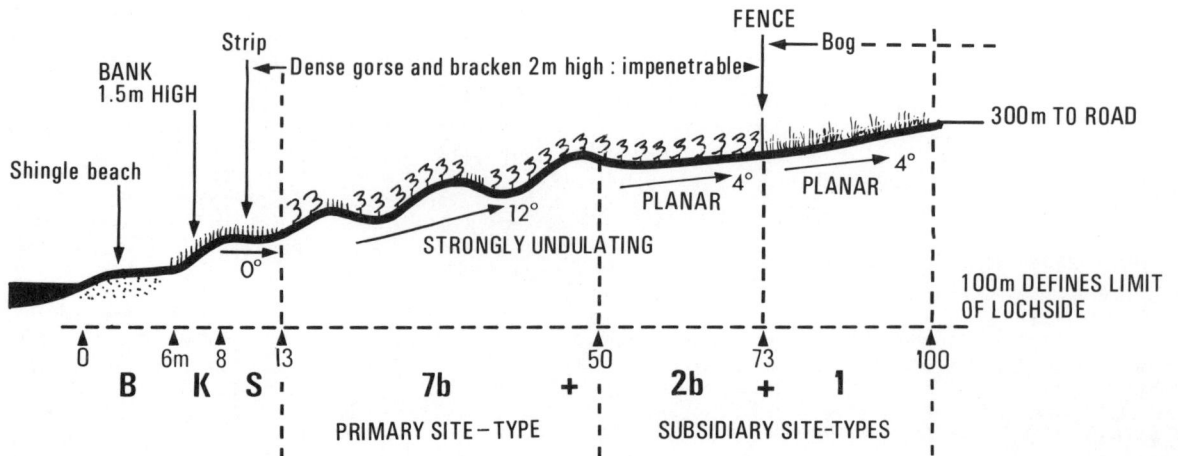

Fig. 42 Examples of annotated lochside profiles

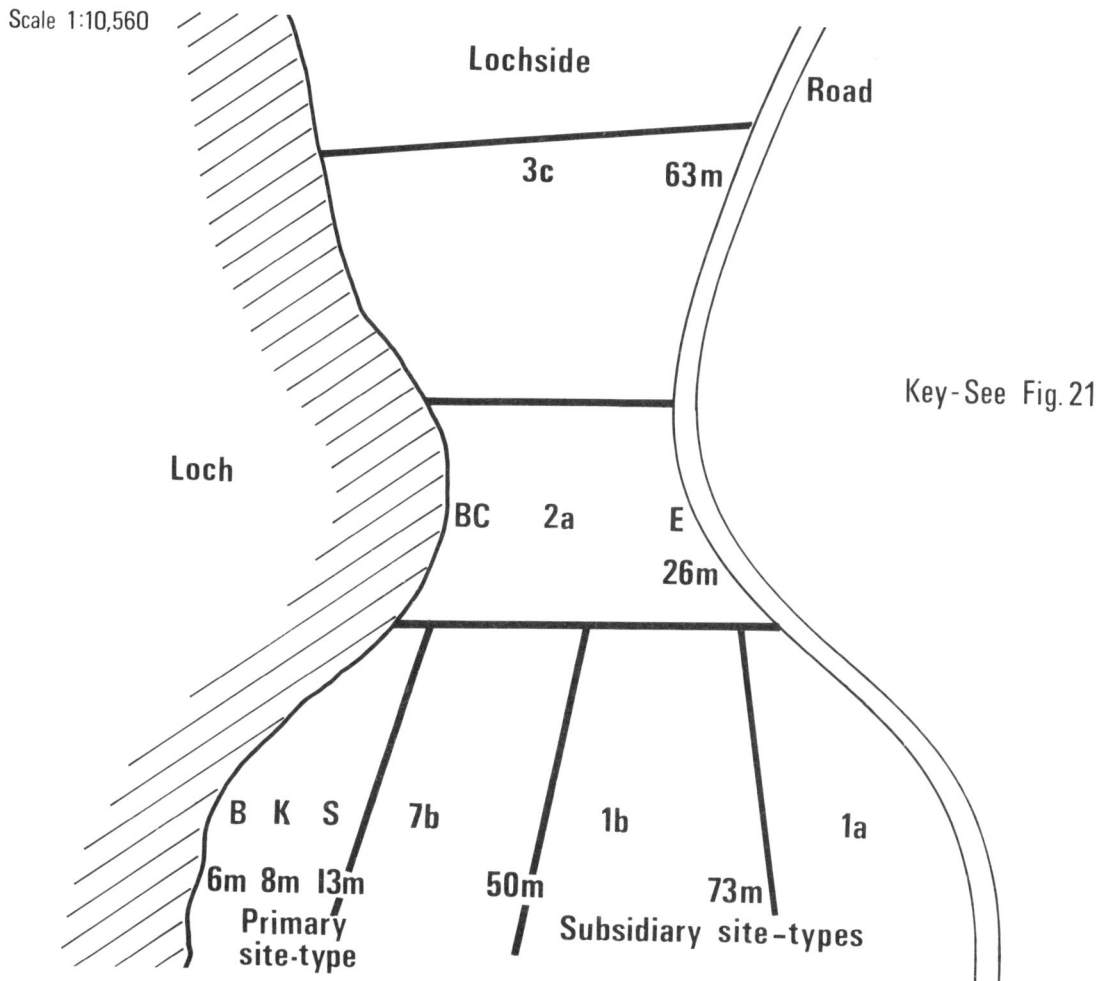

Fig. 43   Base map with data from annotated profiles

2.4   Significance of features in relation to their location on the profile

The significance of land-form and vegetation characteristics decreases the further one moves away from the shore. For example, a bank situated by the water's edge would form an obstacle to launching boats, but if situated within a site-type of gentle slope, 50m from the shore, its significance is diminished. Similarly, a narrow belt of dense scrub near the shore would provide a major limitation for shore-based activities, but would not be significant if located near the landward limit of the lochside.

2.5   Use of boat and aerial photographs

In some cases, it may be quicker to carry out the reconnaissance from a boat, particularly where there is a high proportion of marshy ground or dense woodland on the lochside. However, it is advised that the lochside is viewed from the land until some experience has been gained. Aerial photographs at a scale of 1:10,000 or larger may help in locating the boundaries of site-types on the map.

2.6   Lochside type description

This is arrived at by measuring, on the six-inch base map, the length of shoreline occupied by each of the primary site-types (using a graduated length of non-stretching thread, which is more accurate than a wheel for indented shorelines). The length of shoreline occupied by each site-type is summed and expressed as a percentage of total shorelength. The percentage of shoreline occupied by the dominant site-type defines the loch-type. The loch is then fitted into one of the following categories:

0 - 20% shoreline occupied by dominant site-type

21 - 40% shoreline occupied by dominant site-type

41 - 60% shoreline occupied by dominant site-type

61 - 80% shoreline occupied by dominant site-type

81 - 100% shoreline occupied by dominant site-type

This may be calculated either for the whole lochside, if it is within 100m of the water's edge, or, if only part of the lochside is within this distance, for the proximal parts only.

3. General Assessment of Suitability

For this purpose activities fall into three groups:

(a) those which extend from the road onto the backshore only such as camping, caravanning, car-based activities, picnicking, walking, orienteering, nature study, nature trails, pony trekking;

(b) those which extend across the backshore to use the shore such as bank-fishing, picnicking, walking;

(c) those which extend across the backshore and shore to use the water body (see Table 59) such as all boat-based activities, sub-aqua and swimming.

These categories are not mutually exclusive as, for instance, picnicking which may take place on the backshore or the shore. For activities in Group (a), the backshore is assessed for its suitability for the particular activity. For Group (b), the shore is assessed for its suitability for the activity, the backshore being important only with regard to access from the road to the shore. For Group (c), the water body is assessed for its suitability for the activity, and the shore and backshore are important with respect to access to the water body. In the case of boating activities, the shore access influences the method of boat-launching.

Three classes of suitability are recognised at this stage:

A the lochside component is highly suitable for the activity in its present state;

B the lochside component is moderately suitable for the activity in its present state;

C the lochside component is unsuitable for the activity in its present state.

The procedure for the suitability assessment varies according to the type of activity being considered.

3.1 Land-based activities

The suitability of the relevant components of the lochside for the particular activity is determined from Table 60. As well as the suitability of the land, it is necessary to consider whether the area of lochside is sufficient to meet the minimum length and area requirements for the activity (see Table 44).

3.2 Activities dependent on a biotic resource

This group includes nature study, bank-fishing, wildfowling and boat-fishing. The suitability assessment involves two stages:

(a) an evaluation of the relevant wildlife base (using local knowledge, and consulting organisations such as the Nature Conservancy Council, and the Royal Society for the Protection of Birds);

(b) an assessment, from Table 60, of the suitability of the relevant lochside components for access and pursuit of the activity.

3.3 Water-based Activities

This group includes rowing, canoeing, sailing, power-boating,

Table 59

Matrix of lochside activities and their use of shore components

Activity	Water body used for pursuit of activity	Shore used for access	Shore used for pursuit or activity	Backshore used for access	Backshore used for pursuit of activity
Camping	—	—	—	—	X
Caravanning	—	—	—	—	X
Car parking	—	—	—	—	X
Orienteering	—	—	—	—	X
Pony trekking	—	—	—	—	X
Nature trail	—	—	—	—	X
Nature study	—	—	—	—	X
Picnicking	—	—	E	X	E
Walking	—	—	E	X	E
Bank-fishing	B	—	B	X	—
Swimming	B	—	B	X	—
Sub-aqua	X	X	—	X	—
Boat-based activities	X	X	—	X	—

Camping through Picnicking: Group (a)
Walking through Swimming: Group (b)
Sub-aqua through Boat-based activities: Group (c)

— component not used for the activity
X component is used for the activity
E either component may be used
B both components are required
Group (a) Activities based on the backshore
Group (b) Activities based on the shore, using the backshore for access
Group (c) Activities based on the water body, using the backshore and shore for access

Table 60

General assessment of suitability of lochside components for recreational activities

USE of SHORE COMPONENT / ACTIVITY		OFFSHORE GRADIENT FOR ACCESS INTO WATER BODY	SHORE FOR ACCESS		SHORE FOR PURSUIT OF ACTIVITY		BACKSHORE FOR ACCESS		BACKSHORE FOR PURSUIT OF ACTIVITY
CAMPING •, CARAVANNING •, CAR-BASED PICNIC ON BACKSHORE (minimum area 9m²/pitch)	A								2a 2c 3a 3c
	B								4a 4c 7a 7c
	C								1 2b 3b 4b 5a 5b 5c 6a 6b 6c 7b 8 9 10 11
WALKING • } BACK PONY TREKKING } SHORE, NATURE TRAIL, NATURE STUDY • (minimum length ½ mile)	A								2a 2c 3a 3c 4a 4c 7a 7c
	B								5a 5c 6a 6c
	C								1 2b 3b 4b 5b 6b 8 9 10 11
ORIENTEERING	A								2b 2c 3b 3c 4b 4c 5b 5c 6b 6c 7b 7c
	B								7a
	C								2a 3a 4a 5a 6a 8 9 10 11
WALKING • } SHORE PONY TREKKING • } (minimum length ½ mile)	A				B Sa Sc		2a 2c 3a 3c 4a 4c 5a 5c		
	B				K		6a 6c 7a 7c		
	C				WF R C Sb		1 2b 3b 4b 5b 6b		
PICNICKING SHORE	A				Sa Sc R B K C				
	B						,,		
	C				WF Sb				
BANK FISHING	A	moderate or steeper off-shore gradient			WF C K R				
	B				Sa Sb Sc B		,,		
	C								
SWIMMING, SUB-AQUA (without boat)	A	shallow, non-shelving off-shore gradient	R B						
	B		C K Sa Sc				,,		
	C		WF Sb						
BOAT LAUNCHING	A	shallow or moderate off-shore gradient which shelves	B						
	B		K C Sa Sc				,,		
	C		WF R Sb						

KEY : A Lochside component is HIGHLY SUITABLE, in its present state, for the activity.
　　　 B　　　　,,　　　　MODERATELY   ,,　　　　,,　　　　,,
　　　 C　　　　,,　　　　UNSUITABLE　　　,,　　　　,,

water-skiing, sub-aqua, and boat-fishing; all of which may be carried out on an informal or formal basis. The suitability of a loch for these activities has three components:

(a) the suitability of the water-body itself;

(b) the suitability of the shore for boat-launching;

(c) the suitability of the lochside for car-parking and access to the shore.

The first stage is to determine whether or not the water-body satisfies the minimum threshold conditions, such as area and depth, for the particular activity(ies). This is achieved using the table of user requirements for water-based activities (see Tables 61 a and b) along with information on the physical character of the water-body and the surrounding landscape (to be acquired from the data bank, if available, or by observation or from local knowledge). It may be necessary to divide the loch into a number of parts if the physical conditions vary; for example, in many

Table 61a

User requirements for water-based recreational activities — canoeing and sailing

ACTIVITY USER REQUIREMENT	CANOEING			SAILING			
	casual	long distance	sprint	casual dinghy	compet. dinghy	casual keel	compet. keel
AREA		★TO ACCOMODATE CIRCUIT OF 6×154m	★ 1000 x 81m	★ > 10a FOR SMALL DINGHYS	★ > 20a	★ > 20a	★ > 20a
SHAPE		★TO ACCOMODATE CIRCUIT OF 6×154m	★STRAIGHT RUN	NOT NARROW	NOT NARROW	NOT NARROW	NOT NARROW
DEPTH	★ > 25cm	★ > 3m	★ > 3m	★ > 3.5m OR MAST HEIGHT	★ > 3.5m OR MAST HEIGHT	★ > MAST HEIGHT	★ > MAST HEIGHT
OBSTACLES	NONE UNMARKED	NONE UNMARKED	NONE UNMARKED	NONE UNMARKED	NONE UNMARKED	NONE UNMARKED	NONE UNMARKED
WATER CURRENTS	ABSENCE OF STRONG CURRENTS	NONE	NONE	NOT STRONG	NOT STRONG	NOT STRONG	NOT STRONG
WIND	⊁ FORCE 3	IF PRESENT HEAD OR FOLLOWING	IF PRESENT HEAD OR FOLLOWING	ACROSS LONG AXIS OF LOCH	ACROSS LONG AXIS OF LOCH.NOT GUSTY. SHELTERED AREA FOR MOORING	ACROSS LONG AXIS OF LOCH NOT GUSTY	ACROSS LONG AXIS OF LOCH. NOT GUSTY. SHELTERED AREA FOR MOORING
WATER POLLUTION	PREFERABLY UNPOLLUTED	PREFERABLY UNPOLLUTED	PREFERABLY UNPOLLUTED	PREFERABLY UNPOLLUTED	PREFERABLY UNPOLLUTED	PREFERABLY UNPOLLUTED	PREFERABLY UNPOLLUTED
WATER CLARITY							
WATER TEMP.	NOT FROZEN IN WINTER			NOT FROZEN IN WINTER		NOT FROZEN IN WINTER	
BOTTOM MATERIAL					NOT SOLID ROCK FOR MOORINGS		
NATURE OF SURR. LAND				OPEN TO PREVENT WIND FUNELLING	OPEN TO PREVENT WIND FUNELLING	OPEN TO PREVENT WIND FUNELLING	OPEN TO PREVENT WIND FUNELLING

KEY : ★ Essential condition which must be satisfied for the activity to be pursued.

  > IS greater than

  ⊁ NOT greater than

cases there is an area of very shallow water at the head of the loch which would not satisfy the depth threshold. Some lochs may be completely eliminated at this stage. If the loch is found to satisfy the threshold conditions the next stage is to assess the relative suitability of the shore features of the site-type for boat-launching (see Table 60). The third consideration is that of physical access across the backshore (see Table 60).

4. Detailed Assessment of Suitability

This method is applicable when detailed evaluation is required to assess the suitability of a particular site for one or more activities. The procedure is:

(a) to determine which categories of field data are required for the particular activity using the matrix of activities against field data (see Table 62);

(b) to collect the field data using the instructions given below;

(c) to determine the suitability class of the site using the suitability matrix for the activity concerned.

4.1 Collection of field data

Collection of field data is necessary for detailed suitability ranking of the lochside components for activities. The data is collected along three transects for each component, extending from the water's edge to the landward limit of the lochside. If there are subsidiary site-types, which are not being considered for the activity, they must be assessed for access across the backshore. Transects should be spaced at 20m intervals, and be located away from the site boundaries, preferably near the centre of the site (see Fig. 44).

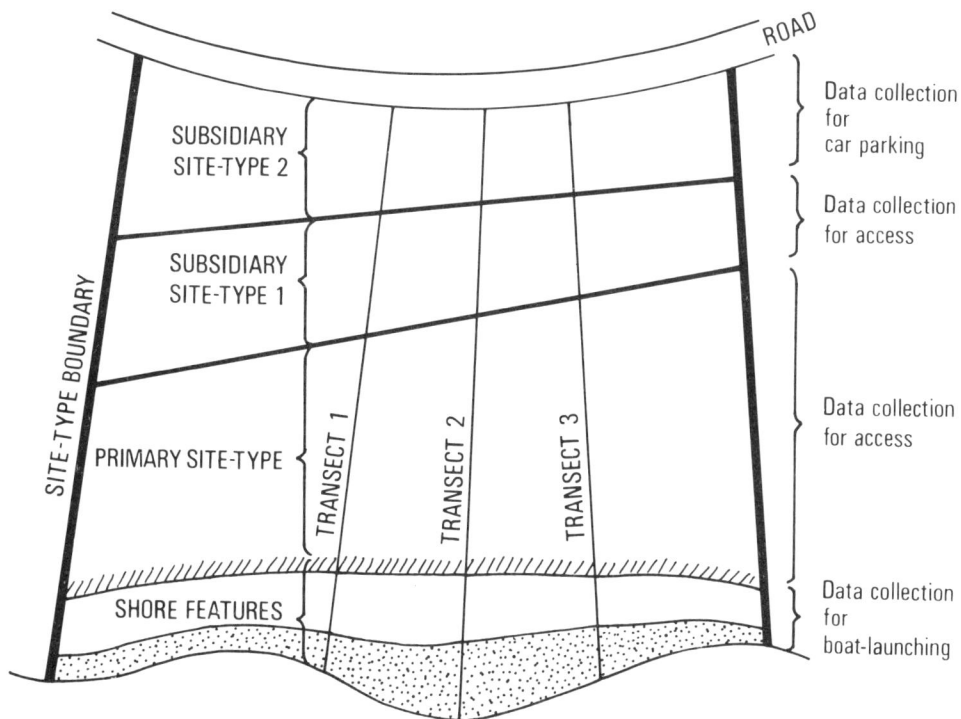

Fig. 44   Procedure for field data collection

4.2   *General procedure*

Equipment includes a clinometer, rule, metre-tape, Field Handbook (Soil Survey, 1960), six-inch to the mile base map, and field notebook. Then:

(a) using Table 61 determine which data categories are required for the activity(ies) under consideration;

(b) consult the relevant capability evaluation sheet, Tables 63-69 at end of this appendix;

(c) collect data as described in Section 4.3.

4.3   *Procedure for site as a whole*

1. Length of site
   (a) Measure average distance between site boundaries by pacing.
   (b) Determine the length of pace using the surveyor's tape.

2. Width of site
   (a) Measure average width (distance between water's edge and limit of lochside) by pacing.

3. Aspect
   (a) Record as N/W/S/E/NW/SW/SE or NE.

4. Flood hazard
   (a) Record any evidence such as presence and location of a trash-line.

5. Shelter
   (a) Record whether the site is sheltered or exposed.

6. Obstacles in water
   (a) State the nature of obstacles such as boulders, vegetation.
   (b) Record density as sparse/moderately dense/very dense.

7. Obstacles to shore access
   (a) Measure maximum gradient occurring along the whole length of the site, between the primary site-type and the landward limit of the lochside.
   (b) Describe the most inaccessible vegetation form occurring along the whole length of the site between the primary site-type and the landward limit of the lochside as:
      (i) wetland;
      (ii) arable;
      (iii) short grass (less than 10cm high);
      (iv) tall forbs and grasses (10cm - 1m high);
      (v) moor grass;
      (vi) low shrubs (less than 1m high such as bracken and heather);
      (vii) shrubs/dense woodland, and
      (viii) open woodland.

8. Alongshore slope
   (a) Estimate average alongshore slope.
   (b) Measure maximum prolonged alongshore slope.

9. Presence of shore features
   Record wetland fringe, rocky shore, beach, cliff, bank, strip, embankment.

Table 61b
User requirements for water-based recreational activities – rowing, water-skiing, sub-aqua, swimming and power-boating

Activity / User Requirement	ROWING casual	ROWING compet.	WATER-SKIING casual	WATER-SKIING slalom	SUB-AQUA novice	SUB-AQUA experienced	Swimming	Power-Boating
AREA		★INTERNAT COMPET 2000×140m	★10a per BOAT	★>25a per BOAT				★MINIMUM 16a – casual 80a-for racing
SHAPE		★STRAIGHT RUN	NOT TOO INDENTED SHORELINE IF SMALL LOCH	NOT TOO INDENTED SHORELINE IF SMALL LOCH				
DEPTH	★>1m	★>3.25m	★>2m	★>2m	★<10m	★10–30m	★≯1.5m	★>2m
OBSTACLES	NONE UNMARKED	NONE UNMARKED	NONE UNMARKED	NONE UNMARKED	NOT WEEDS	NOT WEEDS	NOT WEEDS	NONE UNMARKED
WATER CURRENTS	NOT STRONG	NONE			NOT STRONG	<1 KNOT		
WIND	≯FORCE 3	IDEALLY NONE, HEAD OR FOLLOWING IF PRESENT	ABSENCE OF WAVES	ABSENCE OF WAVES	NOT STRONG	NOT STRONG	NOT STRONG	
WATER POLLUTION	PREFERABLY UNPOLLUTED	PREFERABLY UNPOLLUTED	MUST BE UNPOLLUTED	MUST BE UNPOLLUTED	MUST BE UNPOLLUTED	MUST BE UNPOLLUTED	MUST BE UNPOLLUTED	PREFERABLY UNPOLLUTED
WATER CLARITY					★CLEAR WATER ESSENTIAL	★CLEAR WATER ESSENTIAL		
WATER TEMP.			NOT COLD	NOT COLD	NOT COLD	NOT COLD	NOT COLD	
BOTTOM MATERIAL					MUST HAVE FEATURES OF INTEREST	ROUNDED SAND TO SMALL STONES		
NATURE OF SURR. LAND								

Key ★ Essential condition which must be satisfied for the activity to be pursued
> IS greater than
≯ NOT greater than

4.4 *Procedure for each traverse*
10. Offshore gradient
   (a) Measure water depth at 2m from water's edge.
   (b) Describe form of offshore slope as planar, steeply shelving etc.
11. Beach width
   Measure by pacing along the traverse.
12. Beach gradient
   Measure along traverse, using clinometer.
13. Beach material
   (a) Record as percentages of:
      (i) sand (less than 2mm diameter);
      (ii) stones (see category 25 for size classes);
      (iii) boulders (over 60cm), and
      (iv) rock *in situ*.
   (b) Record shape of material as rounded, angular or sub-angular.
14. Beach drainage
   (a) Record drainage as free, imperfect, poor, waterlogged.

4.5 *Procedure for each site-type*
15. Width
   (a) Measure by pacing.
16. Gradient
   (a) Measure with clinometer.
17. Macro-relief
   (a) Describe as planar, undulating, terraced, hummocky etc.

18. Micro-relief
    (a) This refers to minor changes in relief, where the vertical height variation (ΔH) is less than 1m. Describe the form of the relief as ridges, hummocky etc.
    (b) Estimate ΔH.

19. Vegetation form
    (a) Describe as in 7(b).
    (b) Record average tree-spacing in metres.

20. Unvegetated ground
    (a) Estimate per cent ground without vegetation (but excluding beach already covered in 9).
    (b) Describe nature of the ground surface as rock *in situ*, boulders, stones, bare soil etc.

4.6 *Data from soil pits*

   (a) Dug to at least spade-depth, 25cm, and staggered as shown in Figure 43.

21. Depth organic layer
    (a) Record depth of organic layer if present.

22. Soil texture
    (a) Describe the nature of the mineral horizon in following terms: sand; loam; silt; loamy sand; sandy loam; clay; clay loam; sandy clay; silty loam, and other combinations.

23. Soil drainage
    (a) Describe as:
        (i) excessive (loose, powdery soil);
        (ii) free (no mottling in top 50 cm);
        (iii) imperfect (some mottling along root channels);
        (iv) poor (grey colour due to gleying), and
        (v) very poor (grey, with highly humose surface layer, for example, peat).

24. Soil depth
    (a) Record as:
        (i) less than 5cm;
        (ii) 6 - 10cm;
        (iii) 11 - 15cm;
        (iv) 16 - 20cm;
        (v) 21 - 25cm, and
        (vi) over 26cm

25. Percentage stones in soil
    (a) Record as:
        (i) 0 - 10%;
        (ii) 11 - 20%;
        (iii) 21 - 30%;
        (iv) 31 - 50%, and
        (v) over 51%.

26. Stone size
    (a) Record as:
        (i) very small (2 - 6mm mean diameter);
        (ii) small (6mm - 2cm mean diameter);
        (iii) medium (2 - 6cm mean diameter);
        (iv) large (6 - 20cm mean diameter);
        (v) very large (20 - 60cm mean diameter).

5. Suitability Ranking

The objective is to rank sites according to their relative suitability for a particular activity. The method used here is based largely on the physical limitations of the site, together with a consideration of enhancing factors where relevant for example a very sheltered site for camping and/or caravanning. Each site is allocated a score expressing the extent of limiting factors plus the extent of enhancing factors. A management input may considerably lessen the extent of the limitations on a site, for example drainage and artificial surfacing would make a wet, hummocky site more suitable for car parking. Since the level of management input varies considerably, the suitability rankings are based on the present condition of the site; management is only taken into account when assessing the value of the multiplier, which is discussed below.

5.1 *Procedure for ranking*

1. Determine the average value of each data category for each lochside component being considered on the site.

2. Using the relevant suitability evaluation table (see Tables 63 - 73), scores are allocated to the data categories; these are then multiplied by the value of the multiplier (which reflects the relative importance of data categories for the particular activity). It may be necessary to alter the values of multipliers if there is to be a high management input, for example if the site is to be drained, it would mean that soil drainage could be excluded from the ranking.

3. The scores are summed for all categories and the suitability class to which the site belongs is determined using Table 74.

When interpreting the results, it must be remembered that the suitability class represents the relative degree of limitation of a site for a particular activity; there is no direct comparison between the suitability classes for different activities.

Table 62

Matrix of recreational activities v. field data categories

Data category / activity	1	2	3	4	5	6	7	8	9	10	11	12	13	14	15	16	17	18	19	20	21	22	23	24	25	26
Camping and caravanning			✓	✓	✓		✓								✓	✓	✓	✓	✓	✓	✓	✓	✓	✓	✓	✓
Car parking				✓			✓								✓	✓	✓	✓	✓	✓	✓	✓				
Picnicking (a) Backshore			✓	✓	✓										✓	✓	✓	✓	✓	✓	✓	✓	✓			
(b) Shore				✓					✓				✓	✓												
Walking (a) Backshore	✓			✓				✓							✓			✓	✓			✓	✓			
(b) Shore				✓					✓				✓	✓												
Pony trekking (a) Backshore	✓			✓				✓										✓			✓	✓		✓	✓	✓
(b) Shore				✓					✓				✓	✓												
Orienteering	✓		✓			✓	✓		✓			✓	✓			✓	✓		✓			✓	✓			
Nature study			✓	✓		✓			✓	✓	✓	✓	✓	✓		✓		✓	✓			✓	✓			
Bank-fishing		✓																				✓				
Boat-launching		✓		✓			✓									✓			✓							
Access across backshore																						✓	✓			
Swimming		✓	✓		✓	✓			✓	✓	✓		✓	✓	✓											

Table 63

Suitability evaluation for camping and caravanning

Data category		+1	0	−1	−2	−3	−4	−5	Multiplier
3	Aspect		S/SW/SE	W/E		NW/NE		N	3
4	Flood hazard		Never	Occasional			Frequent		4
5	Shelter	Very sheltered						Very exposed	1
7	Max. gradient		0 - 10°	11 - 15°	16 - 20°	21 - 25°	26 - 30°	>31°	1
7	Inhib. vegn.		Short grass/ Woodland	Tall grass/Moor grass		Low shrub	Arable	Wetland/ Shrub	1
15	Width facet		>31m	26 - 30m	21 - 25m	16 - 20m	11 - 15m	<10m	3
16	Gradient		0 - 4°	5 - 6°	7 - 8°	9 - 10°	11 - 12°	>13°	4
17	Macrorelief		Planar	Gently undulating		Undulating		Strongly und./ narrow terraced	1
18	Microrelief		ΔH=0 - 5cm	6 - 10cm	11 - 15cm	16 - 20cm	21 - 25cm	>26cm	3
19	Veg. form		Short grass	Woodland/Tall grass Moor grass		Low shrub		Arable/Shrub/ Wetland	3
19	Treespacing (av)		>11m	8 - 10m	5 - 8m	4 - 6m	2 - 4m	0 - 1m	1
20	% Rock/stones		0 - 10%	11 - 15%	16 - 20%	21 - 25%	26 - 30%	>31%	2
21	Depth org. layer		1 - 4cm	5 - 6cm		7 - 8cm		>9cm <1cm	1
22	Soil texture		L/SCL/SL	CL/SCL/SiCL/LS/S (not loose)		SC/SiC/Loose - S		O/C*	2
23	Soil drainage		Excessive/Free	Imperfect		Poor		Very poor	4
24	Soil depth		>26cm	21 - 25cm	16 - 20cm	11 - 15cm	6 - 10cm	<5cm	2
25	% Stones		0- 10%	11 - 20%	21 - 30%	31 - 40%	41 - 50	>51%	½
26	Size stones		Very small	Small		Medium		Large	½

* L = loam; Si = silt; C = clay; S = sand; O = organic material

L

Table 64

Suitability evaluation for car parking

Data category	Score +1	0	−1	−2	−3	−4	−5	Multi-plier
4 Flood hazard		Never	Occasional			Frequent		3
7 Max. gradient		0 - 10°	11 - 15°	16 - 20°	21 - 25°	26 - 30°	>31°	1
7 Inhib. vegn.		Short grass/ woodland	Tall grass/Moor grass		Low shrub	Arable	Wetland/ shrub	1
15 Width site-type		>21m	17 - 20m	13 - 16m	9 - 12m	5 - 8m	<4m	2
16 Gradient		0 - 4°	5 - 6°	7 - 8°	9 - 10°	11 - 12°	>13°	3
17 Macrorelief		Planar	Gently undulating		Undulating		Strongly undulating	2
18 Microrelief		0 - 5cm	6 - 10cm	11 - 15cm	16 - 20cm	21 - 25cm	>26cm	1
19 Veg. form		Short grass/ Woodland	Tall grass/Moor grass		Low shrub		Arable/Shrub/ Wetland	2
19 Treespacing (av)		>11m	8 - 10m	3 - 5m	5 - 6m	2 - 4m	0 - 1m	1
20 Rock/boulders %		0 - 5%	6 - 10%	11 - 15%	16 - 20%	21 - 25%	>26%	1
22 Soil texture		L/Si, L/Si	CL/SCL/SiCL/LS/S (not loose)		SC/SLC/Loose - S		O/C*	1
23 Soil drainage		Excessive/Free	Imperfect		Poor		Very poor	1

* See **Table 63** for Soil Abbreviations

Table 65a

Suitability evaluation for picnicking on backshore

Data category \ Score	+1	0	−1	−2	−3	−4	−5	Multiplier
3 Aspect		S/SW/SE	W/E		NW/NE		N	2
4 Flood hazard		Never	Occasional			Frequent		3
5 Shelter	Very sheltered						Very exposed	1
15 Width facet		>21m	17 - 20m	13 - 16m	9 - 12m	5 - 8m	<4m	2
16 Gradient		0 - 4°	5 - 6°	7 - 8°	9 - 10°	11 - 12°	>13°	3
17 Macrorelief		Planar	Gently undulating		Undulating		Strongly undulating	2
18 Microrelief		0 - 5cm	6 - 10cm	11 - 15cm	16 - 20cm	21 - 25cm	>26cm	1
19 Veg. form		Short grass/ Woodland	Tall grass/Moor grass		Low shrub		Arable/Shrub/ Wetland	2
19 Treespacing (av)		>11m	9 - 10m	7 - 8m	5 - 6m	2 - 4m	0 - 1m	1
20 Rock/boulders %		0 - 10%	10 - 20%	21 - 25%	26 - 30%	31 - 35%	>36%	1
22 Soil texture		L/SiL/Si	CL/SCL/SiCL/LS/Si (not loose)		SC/SiC/Loose - S		O/C*	1
23 Soil drainage		Excessive/Free	Imperfect		Poor		Very poor	1

Table 65b

Suitability evaluation for picnicking on shore

4 Flood hazard		Never	Occasional		Frequent		2
×9 Shore features	Beach/Cliff	Rocky shore/ Strip		Bank		Embankment/ Wet. fringe	10
13 Beach material		Sand-Medium stones	Medium – Large stones		Boulders	Clay/Mud	2
14 Beach drainage		Free	Imperfect		Poor	Waterlogged	2

× If shore feature is bank or strip assess as for picnicking on backshore
* See **Table 63** for Soil Abbreviations

Table 66a

Suitability evaluation for walking on backshore

Data category	Score +1	0	−1	−2	−3	−4	−5	Multiplier
1 Length site		>51m	41 - 50m	31 - 40m	21 - 30m	11 - 20m	0 - 10m	1
4 Flood hazard		Never	Occasional			Frequent		1
8 A'shore slope (av)		0 - 5°	6 - 8°	9 - 11°	12 - 14°	15 - 17°	>18°	2
8 A'shore slope (max)		0 - 5°	6 - 10°	11 - 15°	16 - 20°	21 - 25°	>26°	1
18 Microrelief		0 - 5cm	6 - 15cm	16 - 25cm	26 - 35cm	36 - 45cm	>4cm	1
19 Veg. form		Short grass/ Woodland	Tall grass/Moor grass		Low shrub		Arable/Shrub/ Wetland	1
19 Treespacing (av)		>7m	5 - 6m	4 - 5m	3 - 4m	2 - 3m	0 - 1m	1
20 % Boulders		0 - 10%	11 - 20%	21 - 30%	31 - 40%	41 - 50%	>51%	1
22 Soil texture		L/SiL/Si	CL/SCL/SiCL/LS/S (not loose)		SC/SiC/Loose - S		O/C*	1
23 Soil drainage		Excessive/Free	Imperfect		Poor		Very poor	2

Table 66b

Suitability evaluation for walking on shore

Data category	Score +1	0	−1	−2	−3	−4	−5	Multiplier
4 Flood hazard		Never	Occasional			Frequent		2
X9 Shore features	Beach Strip	Cliff				Bank	Embankment	10
13 Beach material		Sand-Medium stones	Medium – Large stones		Boulders		Clay/Mud	2
14 Beach drainage		Free	Imperfect			Poor	Waterlogged	2

X If shore feature is bank/strip access as for walking on backshore
* See **Table 63** for Soil Abbreviations

Method for evaluation of the suitability of lochsides for recreational activities

Table 67a

Suitability evaluation for pony trekking on backshore

Data category	+1	0	−1	−2	−3	−4	−5	Multiplier
1 Length site		>51m	41 - 50m	31 - 40m	21 - 30m	11 - 20m	0 - 10m	1
4 Flood hazard		Never	Occasional			Frequent		2
8 A'shore slope (av)		0 - 4°	5 - 10°		11 - 18°		>19°	2
8 A'shore slope (max)		0 - 10°	11 - 20°		21 - 30°		>31°	1
18 Microrelief		0 - 5cm	6 - 10cm	11 - 15cm	16 - 20cm	21 - 25cm	>26cm	1
19 Veg. form		Short grass/ Woodland	Tall grass/Moor grass		Low shrub		Arable/Shrub/ Wetland	2
19 Treespacing (av)		>7m	5 - 6m	4 - 5m	3 - 4m	2 - 3m	0 - 1m	1
20 % Rock/Boulders/Stones 20mm		0 - 5%	6 - 10%	11 - 15%	16 - 20%	21 - 25%	>26%	1
21 Depth org. layer		0 - 5cm	6 - 10cm		10 - 15cm		>16cm	1
22 Soil texture		L/SCL/Si	CL/SCL/SiCL/LS/S (not loose)		SC/Si/C Loose - S		O/C*	2
24 Soil drainage		Excessive/Free	Imperfect		Poor		Very poor	2
25 Soil stones: %		0 - 5%	6 - 10%	11 - 15%	16 - 20%	21 - 25%	>26%	1
26 Soil stones: size		Very small	Small		Medium		Large	1
x Track ashore	Present							3

Table 67b

Suitability evaluation for pony trekking on shore

Data category	+1	0	−1	−2	−3	−4	−5	Multiplier
4 Flood hazard		Never	Occasional			Frequent		2
ˣ9 Shore features	Beach/Strip	Cliff				Bank	Embankment/ Wet. fringe	10
13 Beach material		Sand-medium rounded stones	Medium – Large rounded stones		Boulders/Angular stones		Clay/Mud	2
14 Beach drainage		Free	Imperfect			Poor	Waterlogged	2

ˣ If shore feature is a bank/strip assess as for backshore
* See **Table 63** for Soil Abbreviations

Table 68

Suitability evaluation for orienteering

Data category \ Score	+1	0	−1	−2	−3	−4	−5	Multi-plier
1 Length site		>51m	41 - 50m	31 - 40m	21 - 30m	11 - 20m	0 - 10m	1
15 Width facet		>51m	41 - 50m	31 - 40m	21 - 30m	11 - 20m	0 - 10m	1
16 Gradient		10 - 20°	21 - 30°	31 - 40°	41 - 50°	51 - 60°	>61° <9°	1
17 Macrorelief		Strongly undulating	Undulating	Gently undulating	Planar			2
19 Veg. form		Shrub	Woodland	Low shrub	Tall grass/Moor grass		Short grass/ Arable/Wetland	2
19 Treespacing (av)		2 - 3m	3 - 4m	4 - 5m	5 - 6m	6 - 8m	<1m >9m	1
22 Soil texture		L/SiL/SL	CL/SCL/SiCL/LS/S/SC/SiC				C/O*	1
23 Soil drainage		Excessive/Free	Imperfect		Poor		Very poor	1

* See **Table 63** for Soil Abbreviations

Table 69

Suitability evaluation for nature study

Data category \ Score	+1	0	−1	−2	−3	−4	−5	Multi-plier
4 Flood hazard		Never	Occasional			Frequent		1
7 Max. gradient		0 - 10°	11 - 15°	16 - 20°	21 - 25°	26 - 30°	>31°	1
7 Inhib. vegn.		Short grass/ Woodland	Tall grass/Moor grass		Low Shrub/Arable		Wetland/ Shrub	1
16 Gradient		0 - 7°	8 - 14°	15 - 20°	21 - 27°	28 - 33°	>34°	1
18 Microrelief		0 - 10cm	11 - 20cm	21 - 30cm	31 - 40cm	41 - 50cm	>51cm	1
19 Veg. form		Rest					Shrub	1
19 Treespacing (av)		>5m	2 - 4m		1 - 2m		<1m	1
22 Soil texture		L/SiL/Si	CL/SCL/SiCL/LS/S/SC				C/O*	1
23 Soil drainage		Excessive/Free	Imperfect		Poor		Very poor	1

* See **Table 63** for Soil Abbreviations

Table 70

Suitability evaluation for bank-fishing

Data category	Score +1	0	−1	−2	−3	−4	−5	Multiplier
2 Width site		>6m	4 - 5m	3 - 4m	2 - 3m	1 - 2m	<1m	1
4 Flood hazard		Never	Occasional			Frequent		1
6 Type obstacles			Boulders		Reeds			1
6 Density obstacles		None	Moderately dense			Very dense		1
9 Shore features	Cliff/Bank/ Rocky shore	Beach/ Embankment	Strip	Wetland fringe				5
12 Beach mat.		Sand/Gravel/ Stones<20cm	Stones > 20cm → Boulders			Mud	Clay	1
13 Beach drainage		Excessive/Free	Imperfect			Poor	Waterlogged	1
22 Soil drainage		Excessive/Free	Imperfect			Poor	Waterlogged	1

Table 71

Suitability evaluation for boat-launching (informal)

Data category	Score +1	0	−1	−2	−3	−4	−5	Multiplier
4 Flood hazard		Never	Occasional			Frequent		1
6 Type obstacles		None	Reeds		Boulders		Rock *in situ*	1
6 Density obstacles		None	Sparse		Moderately dense		Dense	1
9 Shore features	Beach	Strip				Bank/Cliff	Embankment/ Rocky shore	5
10 O.S.G. depth at 2m	17 - 35cm		36 - 99cm	8 - 16cm	0 - 8cm	>99cm		1
10 O.S.G. form		Planar/ Mod. shelving	Steeply shelving				V. steeply shelving	1
11 Beach width		>6m	3 - 5m		<2m			1
12 Beach gradient		0 - 5°	6 - 10°		11 - 15°		>16°	1
13 Beach material		Sand-rounded med. stones	Rounded med. – Lge. stones		Boulders/Angular stones		Clay/Mud	2
14 Beach drainage		Free	Imperfect			Poor	Waterlogged	1

Table 72

Suitability evaluation for access across backshore

Data category		Score +1	0	−1	−2	−3	−4	−5	Multiplier
2	Site width		0 - 10m	11 - 20m	21 - 30m	31 - 40m	41 - 50m	>51m	1
7	Max. gradient		0 - 10°	11 - 15°	16 - 20°	21 - 25°	26 - 30°	>31°	1
7	Inhib. vegn.		Short grass/Woodland	Tall grass/Moor grass		Low shrub/Arable		Wetland/Shrub	1
16	Gradient		0 - 5°	6 - 10°	11 - 15°	16 - 20°	21 - 25°	>26°	2
19	Veg. form		Short grass/Woodland	Tall grass/Moor grass		Low shrub	Arable	Wetland/Shrub	2
19	(av) Treespacing		>6m	5m	4m	3m	1 - 2m	<1m	1
20	Stones/% bare earth		0 - 10%	11 - 20%	21 - 30%	31 - 40%	41 - 50%	>51%	1
22	Soil texture		L/SiL/Si	CL/SCL/SiCL/LS/S (not loose)		SC/SiC/loose S		C/O*	1
23	Soil drainage		Excessive/free	Imperfect		Poor		Very poor	2

* See **Table 63** for Soil Abbreviations

Table 73

Suitability evaluation for swimming

Data category		Score +1	0	−1	−2	−3	−4	−5	Multiplier
3	Aspect		S/SW/SE	W/E		NW/NE		N	1
5	Shelter	Very sheltered						Very exposed	1
6	Type obstacles		None	Reeds		Boulders/Rock *in situ*			2
6	Density obstacles		None	Sparse		Moderately dense		Dense	1
9	Shore features	Beach/Rocky shore	Bank/Cliff/Strip				Embankment	Wet. fringe	5
10	O.S.G. depth at 2m		0 - 8cm	9 - 16cm	17 - 30cm	31cm - 1m		>1m	1
10	O.S.G. form		Planar	Gently shelving		Moderately shelving		Steeply shelving	1
11	Beach width		>6m	5m	4m	3m	1 - 2m	<1m	1
13	Beach material		Sand	V. sm. - sm. rounded stones		Med. - Lge. rounded stones	V. Lge. rounded stones	Boulders/Angular stones	2
14	Beach drainage		Free	Imperfect		Poor		Waterlogged	1

Table 74

Suitability class scores for selected recreation types

Activity	Suitability class 1 High in present condition	2 Moderate	3 Slight limitations	4 Moderate limitations	5 Unsuitable
(a) Camping and caravanning	+1 - 13	14 - 29	30 - 50	50 - 72	73 - 130
(b) Car parking	0 - 10	11 - 22	23 - 29	40 - 53	54 - 96
(c) Picnicking on backshore	0 - 10	11 - 22	23 - 29	40 - 53	54 - 96
(d) Picnicking on shore	+10 - 0	1 - 12	13 - 29	30 - 43	44 - 90
(e) Walking on backshore	0 - 6	7 - 13	14 - 23	15 - 38	39 - 60
(f) Walking on shore	+10 - 0	1 - 12	13 - 29	30 - 43	44 - 90
(g) Pony trekking backshore	+3 - 7	7 - 19	20 - 36	37 - 50	51 - 92
(h) Pony trekking shore	+10 - 0	1 - 12	13 - 29	30 - 43	44 - 90
(i) Orienteering	0 - 5	6 - 11	12 - 19	20 - 31	32 - 48
(j) Nature study	0 - 5	6 - 11	12 - 19	20 - 31	32 - 48
(k) Bank-fishing	+5 - 0	6 - 6	7 - 15	16 - 22	23 - 45
(l) Boat-launching	+5 - 3	4 - 12	13 - 25	26 - 45	46 - 78
(m) Backshore access	0 - 6	7 - 13	14 - 23	15 - 38	39 - 60
(n) Swimming	+5 - 2	3 - 11	12 - 23	24 - 40	41 - 78

N.B. Scores are negative unless indicated positive.

Group 1 Car parks, lay-bys and picnic sites are all associated with the parking of private cars, which are the main means of transportation for visitors to the countryside. A car park is usually a surfaced area with a capacity for a larger number of cars than a lay-by. In rural areas, a picnic site usually consists of a car park interspersed with facilities such as picnic tables, benches, litter bins and possibly fireplaces.

Group 2 Footpaths and guided trails are both associated with the recreational activity of walking. The Countryside Recreation Research Advisory Group (CRRAG, 1970) define a footpath as a "highway over which the public have a right-of-way on foot only". In the lochside context, the number of paths which are rights-of-way are limited, and therefore, a much broader definition is required which includes all paths which are accessible on foot, as well as those which provide short links between car parks and the shore. Longer paths tend to occur along the lochside. A guided trail is defined as a "specifically designed or existing route marked with descriptive numbered posts at points of interest, which may be used in conjunction with written explanatory material" (CRRAG, 1970). These include nature trails, forest walks and way-marked paths.

Group 3 A viewpoint, often located on high ground, provides a vantage point for viewing the surrounding countryside. Features commonly associated with a viewpoint are a cairn, information plaque, seats and a footpath from the nearest road or car park.

Group 4 A camping and/or caravan site is "land on which tents or caravans are stationed for the purposes of human habitation either permanently or temporarily and land used in conjunction with this" (CRRAG, 1970). Types of caravan site are transit, touring and holiday and static. (For full range of definitions see CRRAG, 1970). The number of facilities depends on the level of development of the site. Running water may be all that is provided on temporary sites which farmers, on holdings of more than 2.02h (5 acres), can establish without planning permission for three units for a maximum of 28 days in any one year. Planning requirements for permanent caravan sites are more stringent and facilities usually include litter disposal, toilets, showers and, in some cases, a shop and laundering facilities. There are recommended guidelines concerning the ratio of people to toilets (see Caravan Sites Control and Development Act, 1960). There are no legal requirements for camping sites, but some private organisations such as the Camping Club of Great Britain set their own standards. Camping sites and caravan sites may be exclusively for tents or caravans, but often sites will accommodate both.

Group 5 Accommodation includes youth hostels, hotels, bed and breakfast establishments, outdoor activity centres and field study centres, which may be used as a base from which recreational activities are carried out. The accommodation may not be situated on the lochside itself, but outdoor activity centres in particular may still have a considerable impact in terms of numbers of people using the loch and lochside, for example Glenmore Lodge by Loch Morlich.

Group 6 Toilets and litter bins are basic facilities associated with a number of recreational activities. Litter bins are usually provided at points which are visited by large numbers of people such as car parks and picnic sites, although some site managers encourage visitors to take their litter home, and do not provide litter collection facilities. Toilets are required in similar locations, but because of their high capital cost and the need for maintenance, their provision is limited.

Group 7 Fishing huts and club houses are buildings used as a base by club members. Club houses for water-sports include toilets and showers and may also be used for other functions, such as dances and barbecues.

Group 8 Piers, jetties, slipways, moorings and marinas are all facilities associated with boating activites. Slipways are used for boat launching, piers and jetties for securing a boat, either for a short period as part of the recreation trip, or as a method of mooring. A marina is "a harbour or other sheltered water area providing moorings and/or parking areas for leisure craft, with associated servicing facilities" (CRRAG, 1970).

Group 9 Steamer services and boats for hire are facilities provided on a number of lochs. Steamer services are restricted to larger lochs such as Loch Lomond, Loch Katrine and Loch Ness, whereas small-boat hire is common on all sizes of loch, and includes rowing boats, boats with outboard motors for fishing, canoes and, less frequently, yachts.

Group 10 Hides are shelters from which wildlife (particularly birds) can be observed. Their provision encourages the concentration of the activity, and allows it to be pursued in what might otherwise be adverse weather conditions.

Sources (other than standard dictionaries) of already existing or widely accepted definitions are given where appropriate.

Access	Means of gaining entry or right of approach to the lochside.	*Bog*	Acid peat.

Accessible	Capable of being approached: no physical, legal or land-ownership constraints to approach.
Accessibility	Degree and type of access to a point on the lochside; extent to which a lochside is approachable.
Active recreation (mobile)	Forms of recreation which involve a high degree of physical exercise.
Backshore	The lochside immediately above mean winter water-level, characterised by terrestrial vegetation.
Bank	A land-face of moderate or steeper gradient behind the water's edge/wetland fringe/rocky shore/beach/cliff, of limited width (less than 10m), usually vegetated, and succeeded on the landward side by a site-type of less-steep gradient.
Beach	The zone of accumulated inorganic material (mud to boulders) extending from summer (or lowest sustained) water-level to the highest point reached by storm waves.
Beach profile	A section describing the morphology of the beach surface as surveyed perpendicular to the water-line on the day of survey to the limit of permanent vegetation.
Beach ridges	Lines of unconsolidated material lying above the waterline, and parallel to it, at the location of former loch level(s): they may subsequently be submerged. The result of accumulation due to swash transport, the lochward ridge-face is more commonly flatter than the landward one; or the crest may be indistinct, being replaced by a rounded, convex form.
Beach types	(a) Arc: an area of beach material marked by a more or less abrupt change of shoreline direction away from the loch. (b) Fan: an area of beach material marked by a more or less abrupt change of shoreline direction towards the loch. (c) Line: a more or less unbroken stretch of beach material following the line of the shore.
Biotic resource – based recreation	Recreation which is dependent on living things.
Boat nests	Distinct parallel-sided gaps usually about two metres wide in otherwise continuous reed or sedge beds.

Breaker types	(a) Plunging: the crest is the leading part of the breaker; it eventually becomes unstable as it travels up the beach and falls to the foot of the breaker. The underlying material experiences considerable pressure changes (Galvin, 1972). (b) Spilling: the foot of the breaker leads; the crest, becoming unstable collapses down the breaker face, decaying into wavelets (Galvin, 1972). (c) Surging: though the breaker foot leads, the crest does not collapse, but is carried up the beach by the forward momentum of the breaker, until the energy becomes so low that the entire form collapses into swash. (Galvin, 1972).
Cliff	An extremely steep land-face formed of sand, clay, boulder clay or peat, located at the back of the shore. The height varies between 0.25m and 3m; the face may be planar or undercut, and is usually without vegetation.
Competitive recreation	Forms of recreation which contain an element of competition against oneself or others. Competitive recreation is usually formal.
Concentrated recreation	Forms of recreation which are dependent on facilities or organisations which are fixed in location.
'de facto' access	All land approachable by the general public; land used by the public but not necessarily with the legal right or permission of landowner to be so used.
'de jure' access	Land physically approachable and accessible to general public by right.
Demand	The amount(s) of various recreational activities in which a population will be willing and able to participate. (Coppock, 1975).
Development	Deliberate modification (involving physical changes and capital input) of a site to provide means or to improve conditions, for one or more recreational activities.
Dispersed recreation	Forms of recreation which are not dependent on facilities or organisations which are fixed in location.
Dominant species	Those plants which contribute most to areal vegetation cover and/or total vegetation biomass, usually, but not always, the most abundant.

184

Dominant vegetation forms	See entries for wetland, grass, forbs, shrub, woodland.
Ecocline	A series of plant communities developed under conditions of gradually changing environmental parameters, for example on a lochside the plant community sequence from the backshore margin to increasingly deep water. Precise subdivisions in such a situation are often difficult to establish.
Ecological capacity	Occurs when use causes an unacceptable degree of ecological change away from the ecosystem considered desirable (Burton, 1974).
Embankment	A man-made slope, of moderate or steeper gradient, varying in height from 1 - 10m, composed of unconsolidated material or stone, and usually associated with a road, railway or reservoir.
Emergent vegetation	Tall hydrophytes, rooted under water but with much of the plant standing above the water. Flowers and seeds are produced on the emergent parts, for example reeds and sedge.
Enhancing factor	An element which, although not essential, enhances the potential for developing an area for recreation.
Eutrophic loch	A nutrient-rich loch, highly productive in organic material, often characterised by the species growing there. Excess nutrient input from sewage and agricultural fertilisers can be damaging. Spence (1964) regards nutrient-rich lochs as those which have more than 60ppm Ca CO3.
Exposure	The degree and extent to which any part of the shoreline is open to wind and/or wave action.
Facility	Natural or man-made resources, whose potential for a particular recreational use has been realised by development or other means, or man-made developments provided explicitly for recreational use.
Facility-based recreation	Recreational pursuits which are geared to the provision of user-facilities supplying a specific recreational need or demand.
Fen	An area where the summer water-table is below ground, but liable to flood in winter. Dominant species are partial hydrophytes and helophytes (Spence, 1964).
Fetch	See wave fetch.
Fire scar	Charring or burning of a tree as a result of a fire having been lit at its base.
Fire site	A small area of ground damaged as a result of a fire having been lit in it.
Floating-leaved vegetation	Plants with floating leaves, with or without submerged leaves, frequently with emergent flowers. Most plants are rooted in the substratum, generally in water more than 30cm deep in summer.
Forbs	Non-woody herbaceous flowering plants other than grass species or grass-like species such as sedges and rushes.
Formal recreation	Forms of recreation in which there is organisation of people and/or equipment.
Frequency	See wave frequency; plant species frequency.
Grass and forbs	Non woody plants less than 1m in height, penetrable on foot. (a) Turf: grass which is periodically cut to maintain a low sward. (b) Arable: land supporting grasses and/or forbs, which is bare for some part of the year. (c) Short grasses and forbs: less than 10cm in height, for example permanent pasture. (d) Tall grasses and forbs: between 10cm and 1m in height, for example roadside verges.
Helophyte	A marsh plant with perennating (or over-wintering) buds at the soil surface.
Hydrophyte	A plant which has its perennating buds only or mainly below the water surface. Total hydrophytes are always rooted under water, partial hydrophytes can grow where summer water-table is below the ground.
Impact	Direct physical contact and/or force exerted on land surface (or water body), or a component of it, by recreationists and their equipment.
Informal recreation	Casual, unorganised forms of recreation which do not involve organisation of people and/or equipment.
Irregularity index Ss/Sa	Where Sa = actual shorelength and Ss = smoothest shorelength for loch of given area (loch length + loch breadth).
Land-based recreation	Recreation which is dependent on features of the land.
Landscape capacity	The ability of the landscape to absorb recreational use. (Burton, 1974).
Leisure	The time available to the individual when the disciplines of work, sleep and other basic needs have been met.
Limiting factor	A key element which exerts such a negative influence on the potential for developing a particular kind of recreation area or enterprise, that it poses problems that are difficult or impossible to overcome.
Linear recreation	Forms of recreation which are pursued along linear routes on land or water.

Littoral The zone between extreme low water level and the upper shore.

Lochshore See shore.

Lochside The area of land extending from the water's edge to either
(a) a road track or railway within 100m or
(b) a major obstruction in vegetation or landform or
(c) for a distance of 100m if (a) or (b) do not apply.
(N.B. This is a general guide to be used in a flexible manner).

Lochside profile The surface landform and vegetation-form features of a section of the lochside extending in a perpendicular direction from the water's edge to the outer limit of the lochside.

Lochside types Defined according to the percentage of shoreline occupied by the dominant primary site-type; (a) 0-20%; (b) 21-40%; (c) 41-60%; (d) 61-80%; (e) 81-100%.

Management Input of labour and capital to maintain a formal or informal recreation site at a given level of development, condition and/or quality.

Marsh An area of inorganic soil where the water-table is permanently or seasonally at or near ground-level.

Mesotrophic loch A loch with nutrient status intermediate between an oligotrophic loch and a eutrophic loch: 16-60ppm Ca CO_3.

Minimum user requirements The threshold of physical conditions, which, if not satisfied, preclude the pursuit of a particular recreational activity by the average user.

Oligotrophic loch A loch of low nutrient status, poor in producing organic material, with less than 15ppm Ca CO_3.

Optimum user requirements The ideal physical conditions required for the pursuit of a particular recreational activity by the average user.

Particle-size classes (International scale)

Class		Particle size (dia.mm)
Gravel:	medium	19.0 - 6.35
	fine	6.35 - 2.0
Sand:	coarse	2.0 - 0.60
	medium	0.60 - 0.20
	fine	0.20 - 0.06
Silt:	coarse	0.06 - 0.02
	medium	0.02 - 0.006
	fine	less than 0.006

Passive recreation (static) Forms of recreation which do not involve a high degree of physical exercise. Passive recreation is also informal.

Perceptual capacity That level of recreational use at which a significant proportion of visitors begin to perceive a site as crowded.

Physical capacity The maximum number of user units that a recreation site can satisfactorily accommodate for a given period.

Plant community A group or assemblage composed of two or more species populations, occupying a specified habitat.

Plant cover The proportion of ground covered by any given species.

Plant species frequency The percentage of individual species scores (using point quadrats), expressed as a proportion of total hits. This has been used as a measure of abundance, or dominance.

Potentiality (for recreation) The suitability of an area to accommodate a recreational activity, allowing for a future management input.

Pressure Level of demand for, or intensity of use of, recreational space or facility.

Primary site-type The site-type nearest to the water's edge.

Profile classification
(a) Convex: A beach profile in which most of the profile occurs above the theoretical slope line.
(b) Concave: A beach profile in which most of the profile occurs below the theoretical slope line.
(c) Planar: A beach profile which does not depart appreciably from the theoretical slope line.
(d) Convex/concave: A beach profile which displays both convex and concave elements, the former being predominant.
(e) Concave/convex: A beach profile which displays both convex and concave elements, the latter being predominant.
(f) Convex/planar: A beach profile which displays both convex and planar elements, the former being predominant.
(g) Concave/planar: A beach profile which displays both concave and planar elements, the latter being predominant.
(N.B. Other combinations of the initial three categories are possible, depending on which category predominates (in other words accounts for 50% or more of the profile length) in any one profile.)

Proximal (as applied to lochs) Shoreline within 100m of a road or track capable of taking four-wheeled vehicles.

Proximity Nearness, in terms of distance of a lochside to a road or track capable of taking four-wheeled vehicles.

Public access Right of approach/passage/admittance.

Recreation	Leisure-time activities which give refreshment to the individual via some pleasant occurrence, pastime or amusement.
Recreational activity	A leisure pursuit which affords mental and/or physical enjoyment primarily for its own sake rather than for economic gain.
Recreational carrying capacity	The level of recreation use an area can withstand while providing a sustained quality of recreation.
Recreation resource	Natural or man-made features which provide or may provide in the future, opportunities for recreation.
Resource	Natural or man-made feature(s) of the environment which provides, or may provide in the future, opportunities for the satisfaction of human wants.
Resource-based recreation	Recreation pursuits which are geared to or dependent on features of the natural environment or the countryside which in themselves attract users.
Rhizome	Prostrate root-like stem emitting roots.
Rocky shore	A shoreline of rock *in situ*.
Shingle	Unconsolidated stones and/or gravel on the shore, susceptible to movement by waves.
Shingle pavement	Shore where stones or gravel are firmly embedded in an inorganic substratum; constituent stones are not usually susceptible to movement by waves.
Shoaling wave	A wave which, on entering water approximately half as deep as its deep-water wave-length, begins to 'feel bottom' and undergoes a transformation from its deep-water shape to a sharper, more unstable form which eventually becomes over-steepened and breaks (Komar, 1976).
Shore	Of a natural loch: the seasonally submerged zone between lowest summer water-level and mean winter water-level. Of an impounded loch or reservoir: the zone between the lowest drawdown and highest permitted water-level.
Shore-based recreation	Recreation which is dependent on features of a shore.
Shore features	Features found at or near the shoreline: see wetland fringe, rocky shore, beach, cliff, bank, strip, embankment.
Shoreline	The mean winter water-level often marked by a small cliff or by the beginning of terrestrial vegetation.
Shrub	Woody plants, impenetrable by foot,

	including woodland with an impenetrable undergrowth.
	(a) Low shrub: woody plants less than 1m in height, for example heather, bog myrtle, bracken.
	(b) Tall shrub: woody plants between more than 1m in height, or such a density to be impenetrable by foot, for example gorse, dense forest, open woodland with dense undergrowth.
Site-type	An area of lochside in which adjacent profiles are homogenous with respect to the dominant slope, the dominant vegetation form and any shore features present.
Slope categories	Gentle = 0 - 5°; moderate = 6 - 10°; steep = 11 - 15°; very steep = 16 - 20°; extremely steep = greater than 21°.
Solum	That area of land underlying the water body in a loch or reservoir.
Stand	Any area of vegetation treated as a single unit for the purposes of study.
Stilt roots	Exposed roots from which the substratum has been removed, so that they support the base of the tree at a level above the present ground surface.
Strip	A narrow (less than 5m wide) facet of gentle slope located between the water's edge/wetland fringe/rocky shore/beach/cliff and a bank or the primary facet which has a steeper gradient.
Submerged vegetation	Vegetation growing on the loch floor always submerged in winter. It may be submerged all the year, but shallow water communities are usually exposed in summer. Most species tolerate a considerable range of water depth and some will stand occasional drying out.
Subsidiary site-type	A site-type, to the landward side of the primary site-types, defined by a change in the dominant slope and/or vegetation-form occurring along the line of the profile.
Substratum	The medium in which the plants are growing; in the lochside situation this may include stones and peat.
Suitability	The ability of an area in its present state to accommodate a recreational activity.
Swamp	An area adjoining open water, permanently or seasonally submerged and dominated by total or partial hydrophytes. The summer water-table may vary from 30cm above to a few centimetres below ground. (Spence, 1964).

Swash	A disorganised body of water produced by breaker or wave collapse, that travels up the beach until halted by energy loss or percolation into the substratum. Swash that remains on the surface will return downslope as backswash.
User-based recreation	Recreation pursuits which are geared to the needs of the user.
Water-based recreation	Recreation which is dependent on a water body.
Wave fetch	The uninterrupted straight-line distance over which waves must travel to reach the shore.
Wave form	Shape of wave from one trough to the lowest of the adjacent trough.
Wave frequency	The number of wave crests passing a given point in unit time; or, the time taken for the wave form to move a distance of one wavelength.

Wave height	The vertical distance from the crest of a wave to the base of the preceding or succeeding trough.
Wavelength	The distance between two successive wave crests.
Wave steepness	Wave height divided by wavelength (King, 1972).
Wetland	A habitat where the water-table in summer is near, at or above the ground surface; the plants are adapted to waterlogged conditions. Wetland includes bog, marsh, fen and swamp. (a) Low wetland: low sedges, rushes, grasses and forbs, less than 1m in height; (b) Tall wetland: reeds, shrubs or trees which restrict visibility greater than 1m in height.
Wetland fringe	A shoreline fringe of wetland vegetation between 1 and 3m in width.

M

Agriculture Land Service (1968) *Agriculture Land Classification*. Ministry of Agriculture, Fisheries and Food, Technical Report No. 11.

Appleton, J. H. (1975) The theoretical vacuum. In The Evaluation of Landscape. *Trans. Inst. Br. Geogr.*, 66, 120-123.

Bates, G. H. (1935) The vegetation of footpaths, sidewalks, cart tracks and gateways. *J. Ecol.*, 23, 470-487.

Bates, G. H. (1938) Life forms of pasture plants in relation to treading. *J. Ecol.*, 26, 452-454.

Bayfield, N. G. (1971) Some effects of walking and skiing on vegetation at Cairngorm. *The Scientific Management of Animal and Plant Communities for Conservation* (Ed. E. Duffey and A. S. Watt). Blackwell Scientific Publications, 469-485.

Bayfield, N. G. (1973) Use and deterioration of some Scottish hill paths. *J. appl. Ecol.*, 10, 635-644.

Beazley, E. (1969) *Designed for Recreation*. Faber, London.

Bernard, J. M. (1976) The life history and population dynamics of shoots of *Carex rostrata*. *J. Ecol.*, 64, 1045-1048.

Bibby, J. S. and Mackney, D. (1969) *Land Use Capability Classification*. Soil Survey of Scotland Technical Monograph No. 1.

Bidde, D. D. (1965) *Ship Waves in Shoaling Water*. Hydraulics Laboratory Report HEL-12-6, University of California.

Birch, J. W. (1974) *Information requirements for land and water resource management planning*. Symposium on Resource Management, at Annual Meeting Institute of British Geographers, Norwich (*mimeo*).

Bjork, S. (1967) Ecological Investigations of *Phragmites communis. Folia Limnol. Scand.*, 14, Lund.

British Waterworks Association (1972) *Amenity Use of Reservoirs in Scotland: Analaysis of Returns 1971.*

Britton, R. H. (1974) Factors affecting the distribution and productivity of emergent vegetation at Loch Leven, Kinross. *Proc. R. Soc. Edinb.*, Series B., 209-218.

Brooks, A. (1976) *Waterways and Wetlands*. British Trust for Conservation Volunteers Ltd.

Brown, R. and Chapman, V. (1975) *Loch Lomond Recreation Report*. Occasional Paper No. 5. Countryside Commission for Scotland, Perth.

Bruun, P. (1954) *Coast Erosion and the Development of Beach Profiles*. Corps of Engineers, Beach Erosion Board Technical Memo. 44. Washington, D.C.

Burbridge, V. (1974) *Land Classification*. Unpublished Ph.D Thesis, Heriot-Watt University.

Burden, R. F. and Randerson, P. F. (1972) Quantitative studies of the effects of human trampling on vegetation as an aid to management of semi-natural areas. *J. appl. Ecol.*, 9, 439-457.

Burton, R. C. J. (1974) *The Recreational Carrying Capacity of the Countryside*, Keele University Library, Occasional Publication No. 11.

Canada Land Inventory (1969) *Land Capability Classification for Outdoor Recreation*. Report No. 6 Ottawa.

Central Water Planning Unit (1976) *Some empirical information on demands for water-based recreation for water-based activities*. Technical Note No. 13. Reading.

Chappell, H. G. *et al* (1971) The effect of trampling on a chalk grassland ecosystem. *J. appl. Ecol.*, 8, 868-882.

Chubb, M. and Banmann, E. G. (1976) *Assessing the recreational potential of rivers*. Paper presented at 72nd Annual Meeting of the Association of American Geographers, New York (*mimeo*).

Clapham, A. R., Tutin, T. G. and Warburg, E. F. (1962) *Flora of the British Isles*. University Press, Cambridge.

Clawson, M. (1963) *Land and Water for Recreation*. Rand McNally, Chicago.

Clay, D. P. (1973) *Recreation and Planning with Special Reference to the Use and Capacity of Waterways*. M.Sc. Thesis, University of Strathclyde.

Coppock, J. T. (1966) Recreational use of land and water in rural Britain. *Tijdschr. econ. soc. Geogr.*, 57(3), 81-96.

Coppock, J. T. and Duffield, B. S. (1975) *Recreation in the Countryside: A Spatial Analysis*. MacMillan Press, London.

Countryside Commission (1972) *The Use of Aerial Photography in Countryside Research*. Conference Report, London.

Countryside Commission (1973) *Recreation at Reservoirs*. CCP 60.

Countryside Commission for Scotland and Scottish Civic Trust (1971) *The Caravan in Scotland: Chaos or Compatibility*. Report of a one day conference (21.10.71), Countryside Commission for Scotland, Perth.

Countryside Recreation Research Advisory Group (1970) *Countryside Recreation Glossary*. Countryside Commission, London.

Cressman, E. M. (1968) *Methodology for Ontario Land Inventory*. Ontario Department of Lands and Forests, Toronto.

Dartington Amenity Research Trust (1973) *Langorse Lake Study: A Recreation Survey*. Report to Brecon County Council, Sports Council and Countryside Commission.

Davidson, D. A. (1976) Terrain evaluation: testing a technique in Lower Deeside. *Scott. geogr. Mag.*, 92(2), 109-119.

Davis, R. A. and Fox, W. T. (1971) *Beach and nearshore dynamics in eastern Lake Michigan*. Ohio Natural Resources Dept. Technical Rept. No. 4, Contract 338-092.

Deakin, W. H. (1975) *Kent County Structure Plan*. Kent County Council.

Dean, R .G. (1976) Beach erosion: causes, processes and possible remedial measures. CRC: *Crit. Rev. Environ. Contr.*, 6(3), 259-296.

Department of Trade and Industry (1974) *Seaway Code: A Guide for Small Boat Users*. H.M.S.O. London.

Douglas, R. W. (1969) *Forest Recreation*. Pergamon Press, Oxford.

Duffey, E. (Ed.) (1971) *The Scientific Management of Animal and Plant Communities for Conservation*. British Ecological Society, Blackwell.

Duffield, B. S. and Owen, M. L. (1970) *Leisure and Countryside. A Geographical Appraisal of Countryside Recreation in Lanarkshire*. Edinburgh University.

Floyd, W. E. (1974) *Selection and Evaluation of Land Resources for Outdoor Recreation Areas Adjacent to a Proposed Reservoir*. M.Sc. Thesis. Stephen Austin State University, Texas.

Folk, R. L. (1966) A review of grain size parameters. *Sedimentology*, 6, 73-93.

Forestry Commission (1975) *Glenmore Forest Park, Cairngorms*. Forestry Commission Guide, H.M.S.O.

Galvin, C. J. (1968) Breaker type classification on three laboratory beaches. *J. geophys. Res.*, 73, 3651.

Galvin, C. J. (1972) Waves Breaking. In *Waves on Beaches* (Ed. Mayer). Academic Press, London and New York.

Gill, C. J. and Bradshaw, A. D. (1971) Some aspects of the colonisation of upland reservoir margins. *J. Instn. Wat. Engrs.*, 25(3), 165-171.

Gill, C. J. (1977) Some aspects of the design and management of reservoir margins for multiple use. *J. appl. Biol.*, 2, 129-182.

Gittens, J. W. (1976) *Recreation and conservation in the Welsh wetlands*. Paper read at Royal Society of Arts Conference on Recreation and Conservation (11.11.76), London (*mimeo*).

Goldsmith, F. B. (1974) Ecological Effects of Visitors in the Countryside, 217-231. In *Conservation in Practice*. (Eds.) Warren, A. and Goldsmith, F. B. Wiley, London/New York.

Goodall, B. and Whittow, J. B. (1973) *The Recreational Potential of Forestry Commission Holdings*. A Report to the Forestry Commission, University of Reading.

Greig-Smith, P. (1964) *Quantitative Plant Ecology*. Butterworths, London.

Gunton, T. (1974) *An Application and Evaluation of Selected Resource Analysis Techniques*. M.A. Thesis, University of Waterloo, Ontario.

Hackett, B. (1976) *Creating new wetland resources*. Paper read at Royal Society of Arts Conference on Recreation and Conservation (11.11.76), London (*mimeo*).

Hakansson, L. (1977) On lake form, lake volume and lake hypsographic survey. *Geogr. Annlr.*, Series A. Physical Geography, 59A (1-2), 1-28.

Harvard University (1967) *Three Approaches to Environmental Resource Analysis*. Graduate School of Design, Harvard University. Washington Conservation Foundation.

Haslam, S. M. (1972) *The Reed*. (2nd Edn.), Norfolk Reed Growers Association.

Haslam, S. M. (1973a) The management of British wetlands. Part I: Economic and amenity use. *J. Envir. Mgmt.*, 1, 303-320.

Haslam, S. M. (1973b) The management of British wetlands. Part II: Conservation. *J. Envir. Mgmt.*, 1, 345-361.

Hills, G. A. (1966) *Definition of Capability Classes and Bench Mark Sites for the Recreational Land Inventory*. Research Branch, Ontario Dept. of Lands and Forests, Maple, Ontario (*mimeo*).

Hills, G. A. *et al* (1970) *Developing a Better Environment*. Ontario Economic Council, Toronto.

Hockin, R., Goodall, B. and Whittow, J. B. (1977) *The Site Requirements and Planning of Outdoor Recreation Activities*. Reading University.

Howard, J. A. (1970) *Aerial Photo-Ecology*. Faber & Faber, London.

Huxley, T. (1970) *Footpaths in the Countryside*. Countryside Commission for Scotland, Perth.

Inman, D. L. (1952) Measures for describing the size distribution of sediments. *J. sedim. Petrol.*, 22, 125-145.

I.U.C.N. (1962) Project marsh conservation of temperate marshes, bogs and other wetlands. *I.U.C.N. Proeedings*, No. 3.

I.U.C.N. (1967) Towards a new relationship of man and nature in temperate lands. Pt. I. Ecological impact of recreation and tourism on temperate environments. *I.U.C.N. Proceedings*, No. 7.

Jaakson, R. (1970) Planning for the capacity of lakes to accommodate water-oriented recreation. *Planning*, 10(3), 29-40.

Jaakson, R. (1973a) *Shoreline Recreation Planning: A Systems View*. Occasional Paper No. 7. Faculty of Environmental Sciences, University of Waterloo, Ontario.

Jaakson, R. (1973b) Reservoir operation for recreation usability. *J. Hydraul. Div. Am. Soc. Civ. Engrs.*, 1813-1822.

Jeffreys, H. (1917) On the vegetation of four Durham coal measure fells. Parts III and IV. *J. Ecol.*, 5, 129-154.

Jermy, A. C. and Tutin, T. G. (1968) *British Sedges*. Botanical Society of the British Isles, London.

Johnson, J. W. (1948) The characteristics of wind waves on lakes and protected bays. *Trans. Am. geophys. Un.*, 29, 671.

Johnson, J. W. (1969) Ship waves at recreation beaches. *Shore Beach*, 37(1), 11-15.

Johnstone, M. (1979) *The Assessment of the Capability of Scottish Lochsides for Recreation: A Methodology*. Ph.D. Thesis, University of Glasgow.

Jupp, B. P., Spence, D. H. N. and Britton, R. H. (1974) Distribution and production of submerged macrophytes in Loch Leven, Kinross. *Proc. R. Soc. Edin.*, Series B, 195-208.

King, C. A. M. (1972) *Beaches and Coasts* (2nd Edn.), Edward Arnold.

Klingebiel, A. A. and Montgomery, P. H. (1961)   *Land capability Classification.* United States Dept. Agriculture Soil Conservation Service. Agriculture Handbook 210.

Klotzli, F. (1973)   Uber Belastbarkeit und Produktion in Schilfrohrichten Sonderdruck. *Verhandlungen der Gesellschaft fur Okologie*, Saarbrucken (English Abstract).

Komar, P. D. (1976)   *Beach Processes and Sedimentation.* Prentice Hall, Inc., N.J.

Krumbein, W. C. (1939)   *Graphic Presentation and Statistical Analysis of Sedimentary Data.* Report of Southwest Symposium American Association Petrology and Geology, Tulsa, Oka.

Laing, A. I. (1974)   *An Evaluation of the Capability of the Loch Carron Area for Shore-based Recreational Activities.* Undergraduate Dissertation, Department of Geography, University of Glasgow.

La Page, W. F. (1967)   *Some Observations on Camp Ground Trampling and Ground Cover Response.* United States Forest Survey Research Paper NE 68.

Larson, J. S. (Ed.) (1976)   *Models for Evaluation of Freshwater Wetlands.* University of Massachusetts.

Leney, F. (1974)   *The Ecological Effects of Public Pressure on Picnic Sites.* Ph.D. Thesis, University of Aberdeen.

Liddle, M. J. (1973)   *The Effects of Trampling and Vehicles on Natural Vegetation.* Ph.D. Thesis, University College of North Wales, Bangor.

Liddle, M. J. (1975a)   A selective review of the ecological effects of human trampling on natural ecosystems. *Biol. Conserv.*, 7, 17-36.

Liddle, M. J. (1975b)   A theoretical relationship between the primary productivity of vegetation and its ability to tolerate trampling. *Biol. Conserv.*, 8, 251-255.

Liddle, M. J. (1976)   An approach to objective collection and analysis of data for comparison of landscape character. *Reg. Studies*, 10(2), 173-181.

Liddle, M. J. (undated a)   *Outdoor recreation: a selective summary of pertinent ecological research.* School of Plant Biology, University College of North Wales, Bangor (*mimeo*).

Liddle, M. J. (undated b)   *A survey of twelve lakes in the Gwdyr Forest Region of Snowdonia.* School of Plant Biology, University College of North Wales, Bangor (*mimeo*).

Liddle, M. J. and Greig-Smith, P. (1975)   A survey of tracks and paths in a sand dune ecosystem. I. Soils; II. Vegetation. *J. appl. Ecol.*, 12, 893-930.

Lothian Regional Council, Department of Recreation and Leisure (undated)   *A Guide to Freshwater Fishing in the Lothian Region.*

Macan, T. T. (1974)   *Freshwater Ecology* (2nd Edn.). Longmans, London.

Macan, T. T. and Worthington, E. B. (1972)   *Life in Lakes and Rivers.* The New Naturalist, Collins, London.

MacConnell, W. P. and Stoll, P. (1969)   Evaluating recreation resources of the Connecticut River. *J. Photogrammetric Eng.*, 35, 686-692.

McErlean, M. (1970)   *The effects of recreational activities on the lochshore vegetation between Balmaha and Rowardennan.* Hons. B.Sc., undergraduate thesis in Biogeography, Department of Geography, Glasgow University.

McLaren, M. and Currie, W. B. (1972)   *The Fishing Waters of Scotland.* Murray, London.

McLaughlin, B. P. (1971)   *Recreation: A Planning Methodology with Special Reference to Loch Earn.* M.Sc. Thesis, Heriot Watt University, Edinburgh.

McVean, D. N. (1955-56)   Ecology of *Alnus glutinosa* (L) Gaertn. I. Fruit Formation *J. Ecol.*, 43, 46-60, II. Seed Distribution *J. Ecol.*, 43, 61-71, III. Seedling Establishment *J. Ecol.*, 44, 145-218, IV. Root System *J. Ecol.*, 44, 219-225.

McVean, D. N. (1956)   Ecology of *Alnus glutinosa* (L) Gaertn. V. Notes on some British Alder Populations. *J. Ecol.* 44, 321-330.

McVean, D. N. (1959)   Account of *Alnus glutinosa* (L) Gaertn. for the Biological Flora of the British Isles. *J. Ecol.*, 41, 447-460.

Mather, A. S. and Ritchie, W. (1978)   *The Beaches of the Highlands and Islands of Scotland.* Countryside Commission for Scotland, Perth.

Meyer, R. E. (Ed.) (1972)   *Waves on Beaches and Resulting Sediment Transport.* Academic Press, New York and London.

Millman, R. (1970)   *Outdoor Recreation in the Highland Countryside.* Ph.D. Thesis, University of Aberdeen.

Ministry of Housing and Local Government (1963)   *Caravan Parks: Location, Layout and Design.* H.M.S.O., London.

Mitchell, C. W. (1973)   *Terrain Evaluation.* Longmans, London.

Mithen, Dallas (1975)   The Public Forest. In *Land Resources for Recreation in Scotland.* Landscape Research Group in association with Countryside Commission for Scotland (Eds.) Tivy, J. and Dickinson, G., Department of Geography, University of Glasgow.

Moeller, G. H. and Engellun, J. H. (1972)   What fishermen look for in a fishing experience. *J. Wildl. Mgmt.*, 36, 1253-1257.

Morgan, S. M. (1977)   *The Recreational Capability of Loch Ken and the River Dee.* Undergraduate Dissertation. Department of Social Science, Middlesex Polytechnic, Enfield.

Morzer-Bruijns, M. F. (1967)   The influence of recreational activities on aquatic biocenoses. *I.U.C.N. Proceedings*, No. 7.

Moss, B. (1977)   *Development and change in broadland freshwater ecosystems.* Paper presented to Recreation Ecology Research Group Meeting, Norwich.

Murray, J. and Pullar, L. (1910)   *Bathymetrical Survey of the Scottish Freshwater Lochs.* 6 Vols. Edinburgh, Challenger Office.

Nature Conservancy *(1967) Biotic Effects of Public Pressure on the Environment.* Symposium Papers. Monks Wood Experimental Station, Huntingdonshire.

192

Nature Conservancy (1970)   *The application of air photography to the work of the Nature Conservancy.* Nature Conservancy Staff Seminars (*mimeo*).

Nicholls, D. C. (1968)   *Recreation and Tourism in the Loch Lomond Area.* Report to Dunbarton County Council prepared in Department of Social and Economic Research, University of Glasgow and Scottish Tourist Board, Edinburgh.

Norrman, J. D. (1964)   Lake Vattern: investigations on shore and bottom morphology. *Geogr. Annlr.*, 46(1-2), 1-238.

Ogilvie, A. G. (1944)   Debateable land in Scotland. *Scott. geogr. Mag.*, 60(2), 42-43.

Ontario Ministry of Natural Resources (1972) *Lake Alert Phase 2.* Ontario Ministry of Natural Resources.

Ontario Ministry of Natural Resources, Land Use Co-ordination Branch (1977)   *Interim Lake Planning Guidelines.* For use within the Ministry of Natural Resources, Toronto.

Osborne, C. (1975)   *Water-based recreation in the United Kingdom. A review of literature.* University of Leeds Geography Department, Working Paper 108.

Outdoor Recreation Resources Review Commission (1962)   *Outdoor Recreation for America.* Washington, D.C.

Patmore, J. A. (1970)   *Land and Leisure.* David and Charles, Newton Abbot.

Penning-Rowsell, E. C. (1973)   *Alternative Approaches to Landscape Appraisal and Evaluation.* Middlesex Polytechnic Planning Research Group. Report No. 11, Hendon, Middlesex.

Potter, S. (1976)   *The Capability of Lochshores for Recreation.* Undergraduate Dissertation. Department of Social Science, Middlesex Polytechnic, Enfield.

Rees, J. R. (in press)   A people-counter for unsurfaced wetland footpaths. *Environ. Conserv.*

Rees, J. R. and Tivy, J. (1977)   Recreational impact on lochshore vegetation. *J. Scott. Ass. geogr. Teach.*, 6, 8-24.

Rees, J. R. and Tivy, J. (1978)   Recreational impact on Scottish lochshore wetlands. *J. Biogeog.*, 5, 93-108.

Reid, G. K. (1961)   *Ecology of Inland Waters and Estuaries.* Van Nostrand Reinhold.

Ripley, T. H. (1962)   *Recreational Impact on Southern Appalachian Camp Grounds and Picnic Sites.* United States Department of Agriculture South-East Forest Experimental Station, Paper No. 153.

Schneberger, E. and Threinen, C. W. (1964)   Lake Management for recreational uses. *Wis. Acad. Trans.*, 53(A): 49-55.

Scott, T. (1954)   *Sand Movement by Waves.* United States Beach Erosion Board Technical Memo. 48.

Scottish Orienteering Association (undated)   *A Short Guide to Orienteering.*

Scottish Tourist Consultative Council (1973)   *Report by the Working Party on the Caravan in the Environment* (*mimeo*).

Seddon, B. (1972)   Aquatic macrophytes as limnological indications. *Freshwater Biol.*, 2, 107-130.

Sheldon, A. (1972)   A quantitative approach to the classification of inland waters. In Krutilla, J.V. *Natural Environments: Studies in Theoretical and Applied Analysis.* John Hopkins University Press, Baltimore and London.

Shepherd, F. P. (1976)   Coastal classification and changing coastlines. In *Coastal Research* (*Geoscience and Man* Vol. 14) (Ed.) Walker, H. J., Louisiana State University, Baton Rouge.

Simmons, I. G. (1975)   *Rural Recreation in the Industrial World.* Arnold, London.

Slater, F. M. and Agnew, A. D. (1977)   Observations on a peat bog's ability to withstand increasing public pressure. *Biol. Cons.*, 11, 21-27.

Smart, C. W. W. (1973)   Research associated with water recreation. *Landscape Design,* 104, 33-35.

Smith, I. R. and Sinclair, I. J. (1972)   Deep water waves in lakes. *Freshwater Biol.*, 2, 387-399.

Soil Survey (1960)   *Field Handbook.* Soil Survey of England and Wales.

Sorensen, J. C. (1971)   *A Framework for Identification and Control of Resource Degradation and Conflict in the Multiple Use of the Coastal Zone.* Dept. Landscape Architecture, College of Environmental Design, University of California, Berkeley.

Sorensen, R. M. (1960)   *Investigation of Ship-generated Waves.* Hydraulic Engineering Lab. College of Engineering, University of California, Berkeley.

Sorensen, R. M. (1967)   Investigations of ship-generated waves. *J. Wat. Ways Harb. Div. Am. Soc. Civ. Engrs.*, 85-99.

Spence, D. H. N. (1964)   The macrophytic vegetation of freshwater lakes, swamps and associated fens. In Burnett, J. H. (Ed.). *The Vegetation of Scotland.* Oliver and Boyd, Edinburgh.

Spence, D. H. N. (1967)   Factors controlling the distribution of freshwater macrophytes with particular reference to the lochs of Scotland. *J. Ecol.*, 55, 147-170.

Stewart, G. A. (Ed.) (1968) *Land Evaluation.* Papers of a CSIRO symposium. Macmillan of Australia, Melbourne.

Stiegler, S. E. (Ed.) (1976)   *A Dictionary of Earth Sciences.* Macmillan, London.

Streeter, D. T. (1971)   The effect of public pressure on the vegetation of chalk downland on Boxhill, Surrey. In *The Scientific Management of Animal and Plant Communities for Conservation.* (Ed.) Duffey, E. and Watt, A. S., 459-468. Blackwell Scientific Publications, Oxford.

Sukopp, H. (1971)   Effects of man, especially recreational activities on littoral macrophytes. *Hidrobiologia*, 12, 331-340.

Sukopp, H. (1973)   Conservation of wetlands in Central Europe. *Polskie Archium Hydrobiol.*, 20, 223-228.

Tanner, M. F. (1973)   *Water Resources and Recreation.* Sports Council Water Recreation Series, Study 3, London.

Thompson, W. C. and Harlett, J. C. (1967)   The effects of waves on the profile of a natural beach. In *Proceedings 10th Conference on Coastal Engineering of the American Society of Civil Engineers.*

Tivy, J. (1972)   *The Concept and Determination of Carrying Capacity of Recreational Land in the U.S.A.* Countryside Commission for Scotland Occasional Paper No. 3, Perth.

Tivy, J. (1974)   *Loch Lomond Report: Intensity of shoreside environmental damage and management recommendations.* Unpublished Report to Countryside Commission for Scotland (typescript).

T.R.R.U. (1976)   *STARS SERIES No. 2: Summary Report.* Tourism and Recreation Research Unit, Edinburgh University.

United States Department of Agriculture (1968)   *Soils Surveys: soil interpretations for recreation.* Soil Conservation service, Washington, D.C.

University of Toronto (1971) *Lakeshore Capacity: Bibliography.*

Venables, B. (1967)   *Freshwater Fishing,* Jenkins, London.

Wagar, J. A. (1961)   How to predict which vegetated areas will stand up best under active recreation. *Am. Recreat. J.*, 7, 20-21.

Water Ski Study Group (undated)   *Water-skiing in the Eastern Region.* Report of Eastern Sports Council, Water Recreation Committee.

Weir, T. (1970)   *The Scottish Lochs* Vol. I. David and Charles, Newton Abbott.

Weir, T. (1972)   *The Scottish Lochs* Vol. II. David and Charles, Newton Abbott.

Westhoff, V. (1966)   Ecological impact of pedestrians, equestrians and vehicular traffic on vegetation. *I.U.C.N. Proceedings*, No. 10, 218-223.

Westlake, D. F. (1971)   Population Dynamics of *Glyceria maxima. Hidrobiologia,* 12, 133-4.

Wiegel, R. L. (1964)   *Oceanographical Engineering.* Prentice-Hall, Inc., N.J.

194

INDEX

Abbreviations: def. = definition; fig. = figure; L. = Loch; Res. = Reservoir; tab. = table